Proceedings in Life Sciences

The Physiological Properties of Plant Protoplasts

Edited by Paul-Emile Pilet

With 82 Figures and 35 Tables

Springer-Verlag
Berlin Heidelberg New York Tokyo

Professor Dr. PAUL-EMILE PILET
Director of the
Institut de Biologie et de Physiologie Végétales
Bâtiment de Biologie de l'Université
1015-Lausanne, Switzerland

ISBN-13: 978-3-642-70146-7 e-ISBN-13: 978-3-642-70144-3
DOI: 10.1007/ 978-3-642-70144-3

Library of Congress Cataloging in Publication Data. Main entry under title: The physiological properties of plant protoplasts. (Proceedings in life sciences). Bibliography: p. Includes index. 1. Plant protoplasts. I. Pilet, P.-E. (Paul-Emile), 1927–. II. Series. QK725.P467 1985 581.87 84-26689

Softcover reprint of the hardcover 1st edition 1985

2131/3130-543210

Preface

The idea for the present book arose from a 3-day *seminar* which I organized in March 1984 for young research workers in plant physiology. Participants came from several universities of the French-speaking part of Switzerland and speakers from Basel, München, Nottingham, Perpignan, Regensburg, Sheffield, Toulouse, Yale, Zürich ... and Lausannne.

The theme of the seminar was chosen from the range of research fields of our Institute. Meanwhile, feeling it was important to bear in mind that some of our hearers were not specialists in the chosen topic, I wished to provide a subject that would be of scientific, methodological and epistemological interest.

The critical analysis of the *structural and functional characteristics of plant protoplasts* exactly met these criteria. There exists ample material for discussion of the techniques of protoplast preparation, methods used in morphological, biological and biochemical studies, and for the comparison of protoplasts with the cells from which they are obtained.

When I was organizing the programme of our seminar at the end of 1983, I contacted about 50 scientists whose research corresponds to the theme adopted. Almost all replied positively and with enthusiasm, confirming the real interest shown by the participants in the topics which were later selected for this book. *Springer* immediately supported my suggestion to publish an enlarged version of our conference proceedings comprising oral presentation and additional reports from biologists who cannot be present at the Lausanne seminar. This project would have been impossible without the support of the *Fondation Herbette* (Faculté des Sciences de l'Université de Lausanne).

More and more papers and reviews are being published relating to the studies of protoplasts as an experimental tool for plant genetics (fusion between protoplasts, somatic hybrids, genetically transformed cells, plant propagation, etc.). It seemed appropriate to gather some research results devoted to another aspect, namely the "physiological properties of plant protoplasts".

Without overestimating the importance of this topic, I really feel that progress in our knowledge concerning the physiology of protoplasts themselves is essential for the advance of research in genetic engineering in plants and, in a general way, in plant physiology.

Lausanne, Spring 1985 Paul-Emile Pilet

Contents

Initially, my intention was to group together the contributions under different chapter headings. However, many of the texts were found to cover several distinct but complementary themes. The reports have therefore been arranged one after another without chapter divisions. The first deal primarily with problems of technique followed by those concerned with structure and finally functions.

The Editor

List of Contributors

Alibert, G. Centre de Physiologie Végétale de l'Université Paul Sabatier, Unité associée au CNRS No 241, 118, Route de Narbonne, 31062 Toulouse cedex, France

Blaschek, W. Fakultät Chemie/Pharmazie der Universität, Universitätsstraße 31, 8400 Regensburg, FRG

Boller, T. Botanisches Institut der Universität, Schönbeinstraße 6, 4056 Basel, Switzerland

Bornman, C.H. Cell and Tissue Culture, Hilleskög Research AB, Box 302, 261 23 Landskrona, Sweden

Bornman, J.F. Department of Plant Physiology of the University, Box 7007, 222 07 Lund, Sweden

Boudet, A.M. Centre de Physiologie Végétale, Université Paul Sabatier, 118, Route de Narbonne, 31062 Toulouse, France

Canut, H. Centre de Physiologie Végétale, Université Paul Sabatier, 118, Route de Narbonne, 31062 Toulouse, France

Cocking, E.C. Department of Botany of the University, University Park, Nottingham, NG7 2RD, United Kingdom

Crozier, A. Department of Botany of the University, Glasgow G12 8QQ, Scotland

Edwards, G. Department Botany of the Washington State University, Pullman Washington, 99164-4230, USA

Elliott, D.C. School of Biological Sciences, The Flinders University of South Australia, Bedford Park, S.A. 5042, Australia

Fevre, M. Laboratoire de Mycologie associé au CNRS de l'Université de Lyon I, 43, Bd 11 Novembre 1918, 69622 Villeurbanne, France

Fitzsimons, P.J. Department of Biological Sciences of the University, Dundee, DD1 4HN, Tayside, United Kingdom

Fowke, L.C. Department of Biology of the University of Saskatchewan, Saskatoon, S7N OWO, Canada

Franz, G. Fakultät Chemie/Pharmazie der Universität, Universitätsstraße 31, 8400 Regensburg, FRG

Girard, V. Laboratoire de Mycologie associé au CNRS de l'Université de Lyon I, 43, Bd 11 Novembre 1918, 69622 Villeurbanne, Frace

Griesbach, R. USDA, ARS, Florist and Nursery Crops Laboratory, BARC-West, Beltsville, MD 20705, USA

Griffing, L.R. Department of Biology of the University of Saskatchewan, Saskatoon, S7N 0W0, Canada

Hampp, R. Institut für Biologie I der Universität, Auf der Morgenstelle 1, 7400 Tübingen, FRG

Harkes, P. Department of Plant Molecular Biology of the University, Nonnensteeg 3, 2311 VJ Leiden, Netherlands

Hartmann, E. Institut für Allgemeine Botanik der Universität, Saarstraße 21, 6500 Mainz, FRG

Henry, H. Institute de Biologie et de Physiologie végétales, Bâtiment de Biologie de Université, 1015 Lausanne, Switzerland

Hinderer, W. Institut für Biologie I der Universität, Auf der Morgenstelle 1, 7400 Tübingen, FRG

Hock, K. Institut für Allgemeine Botanik der Universität, Saarstraße 21, 6500 Mainz, FRG

Hopp, W. Institut für Biologie I der Universität, Auf der Morgenstelle 1, 7400 Tübingen, FRG

Iversen, T.-H. Department of Botany of the University, 7055 Dragvoll, Norway

Jollès, C. Institut de Biologie et de Physiologie végétales, Bâtiment de Biologie de l'Université, 1015 Lausanne, Switzerland

Keller, F. Department of General Botany, Swiss Federal Institute of Technology, Sonneggstraße 5, 8092 Zürich, Switzerland

Lawson, R. USDA, ARS, Florist and Nursery Crops Laboratory, BARC-West, Beltsville, MD 20705, USA

Libbenga, K. Department of Plant Molecular Biology of the University, Nonnensteeg 3, 2311 VJ Leiden, Netherlands

Lin, W. Central Research and Development Department, Experimental Station, E402/2135, E.I. Du Pont de Nemours and Company, Wilmington, Delaware 19898, USA

Maan, A. Department of Plant Molecular Biology of the University, Nonnensteeg 3, 2311 VJ Leiden, Netherlands

Manigault, P. Laboratoire de Physiologie Cellulaire Végétale, Avenue de la Terrasse, 91190 Gif-sur-Yvette, France

De March, G. Résidence Jardin de Cys., 8, Avenue des Cévennes, 91940 Les Ulis, France

Matile, P. Department of General Botany, Swiss Federal Institute of Technology, Sonneggstraße 5, 8092 Zürich, Switzerland

Mennes, A.M. Department of Plante Molecular Biology of the University, Nonnensteeg 3, 2311 VJ Leiden, Netherlands

Mersey, B.G. Plant Biotechnology Institute, National Research Council of Canada, Saskatoon, S7N 0W9, Canada

Meyer, Y. Laboratoire de Physiologie Végétale de l'Université, (UA 565), Avenue de Villeneuve, 66025 Perpignan, France

Morris, P. Wolfson Institute of Biotechnology of the University, Sheffield, S10 2TN, United Kingdom

Pahlich, E. Institut für Allgemeine Botanik und Pflanzenphysiologie der Justus-Liebig-Universität, Heinrich-Buff Ring 54–62, 6300 Giessen, FRG

Petersen, M. Institut für Biologie I der Universität, Auf der Morgenstelle 1, 7400 Tübingen, FRG

Pilet, P.E. Institut de Biologie et de Physiologie végétales, Bâtiment de Biologie de l'Université, 1015 Lausanne, Switzerland

Rataboul, P. Centre de Physiologie végétales, Université Paul Sabatier, 118, Route de Narbonne, 31062 Toulouse, France

Sandberg, G. Department of Forest Genetics and Plant Physiology, The Swedish University of Agricultural Sciences, 901 83 Umeå, Sweden

Schibeci, A. Department of Biochemistry, La Trobe University, Bundoora, Victoria 3083, Australia

Schnabl, H. Institut für Botanik der Technischen Universität, Arcisstraße 21, 8000 München, FRG

Scott, R. Botany Department of the Washington State University, Pullman Washington, 99164-4230, USA

Seitz, H.U. Institut für Biologie I der Universität, Auf der Morgenstelle 1, 7400 Tübingen, FRG

Tremolières, A. Laboratoire du Phytotron, CNRS, 91190 Gif-sur-Yvette, France

Van der Linde, P. Department of Plant Molecular Biology of the University, Nonnensteeeg 3, 2311 VJ Leiden, Netherlands

Weyers, J.D.B. Department of Biological Sciences of the University, Dundee, DD1 4HN, Tayside, United Kingdom

Wiemken, A. Institute of Botany of the University, Schönbeinstraße 6, 4056 Basel, Switzerland

Introduction: The Use of Plant Protoplasts in Physiological Research

P.E. PILET

The potential of plant protoplasts as an experimental system has been exploited more and more over the last 25 years. For some time, protoplasts were used per se and a number of papers were published on some of their cytological and biological properties.

More recently, most of the research on protoplasts has been orientated towards their genetic applications. Several directions of experimentation are now available. One aspect is the entry into the protoplasts of micro-organisms, metaphasic chromosomes and several types of organelles. Other possibilities are fusion between protoplasts, somatic hybridisation, genetic manipulations illustrated by the transformation by exogenous DNA and experimental mutagenesis. The concentration of protoplast research on short- and long-term investigations in plant genetic engineering has tended to lead to a neglect of the interesting question of the physiological characteristics of protoplasts.

Some Physiological Problems

From the earliest observations on protoplasts (Bajaj 1977) it became clear that such plant material can be used for analysing several biochemical and biophysical problems for which cells are not so suitable.

However, the question arises of whether protoplasts are directly comparable with the cells from which they have been obtained. This in turn leads to the question whether cells and the protoplasts which they yield can be analysed in parallel (Pilet 1973, 1981). In order to resolve these specific and fundamental questions particular attention must be paid to the technical questions related to obtaining protoplasts (Gamborg and Wetter 1975). Here again physiological properties of cells and protoplasts are directly concerned.

Such an experimental approach may indicate if protoplasts can really be considered or not as a useful model for plant cells. Some data already suggest (see below) that the plasmalemma of protoplasts may have quite different characteristics from those of the cells from which they are derived; for instance, the ability of protoplasts to fuse together could be a result of the new properties acquired by the plasmalemma during protoplast formation.

Since freshly prepared protoplasts have no wall, they can be used (as will briefly be discussed later) to give direct access to the plasmalemma. Some general problems

The Physiological Properties of Plant Protoplasts
Edited by P. E. Pilet

could therefore be analysed without undue difficulty: the chemical structure of the plasmalemma (lipids, proteins, enzymes etc.), its physical properties (surface charges, proton extrusion, ionic pumps, permeability, pH of different compartments, etc.) (Boller and Galbraith 1983). As protoplasts will regenerate a new wall, ultrastructural conditions (presence of microtubules, coat vesicles, dictyosomes etc.) and biochemical reactions could also perhaps be much more easily studied than in whole cells.

The culture of protoplasts raises many physiological problems (Ahuja 1982) beside the choice of environmental conditions (light, temperature, pH etc.). The mineral and organic nutrients for the culture medium have to be carefully selected. Experiments with vitamines and growth regulators are necessary; they could give some useful information about the biochemical control of wall formation and extension, synthesis of new material for membranes and hyaloplasm. Using protoplasts it may be possible to separate one or two aspects of the regulation of biological processes brought about by endogenous factors (as auxins, gibberellins, cytokinins etc.) which, in intact plants, act at many levels difficult to discriminate.

Some of these physiological reactions need special attention. Because they were crucial for the viability and the division of protoplasts, they have been fully discussed in several recent publications (Cassells and Cocker 1982; Kaiser and Heber 1983).

Obtaining Protoplasts

From cells to protoplasts and from the protoplasts to plants, several steps are necessary (Fig. 1). None is fully understood, however, some of them are technically well defined, whereas others require further testing.

The first step consists in obtaining some "competent" cells (Dale 1983), those able to form active protoplasts. These cells may be taken from several parts of the plants (leaves, roots etc.) or by using organs, tissues or cells cultured in vitro.

However, it is well established that most cells appear to lose their ability to respond to some stimuli and particularly to growth hormones (Wareing 1982) when they are separated from each other. This should not be forgotten when the physiological properties of cells and their protoplasts are compared. Two essential steps have to be first mentioned: plasmolysis and removal of the cell coat and wall by enzyme mixtures. These cell treatments may modify considerably the behaviour of the protoplasts obtained and they could especially alter the integrity of their plasmalemma.

Protoplast Culture

As previously discussed, physical and chemical conditions for culturing protoplasts involve several preliminary experiments. The finality of these assays consists in precising optimal parameters of the environment and of the growing medium.

A concept often used is the *viability* of the protoplasts. It has to be noted that the evaluation of this biological characteristic is indeed subjective. Viability can be appre-

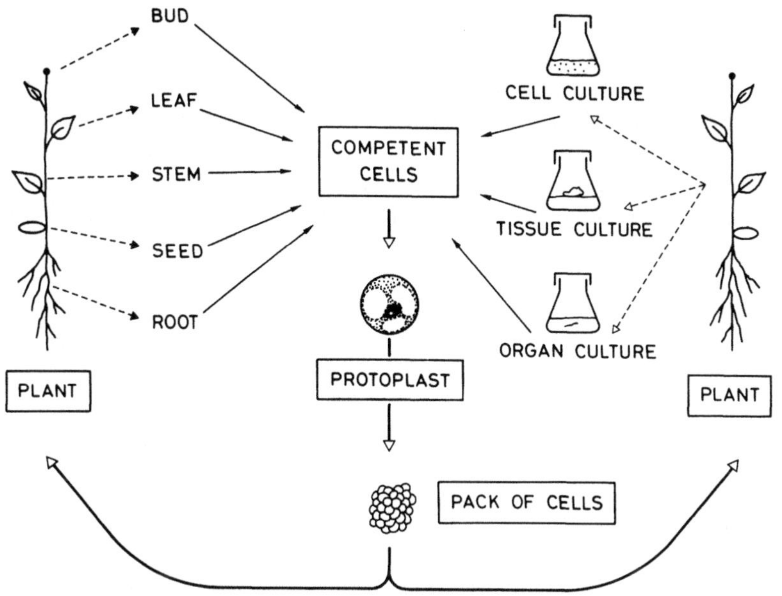

Fig. 1. Schematic representation of the several possible sources of "competent" cells for obtaining plant protoplasts. They may give a "pack" of cells from which, under certain conditions, a complete plant derives

ciated when looking at the form and the cyclosis of the protoplasts or better by vital stains. However, the proportion of surviving protoplasts could very well be a "compartmentalized phenomenon" (Bornman and Bornman 1983) and should preferably be measured quantitatively.

When protoplasts are cultured in optimal conditions, several essential events occur: wall biosynthesis, DNA replication, cell and organelle division. By dedifferentiation and differentiation, a single protoplast may give, after a few days, a "pack" of cells from which, by morphogenesis and organogenesis, a whole plant could be formed. All these steps are regulated by endogenous hormones but, at the moment, nothing is clearly known about their possible effects.

Plasmalemma

Protoplasts, with their exposed plasma membrane, were first used in studies of plasmalemma integrity. They respond to auxin by an increased rate of vacuolation until they ultimately burst (Cocking 1962; Pilet 1971, 1981). Such disruption of the plasmalemma may depend on the endogenous auxin level of the cells from which protoplasts have been obtained (Pilet 1984).

The plasmalemma forms a boundary between the cytoplasm and the periplasm. It regulates the movement of ions, amino acids and glucids into and out of the hyaloplasm. Lectin and specific antigen-binding sites are located on the periplasmic part of the plasmalemma, which consequently may be involved in recognition phenomena (Knox and Clarke 1978; Rubery 1981).

The charges associated with the surface of the plasmalemma can also be characterized (Pilet and Senn 1974; Nagata and Melchers 1978). This is mostly the result of the combined ionization of surface group such as proteins, amino acids, lipids and polysaccharides. These charges expressed by the zeta potential can easily be measured by the electrophoretic mobility of the protoplasts (Grout et al. 1972; Senn and Pilet 1981; Pilet et al. 1984).

However, as shown below, one of the most important functions of the plasmalemma is to participate in wall reformation. During the time of cell coat and wall deposition, precursors move continuously through the plasmalemma. Such transport involves exocytes in association to microtubules, coated vesicles, microfilaments etc. (Fowke et al. 1983).

Wall Biosynthesis

One of the first observations reported on protoplasts was that of C.O. Townsend, in 1897, concerning wall deposition on protoplasts of *Elodea* and *Gaillarda* (Cocking) 1972).

Of course, the wall is involved in overall growth and cell expansion. Both isotropic and anisotropic materials for assembling the new wall have to be channelled through the plasmalemma.

Studies on protoplasts have provided information about the participation of such organelles as microtubules and membrane vesicles as well as dictyosomes, translosomes, reticulum etc. (Gorshkova and Salnikov 1983).

The chemical nature of the regenerated walls was analysed. Now it seems clear that the compounds forming the microfilbrills and the matrix may significantly differ from those in the walls of the cells from which protoplasts have been produced (Pilet et al. 1984).

In conclusion, during the last few years protoplasts have largely been used for genetic experiments and the study of their physiological properties has been pushed into the background. However, the study of some of their biological characteristics is of value in itself. In addition it is obvious that it is necessary to understand better structures and functions of protoplasts for valid interpretations of experiments in genetic engineering.

References

Ahuja MR (1982) Isolation, culture and fusion of protoplasts: problems and prospects. Silvae Genet 31:66–77

Ambrose EJ (1966) Electrophoretic behaviour of plant cells. Prog Biophys Mol Biol 16:243–265

Bajaj YPS (1977) In: Reinert J, Bajaj YPS (eds) Plant Cell, Tissue and Organ Culture. Springer, Berlin Heidelberg New York, pp 467–496

Boller T, Galbraith DW (1983) Plant protoplasts as tools for physiological studies. Exper Suppl (Basel) 46:130–131
Bornman JF, Bornman CH (1983) Protoplast viability: a relative concept. Exper Suppl (Basel) 45:208–209
Cassells AC, Cocker FM (1982) Seasonal and physiological aspects of the isolation of tobacco protoplasts. Physiol Plant 56:69–79
Cocking EC (1962) Action of growth substances, chelating agents and antibiotics on isolated root protoplasts. Nature (Lond) 193:998–999
Cocking EC (1972) Plant cell protoplasts: isolation and development. Annu Rev Plant Physiol 23:29–50
Dale PJ (1983) Protoplast culture and plant regeneration of cereals and other recalcitrant crops. Exper Suppl (Basel) 46:31–41
Fowke LC, Griffing LR, Mersey BG, Van der Valk P (1983) Protoplasts for studies of the plasma membrane and associated cell organelles. Exper Suppl (Basel) 46:101–110
Galbraith DW, Shields BA (1982) The effects of inhibitors of cell wall synthesis on tobacco protoplast development. Physiol Plant 55:25–30
Gamborg OL, Wetter LR (eds) (1975) Plant Tissue Culture Methods. Publ Nat Res Council of Canada, Saskatoon, p 109
Gorshkova TA, Salnikov VV (1983) Ultrastructural dynamics of cell wall regeneration around isolated bean leaves protoplasts. Exper Suppl (Basel) 45:194–195
Grout BWW, Willison JHM, Cocking EC (1972) Interaction at the surface of plant cell protoplasts. Bioenerg 4:585–602
Kaiser G, Heber U (1983) Photosynthesis of leaf cell protoplasts and permeability of the plasmalemma to some solutes. Planta (Berl) 157:462–470
Knox RB, Clarke AE (1978) In: Hall JL (ed) Electron Microscopy and Cytochemistry of Plant Cells. Elsevier/North Holland, Amsterdam, pp 149–185
Nagata T, Melchers G (1978) Surface charge of protoplasts and their significance on cell interaction. Planta (Berl) 142:235–238
Pilet PE (1971) Effet de quelques auxines sur les protoplasts racinaires. CR Acad Sci Paris 273: 2253–2256
Pilet PE (1973) In: Tempé J (ed) Protoplastes et fusion de cellules somatiques végétales. Colloq Int Cent Nat Rech Sci, Paris, pp 99–107
Pilet PE (1981) In: Göring H, Paul E (eds) Pflanzliche Gewebekultur. Humbolt Univ Berlin, pp 35–42
Pilet PE (1984) Auxin effect on protoplasts from gravireacting maize roots. Z Pflanzenphysiol 113:373–376
Pilet PE, Blaschek W, Senn A, Franz G (1984) Comparison between maize root cells and their respective regenerating protoplasts: wall polysaccharides. Plant (Berl) 161:465–469
Pilet PE, Hervé R, Senn A (1984) Zeta potential of protoplasts from gravireacting maize roots. Planta (Berl) 162:17–22
Pilet PE, Senn A (1974) Effet deu CA^{2+} et du K^{+} sur la mobilité électrophorétique des protoplastes. CR Acad Sci Paris 278:269–272
Rubery T (1981) Auxin receptors. Annu Rev Plant Physiol 32:569–596
Senn A, Pilet PE (1981) Electrophoretic mobility, zeta potential and surface charge of maize root protoplasts. Z Pflanzenphysiol 102:19–32
Wareing PF (1982) In: Smith H, Grierson D (eds) The molecular biology of plant development. Blackwell Scient Publ, London, pp 517–541

Applications of Protoplast Technology

E.C. COCKING

Introduction

The use of plant protoplasts for genetic manipulations has been the central theme for most of the applications of protoplast technology. Significant progress has been made in various aspects of genetic manipulations using such protoplasts. This progress will be briefly surveyed. Increasingly, however, it is becoming realised that a better understanding of the physiological properties of protoplasts will greatly enhance our ability in this respect, and these aspects will be highlighted.

Protoplast Culture: Preparatory Studies and the Induction of Sustained Divisions

The application of protoplast technology to the major cereals has been limited by the inability to regenerate plants from isolated properties. One of the central problems in this connection is our present inability to induce protoplasts isolated directly from cereals to undergo sustained divisons. Cell division appears to be blocked very effectively in protoplasts isolated from cereal cells, particularly from those that are highly differentiated. Why this is so is not clear. Protoplasts have to be maintained in a suitable plasmolyticum; and, generally, insufficient attention has been given to the inhibitory effects of increased osmotic pressure on the inhibition of mitosis (Smith et al. 1984). When protoplasts are isolated enzymatically there is always, of course, the possibility that the commercial cell-wall degrading enzymes usually employed contain toxic or undesirable substances; and there is good evidence that in certain species the sustained division of protoplasts is prevented unless some purification is carried out. Partial purification of commercial preparations using Bio Gel P6 is usually employed. Bio Gel purified enzyme was essential for colony formation from *Medicago* mesophyll protoplasts, and for plant recovery from single protoplasts of *Petunia parodii* (Davey and Kumar 1983). Relatively little work has been done on the detailed purification of the various enzyme components, recently, however, significant steps in this direction have been taken by Ishii and Mogi (1983). It is encouraging that suitable enzyme mixtures are now available for the isolation of protoplasts from woody species, and that sustained division has been achieved (Faye and David 1983).

The Physiological Properties of Plant Protoplasts
Edited by P. E. Pilet

Most workers investigating the isolation and sustained division of protoplasts pay little attention to the onset of cell-wall regeneration. Burgess (1983) has highlighted the use of reversible inhibitors of cell-wall regeneration, such as coumarin and 2:6 dichlorobenzonitrile. He has also pointed out these inhibitors are interesting and useful tools in the prolongation of the protoplast state, and the synchronization of protoplast populations.

Interest in the immobilisation of protoplasts, either to enhance division capabilities or to facilitate secretion of secondary products, has been greatly stimulated in recent years by studies on the immobilisation of plant cells in agarose and other matrices. Dix et al. (1983) have suggested that the ability to avoid the use of agar and its associated heat shock could be advantageous; and very significantly Shillito et al. (1983) have shown that not only does culture in agarose improve protoplast plating and proliferation, but that it also permits division in otherwise unresponsive systems. It has also been shown that such immobilisation very significantly delays the onset of senescence in protoplasts (Schnabl et al. 1983). These basic physiological and biochemical studies could greatly help in elucidating factors controlling the induction of cell division.

Protoplast Fusion and Somatic Hybridization

Numerous reviews have been written in recent years on protoplast fusion and the application of this technology to plant improvement (Cocking 1984), and very recently by Gleba and Sytnik (1984). What will be most useful is to identify remaining problems and general difficulties.

Whilst most investigations have emphasized the influence of various treatments on the fusion of protoplasts, including polyethylene glycol, high levels of calcium at high pH and electric field stimulation, there has been a neglect of the influence of the developmental stage of the cells, from which protoplasts are derived, on fusion potential. Early studies, particularly those using pollen tetrad protoplasts, indicated high frequencies of fusion without the presence of an obvious inducer (Bhojwani and Cocking 1972; Ito and Maeda 1973). Only recently, however, have Boss et al. (1983) begun a systematic investigation of the manipulation of membrane structure for fusion potential. They noted that with all the usual treatments (including electric field stimulation) increasing the percentage of fusion beyond a certain threshold decreases the protoplast viability. These workers showed that by varying the culture conditions of wild carrot suspension cultures they could vary the fusogenic state of the plasmamembrane until the membrane became truly fusion-permissive. The fusogenic membrane was characterized by thin fibrillar thread-like protrusions visible only with video-enhanced microscopy. Electron spin resonance identified a uniquely fluid location of the fatty acid spin label associated with fusogenic protoplasts (Boss et al. 1983). These significant observations should be perhaps related to the suggestion of Poste and Pasternak (1978) that fusion resistance of the plasmamembrane may be more apparent than real, and that a low frequency of fusion is determined by some additional constraint. In cell-to-cell fusion, this type of steric hindrance could perhaps be mediated by the various glycoproteins present in the "cell coat" which would provide a "barrier" to a surface projection of one cell readily achieving direct apposition of its layer with that of another cell.

An exciting vista that has recently been discussed (Cocking 1984) is that it may be possible to obtain limited gene transfer, even between sexually isolated species, by protoplast fusion, if one of the fusion partners is suitably irradiated, or treated with chemicals inducing rapid chromosome loss (Roth et al. 1982). Concerted efforts are now required to evaluate the extent of limited gene transfer that can be obtained as a consequence of protoplast fusions.

The cytoplasmic mix obtained from protoplast fusions is novel, with the opportunity for the production of cybrids with the opportunity for the formation of mitochondrial recombinants (Cocking 1981). Important advances have been made in recent years in our knowledge of the segregation of cytoplasmic traits in higher plant somatic fusion hybrids (Evans and Sharpe 1983). Hybrid vigour is well known in sexual hybridization as a consequence of gametic fusion, and it has been suggested (Bingham 1983) that somatic hybridization by somatic protoplast fusion may produce an even greater hybrid vigour.

Plasmid Delivery into Protoplasts and Transformation

Increasingly, plant protoplasts are being utilized in experiments involving the direct interaction of plant cells with plasmid DNA, in experiments involving fusion with bacterial protoplasts and in experiments involving co-cultivation with *Agrobacterium* strains.

There are a limited number of reports of plant protoplast transformation by the uptake of the tumour-inducing (Ti) plasmid of *Agrobacterium* (Davey et al. 1980; Draper et al. 1982; Krens et al. 1982). At present transformation frequencies are low (about 1 in 10^5 protoplasts). Freeman et al. (1984) have recently attempted to quantify plasmid DNA associated with tobacco leaf protoplasts after delivery using several different procedures. In all experiments, the plasmid vector DNA was quantified by Southern blotting, following extraction from intact protoplasts which had been well washed and DNAase-treated. The results indicated that maximum association of vector DNA with protoplasts could be achieved wth plant protoplasts interacted with either naked vector DNA, or vector-containing liposomes, in the presence of 15% w/v polyethylene glycol 6000, or with *E. coli* spheroplasts containing chloramphenicol-amplified plasmid in the presence of 25% w/v PEG 6000. However, it may not always follow that the delivery method giving the greatest uptake of intact plasmid will be the best to achieve transformation. The central challenge is to increase the frequency of transformation when using isolated protoplasts with plasmid DNA. Since so much vector DNA is apparently "wasted", more direct methods such as microinjection (Steinbiss and Broughton 1983) or electrically mediated transfer, electroporation, (Neumann et al. 1982) may prove more efficient for the transformation of protoplasts. This needs evaluation using the Ti plasmid of *Agrobacterium* or suitable chimeric plasmids.

Agrobacterium-mediated transformation of dicotyledons has been achieved by co-culturing protoplasts, regenerating cell walls and undergoing division, with suitable agrobacteria (Davey et al. 1979; Marton et al. 1979). The frequency of transformation is very high and sometimes up to 10% of the protoplasts (depending on the particular species) can be selected. This co-cultivation procedure has also been used to introduce, and have expressed, chimeric genes containing bacterial genes for antibiotic resistance (Herrera-Estrella et al. 1983; Horsch et al. 1984), which are being utilized as dominant selectable markers. Monocotyledons have not, as yet, proved amenable to this co-cultivation procedure.

Whilst co-cultivation which utilises *Agrobacterium* as the delivery system is an attractive procedure for transformation, on-going interest is also being maintained in the direct interaction of DNA with protoplasts; and the fusion of bacterial protoplasts, in which the plasmid has been amplified, with plant protoplasts using the fusion procedures already discussed. Of special interest, therefore, are the factors which define the transformation-competence of cultured plant protoplasts. As discussed by Gould and Ashmore (1982), a consideration of plant cell population kinetics, and more especially of cell cycle distribution, will be essential to the improvement of the efficiency of DNA-mediated transformation of higher plant protoplasts.

General Conclusions

As we have seen, the application of the technology associated with both protoplast fusion and transformation is requiring better characterization of the initial protoplast populations. The solution to these biological problems is often made complex by the cellular heterogenicity of the protoplast populations. In the future it is likely that flow cytometry, which will enable rapid biochemical analysis of single protoplasts, will be increasingly utilized because flow cytometric analysis permits delineation of analytic levels for each cell within a population. As discussed by Kruth (1982) cell sorting allows for recognition and purification of specific cell subpopulations. Multiple and correlated measurements can be obtained for each cell, and cell subpopulations defined by two or more parameters can be characterized.

Overall these cell biological and physiological studies are needed to advance the precision with which genetic manipulations using protoplasts can be undertaken – this application of protoplast technology will be in the forefront of studies in the next decade.

References

Bhojwani SS, Cocking EC (1972) Isolation of protoplasts from pollen tetrads. Nature New Biology 239:29–30

Bingham ET (1983) In: Better crops for food. Ciba Found Symp 97, p 130

Boss WF, Allen NS, Grimes HD (1983) Developmentally regulated fusion of carrot protoplasts. In: Potrykus I, Harms CT, Hinnen A, Hütter R, King PJ, Shillito RD (eds) Protoplasts 1983, Post Proc 6th Int Protoplast Symp Basel, August 1983. Birkhäuser, Basel, p 96

Burgess J (1983) Wall regeneration around isolated protoplasts. In: Giles KL (ed) Plant Protoplasts. Int Rev Cytol Suppl 16. Academic, New York, p 219

Cocking EC (1981) Opportunities from the use of protoplasts. Philos Trans R Soc London Biol Sci 292:5570568

Cocking EC (to be published 1985) Use of protoplasts: Potentials and progress. In: Uses of protoplasts for genetic manipulation. Progress and outlook. 16th Stadler Genet Symp

Davey MR, Kumar A (1983) Higher plant protoplasts – retrospect and prospect. In: Giles KL (ed) Plant Protoplasts. Int Rev Cytol Suppl 16. Academic, New York, p 219

Davey MR, Cocking EC, Freeman J, Draper J, Pearce N, Tudor I, Hernalsteens JP, Beuckeleer M de, Van Montague M, Schell J (1979) The use of plant protoplasts for transformation by *Agrobacterium* and isolated plasmids. In: Advances in protoplast research. Proc 5th Int Protoplast Symp, Szeged 1979

Davey MR, Cocking EC, Freeman JP, Pearce N, Tudor I (1980) Transformation of *Petunia* protoplasts by isolated *Agrobacterium* plasmids. Plant Sci Lett 18:307–313

Dix PJ, Kane EJ, Keane GJ, O'Sullivan MT (1983) Immobilisation of *Nicotiana* cells and protoplasts in alignate beads. In: Potrykus I, Harms CT, Hinnen A, Hütter R, King PJ, Shillito RD (eds) Protoplasts 1983, Poster Proc 6th Int Protoplast Symp Basel, August 1983. Birkhäuser, Basel, p 226

Draper J, Davey MR, Freeman JP, Cocking EC, Cox BJ (1982) Ti plasmid homologous sequences present in tissues from *Agrobacterium* plasmid-transformed *Petunia* protoplasts. Plant Cell Physiol 23:255–263

Evans DA, Sharp WR (1983) Single gene mutations in tomato plants regenerated from tissue culture. Science (Wash DC) 221:949–951

Faye M, David A (1983) Isolation and culture of gymnosperm root protoplasts (*Pinus pinaster*). Physiol Plant 59:359–362

Freeman JP, Draper J, Davey MR, Cocking EC, Gartland KMA, Harding K, Pental D (1984) A comparison of methods for plasmid delivery into plant protoplasts. Plant Cell Physiol (submitted for publication)

Gleba YY, Sytnik KM (1984) In: Shoeman R (ed) Protoplast fusion: genetic engineering in higher plants. Springer, Berlin Heidelberg New York

Gould AR, Ashmore SE (1982) Interaction of purified DNA with plant protoplasts of different cell cycle stage: The concept of a competent phase for plant cell transformation. Theor Appl Genet 64:7–12

Herrera-Estrella L, De Block M, Messens E, Hernalsteens J-P, Van Montagu M, Schell J (1983) Chimaeric genes as dominant selectable markers in plant cells. EMBO J 2:987–995

Horsch RB, Fraley RT, Rogero SG, Sanders RR, Lloyd A, Hoffmann N (1984) Inheritance of functional foreigh genes in plants. Science (Wash DC) 223:496–498

Ishii S, Mogi Y (1983) Isolation and identification of enzymes that are responsible for the isolation of plant protoplasts. In: Potrykus I, Harms CT, Hinnen A, Hütter R, King PJ, Shillito RD (eds) Protoplasts 1983, Poster Proc 6th Int Protoplast Symp Basel, August 1983. Birkhäuser, Basel, p 6

Ito M, Maeda M (1973) Fusion of meiotic protoplasts in liliaceous plants. Exp Cell Res 80: 453–456

Krens FA, Molendijk L, Wullems GJ, Schilperoort RA (1982) In vitro transformation of plant protoplasts with Ti-plasmid DNA. Nature (Lond) 296:72–74

Kruth HS (1982) Flow cytometry: rapid biochemical analysis of single cells. Anal Biochem 125: 225–242

Marton L, Wullems GJ, Molendijk L, Schilperoort RA (1979) In vitro transformation of cultured cells from *Nicotiana tabacum* by *Agrobacterium tumefaciens*. Nature (Lond) 277:129–131

Neumann E, Schaefer-Ridder M, Wang Y, Hofschneider PH (1982) Gene transfer into mouse lyoma cells by electroporation in high electric fields. EMBO J 1(7):841–845

Poste G, Pasternak CA (1978) Virus induced cell fusion. In: Poste G, Pasternak CA (eds) Membrane fusion. Elsevier/North Holland Biomedical Press, Amsterdam, p 305

Roth EJ, Weber G, Lark KG (1982) Use of isopropyl-N (3-chlorophenyl) carbamate (CIPC) to produce partial haploid cells from suspension cultures of soybean (*Glycine max*). Plant Cell Rep 1:205–208

Schnabl H, Youngman RJ, Zimmermann U (1983) Maintenance of plant cell membrane integrity and function by the immobilisation of protoplasts in alginate matrices. Planta (Berl) 158: 392–397

Shillito RD, Paszkowski J, Potrykus I (1983) Culture in agarose improves protoplast plating and proliferation, and permits division in otherwise unresponsive systems. In: Potrykus I, Harms CT, Hinnen A, Hütter R, King PJ, Shillito R (eds) Protoplasts 1983, Poster Proc 6th Int Protoplast Symp Basel, August 1983. Birkhäuser, Basel, p 266

Smith MAL, Palta JP, McCown BH (1984) The measurement of isotonicity and maintenance of osmotic balance in plant protoplast manipulations. Plant Sci Lett 33:249–258

Steinbiss HH, Broughton WJ (1983) Methods and mechanisms of gene uptake in protoplasts. In: Giles KL (ed) Plant Protoplasts. Int Rev Cytol Suppl 16. Academic, New York, p 191

Properties of Some Enzymes Used for Protoplast Isolation

P.J. FITZSIMONS and J.D.B. WEYERS

Introduction

The degree of success achieved in obtaining protoplasts in consistent yields and quality can vary enormously. This can depend on the plant species or tissue used and its age (e.g. Watts et al. 1974) or on the nature of the digestion medium (e.g. Zeiger and Hepler 1976). In parallel with the increase in popularity of protoplasts as experimental tools, more commercial cellulases and pectinases have become available and the use of various non-enzyme components in isolation media has become more common. Unfortunately, our knowledge of the properties of enzyme mixtures in use or of other "additives" has not kept pace, and there is a need for quantitative data on the effects of particular products on protoplast yield and physiology. In this chapter we present the results of a study on the properties of Cellulysin, a commercial product in general use, and describe some effects of various additives to the isolation medium.

The enzyme mixtures used for protoplast isolation are products of undefined composition. Some qualitative information is available (Cocking 1972), but manufacturer's specifications rarely indicate details of the purification procedures used and activity is expressed only as cellulase units. However, many commercial cellulases appear to be widely treated as products of some uniformity, and this may not be justified.

Published protoplast isolation methods can be very simple: good results have been obtained with digestion media consisting of an osmoticum/enzyme mixture alone (e.g. Guy et al. 1978; Randall and Ruesink 1983) or with a similar mixture plus a few salts (e.g. Morris and Thain 1980). Alternatively, protoplasts may be isolated using complex mixtures of enzymes (Lin 1980) and/or other additives, e.g. bovine serum albumin (BSA), dithiothreitol (DTT), potassium dextran sulphate and the protease inhibitors phenyl-methyl sulphonylfluoride (PMSF) and di-isopropyl-fluorophosphate (DIFP). The need for some additives is obvious, for example where a buffer is used, but for a number of chemicals the specific intended function and actual effect on protoplast yield and physiology may not be clear.

Only a few papers (e.g. Von Borner 1973; Gronwald and Leonard 1982) contain data in support of the addition of a particular component to protoplasts isolation media to increase yields. Obtaining such information is complicated by properties of the root and shoot tissue segments normally used in assays of protoplast yield. For these, it is difficult to make accurate estimates of original cell densities, and final yields may be dependent on factors like the degree of mechanical agitation applied to digested tis-

The Physiological Properties of Plant Protoplasts
Edited by P. E. Pilet

sues (Gronwald and Leonard 1982) which may be difficult to standardize. On the other hand, the epidermal tissue of *Commelina communis* represents a convenient system to study protoplast isolation. The tissue is a monolayer of cells (epidermal cells, subsidiary cells and guard cells) and each cell type is present at near-uniform density. Known areas of epidermis can therefore be used to determine accurately yields of epidermal cell (EC) and/or guard cell (GC) protoplasts.

Cellulysin can completely digest the cell walls of the epidermal tissue of *Commelina* (Fitzsimons and Weyers 1983) and many other species (see Weyers and Fitzsimons, Chap. 18, this Vol.) but may also be inactive for protoplast isolation from this or other plant tissues (e.g. Lin 1980). Here, we describe investigations of the cellulase activity of Cellulysin and the effect of various enzyme and non-enzyme additives on its protoplast-releasing characteristics.

Information about the quality of protoplasts obtained is also of importance. We have studied the activity of protease enzymes in Cellulysin and its modulation by various additives and treatments (cf. Lin and Wittenbach 1981). Further, we describe the use of Boyle-van't Hoff analysis (see Weyers and Fitzsimons, Chap. 18, this Vol.) to test protoplast membrane integrity. Other viability assays are also considered and their relative value critically appraised.

Materials and Methods

Plant Material

Commelina communis L. plants were grown as previously described (Fitzsimons and Weyers 1983). The youngest two fully-expanded leaves on the main axis of plants 5–7 weeks old were used to obtain epidermal tissue by the method of Weyers and Travis (1981).

Enzyme Assays

Cellulase activity was determined by measuring the release of glucose from cellulose (Whatman No. 1 filter paper). 100 mm^3 of each enzmye solution (1% w/v) was added to 200 mg of filter paper and 2.40×10^{-6} m^3 50 mol m^{-3} 2-(n-morpholino) ethane sulphonic acid (MES) buffer (pH 5.0, KOH). Incubations were for 60 min at 37°C. 1.0×10^{-6} m^3 of the enzyme/buffer mixture was then added to 1.0×10^{-6} m^3 of 5% (w/v) trichloracetic acid (TCA) and the resulting precipitate removed by centrifugation (500 *g*, 5 min). Depending on the activity of each enzyme tested 50–200 mm^3 of the final supernatant were added to 2.0×10^{-6} m^3 of glucose assay medium (glucose oxidase/phenol-aminophenazone test, E. Merck, Darmstadt, FRG) and the H_2O_2/phenol-aminophenazone dye complex allowed to develop for 15 min before measuring its absorbance at 510 nm. Standards and blanks were assayed and cellulase units computed (one unit releases 1.0 μmol of glucose from cellulose in one hour at 37°C, pH 5.0). Glucose contamination of cellulysin was determined by mixing equal volumes of a 4% (w/v) enzyme solution and 5% TCA and proceeding as above.

Protease activity was determined by measuring the release of dyes from azo-protein complexes (from Sigma Limited, Poole, U.K.) or hide powder azure (HPA, from Calbiochem-Behring, La Jolla, USA). For azo-protease assays all concentrated stock reagents, which contained 50 mol m^{-3} MES buffer (pH 5.0, KOH), were added together to give a final assay volume of 2.5 x 10^{-6} m^3 including 10 mg of azo-protein substrate. After 30 min at 30°C, 2.5 x 10^{-6} m^3 of 5% TCA was added and the precipitate removed as above. Azo-dye in the supernatant was measured spectroscopically at 440 nm. Blanks were prepared for each enzyme and protease activity was defined as the final absorbance attained under the assay conditions described divided by the final concentration of each enzyme employed (w/v, %). The concentrations of each product tested were ajusted so that all assays contained between 0.5–1.0 "units" of protease enzyme. For HPAase assays, 500 mm^3 of a HPAsuspension (1%, w/v) was added to 500 mm^3 of universal buffer (giving constant salt concentration at pH values between 2.5 and 10.0, Dawson et al. 1969) and 500 mm^3 of enzyme solution. After incubation at 30°C for 30 min the reaction was stopped by adding 500 mm^3 40% TCA and the resulting mixture clarified by centrifugation at ca. 10,000 *g* for 30 s. The absorbance of the supernatant was measured at 595 nm and protease units calculated as above.

Protoplast Yield Assays

Known areas of epidermis tissue (500–1000 mm^2) floated for at least 30 min on 300 mol m^{-3} mannitol, were transferred to small (50 mm) Petri dishes containing 3 x 10^{-6} m^3 of enzyme medium (in 300 mol m^{-3} mannitol). The dishes were kept either on an orbital shaker in a controlled temperature room (24°C) or sealed with Nescofilm and floated on a thermostatically-controlled water bath (30°C incubations). To harvest protoplasts, the enzyme medium was removed and the strip remains rinsed twice with fresh 300 mol m^{-3} mannitol, and the rinsings and enzyme medium pooled. Protoplasts were collected by centrifugation (100 *g*, 5 min) and the pellet resuspended in a small volume (500–1500 mm^3) of fresh osmoticum prior to cell density estimation. At least two replicates were prepared for each set of conditions tested. Density estimates were made on 5–10 20 mm^3 aliquots transferred to a haemocytometer (grid area = 0.1 mm^3).

Boyle-van't Hoff analyses and enzyme dialysis procedures were as previously described (Weyers and Fitzsimons 1982). Protein was determined according to Bradford (1976). Cellulysin salt concentrations were determined using an atomic emission spectrophotometer (Corning-EEL, Stone, U.K.) on solutions de-proteinated by treatment with TCA.

Enzyme Suppliers

Cellulysin was obtained from Calbiochem-Behring Corp., La Jolla, USA, Onazuka R-10 and RS from Yakult Honsha Co. Ltd., Nishinomiya, Japan. Other cellulases were supplied by Boehringer Chemicals Ltd., Mannheim, FRG (ex *Trichoderma viride*) and Sigma Chemical Co. Ltd., Poole, U.K. (ex *Aspergillus niger*). Pectolyase Y-23 came from Seishin pharmaceutical Co. Ltd., Tokyo, Japan. Sigma proteases used were Ficin (from fig tree latex), Protease type XIII (from *Aspergillus saitoi*) and Protease type XIV (from *Streptomyces griseus*, "Pronase E").

Results and Discussion

Cellulase Activity: Its Use as a Measure of the Cell-Wall Digesting Capacity of Cellulysin

The manufacturer's specifications for Cellulysin imply that the cellulase activity of each batch is closely related to the ability of this enzyme mixture to digest cell walls. We have determined the cellulase activity (C_1 and C_x, see Cocking 1972) of Cellulysin batches and compared the values obtained with the optimum concentration of each batch for GCP release from *Commelina* epidermal tissue (Table 1). Cellulase activity and guard cell protoplast (GCP) release potential were positively correlated but the relationship was complex. In some cases where cellulase activity was high (ca. 3 units mg^{-1}), low concentrations of Cellulysin gave high yields of GCP's within 4–6 h at 30°C, pH 5.0 (designated "class 1" batches). Of the other batches tested which only had slightly lower cellulase activity (1.2–2.0 units mg^{-1}) some proved to be active in GCP release at high concentrations ("class 2") but others were inactive at up to 4% w/v ("class 3"). Our observations that some enzyme batches exhibited reasonable levels of cellulase activity but were inactive in tissue digestion suggested that these batches were lacking some other enzyme essential for successful GCP isolation.

Figure 1a shows the time course of protoplast release from the epidermal tissue using only an active (Class 1) Cellulysin batch as digestive agent. At the chosen conditions of temperature and pH, yields of GCP's were maximal after 5 h but the viability of the protoplast declined thereafter, reducing the final yield. Cellulysin appears to be toxic to the GCP's only after the cell wall has been digested (see also Morris et al. 1981). Figure 1b shows that GCP release by an inactive (class 3) batch could be improved in a number of ways. Dialysis slightly improved GCP yields from Cellulysin batch 7, a phenomenon not related to any change in its cellulase activity (see below). Even at a lower temperature (25°C) 0.1 mol m^{-3} DTT was effective in increasing GCP yields. In fact, DTT was increasingly effective at higher concentrations, but this was only at the expense of reduced protoplast viability (GCP's became unspherical and/or opaque). The most dramatic effect upon GCP release was brought about by adding low concentrations of Pectolyase Y-23. Since this compound contains essentially only pectinases and a "maceration stimulation factor" (Nagata and Ishii 1979), it seems likely that the absence or inactivation of pectinase enzymes normally present in Cellulysin (Class 1) was the cause of inactivity of Class 3 batches, although we have not yet verified this.

The above information clearly indicates that the cellulase activity of Cellulysin is not always the major factor controlling GCP release from *Commelina* epidermis and low concentrations of Class 3 Cellulysin batches plus 0.005–0.1% Pectolyase Y-23 can thus be used in tissue digestions. Other pectinases (Macerozyme R-10 and Sigma pectinase, both from *Rhizopus* sp.) have been tried by us but even at 4% w/v and

Table 1. The cellulase activity and GCP yields from several batches of Cellulysin and some other cellulases

Enzyme	Cellulase activity		Digestion activity			Class
	Fresh (F) (units mg^{-1})	Dialysed (D) (D/F x 100)	Enzyme concentration (%)	6 (h)	GCP yield (%)	
Cellulysin 1	3.3	93	2 (F)	6	99	
Cellulysin 2	3.1	52	1 (F)	5	86	1
Cellulysin 3	3.1	86	1 (D)	5	85	
Cellulysin 4	2.0	96	4 (D)	5	85	2
Cellulysin 5	1.8	157	4 (F + D)	6	0–5	
Cellulysin 6	1.8	96	4 (F+ D)	6	0–5	3
Cellulysin 7	1.2	84	4 (F + D)	6	0–5	
Onazuka R-10	2.0	–	a	–		
Onazuka R-S	2.5	–	b	–		
Sigma cellulase	0.4	–	–	–		
Boehringer cellulase	0.5	–	–	–		

a In the one batch tested, epidermis digestion rates were similar to class 1 Cellulysin but yields of *viable* GCPs were very low

b Equivalent to Class 3 Cellulysin

– Not determined

30°C, they did not enhance GCP release in conjunction with class 3 Cellulysin. This may be due to batch variation similar to that described here for Cellulysin because Onazuka R-10 and Macerozyme R-10 have been used successfully elsewhere to obtain GCP's (Jewer et al. 1982; Gotow et al. 1982).

Protease Activity: Inhibition of a Potentially Toxic Component of Cellulysin

Cellulysin exhibits a level of azo-albuminase (AAase) activity comparable with a commercial protease (Table 2). The pH sensitivity of Cellulysin (and Onazuka R-10 and RS) proteases suggests that they are serine or alkaline proteases (pH optimum ca. 8.0), typified by Pronase from *Streptomyces griseus* (Fig. 2b). However, Cellulysin still exhibited a greater AAase activity at pH 5 than an acid or carboxyl protease from *Aspergillus saitoi* having an optimum at pH 4 (Fig. 2c). This relationship was substrate specific (Table 2).

Including 0.5% BSA in Cellulysin AAase assays gave levels of inhibition of 50–73%, presumably by acting as a competitive substrate. In Azo-caseinase (ACase) assays Cellulysin exhibited similar levels of protease activity and BSA again brought about some inhibition of this, but to a lesser extent (Table 2). What is the relevance to protoplast isolation of these levels of inhibition of protease activity by BSA? The substrate concentration in a protease assay determines the apparent activity, and the relative proportions of BSA and substrate determine the extent of competitive inhibition. In protoplast isolation experiments, concentrations of exposed plant proteins in the plasmalemma must be very low compared to 0.4% (w/v) in our assays and so 0.5% BSA is likely to have a very strong inhibitory effect upon Cellulysin proteases even though some of them may not be specific for albumin substrates. BSA can thus be expected to protect protoplast membranes from the damage which might otherwise occur from the action of Pronase-type enzymes (Branton 1971; Vreugdenhil et al. 1980).

DTT at low concentrations also alters the protease activity of Cellulysin. At 0.1 and 1.0 mol m^{-3}, DTT inhibited azo-dye release by ca. 10 and 20% respectively (AA ase assays) when using a batch of Cellulysin (Class 3) with an original protease activity of 0.41 "units". (This result was not included in Table 2 because AA batches vary in the degree of azo-dye content: care should be taken in comparative work always to use a single AA or AC batch).

Dialysis: A Partial Purification of Cellulysin

Originally, dialysis was adopted by us in order to standardise experimental conditions when separate batches of Cellulysin were used. In fresh Cellulysin solutions concentrations of salts and sugars were already reasonably low, but dialysis could reduce this even further. In 4% Cellulysin solutions (similar results for 3 batches) the Na^+ and K^+ concentrations were reduced from ca. 7 mol m^{-3} to <1.0 mol m^{-3}. Glucose was present at concentrations less than 4 mol m^{-3} in fresh solutions but undetectable (<0.4 mol m^{-3}) in dialyzed samples. The effect of dialysis on cellulase and protease enzyme activities was also measured. Table 1 shows that the cellulase

Table 2. The protease activity of Cellulysin, Onazuka RS and some commercial proteases: the effects of BSA and dialysis

Enzyme	Protease activity Azo-Albuminase Control	% Inhibition (+ 0.5% BSA)	Dialyzed (% control)	Azo-Caseinase Control	% Inhibition (+ 0.5% BSA)	HPAase
Cellulysin 1[a]	0.66	61	56	0.58	–	0.4[f]
Cellulysin 2	0.65	65	15	0.57	5	–
Cellulysin 3	0.66	59	60	0.62	36	–
Cellulysin 4	0.32	72	38	–	–	–
Cellulysin 5	0.33	59	154	0.35	15	–
Cellulysin 6	0.20	73	73	0.30	–	–
Cellulysin 7	0.34	50	60	0.37	20	–
Aspergillus protease[b]	0.27	55	–	–	–	0.9
Streptomyces protease[cg]	–	–	–	–	–	8.4
Ficin[d]	7.5	63	–	–	–	18.1
Onazuka - RS[e]	–	–	–	–	–	0.6

[a–e] Protein contents (% dry weight): 9, 2, 42, 7 and 16 respectively

[f] 0.9 at pH 8 (see Fig. 2a)

[g] "Pronase"

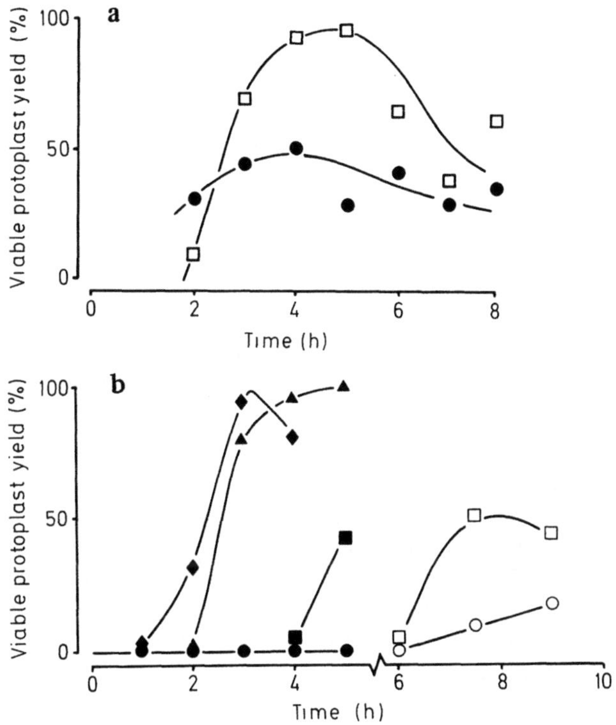

Fig. 1a,b. Protoplast release from *Commelina* epidermal tissue. GCP viability was deduced from overall spherical shape and the possession of good internal organization of cytoplasm and vacuole. Conditions were constant throughout with respect to mannitol (300 mol m^{-3}), BSA (0.5%) and MES (10 mol m^{-3}). pH was adjusted to 5.0 with KOH. **a** 30°C + 2% dialyzed Cellulysin (class 1); □: GCP's, ●: epidermal cell protoplasts. **b** Closed symbols 25°C, 2% fresh Cellulysin (class 3 batch 7); ●control; ■ + 0.1 mol m^{-3} DTT; ▲ + 0.01% Pectolyase Y-23; ◆ + 0.1 mol m^{-3} DTT and 0.01% Pectolyase Y-23. *Open symbols* 30°C, 4% fresh (○) and dialyzed (□) Cellulysin (batch 7)

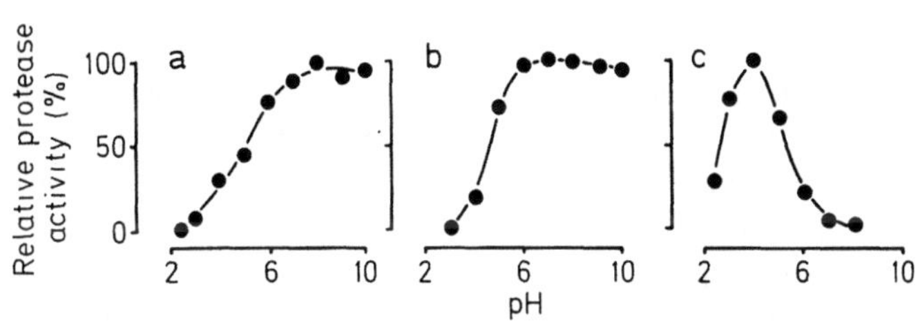

Fig. 2a–c. The pH sensitivity of protease activity in Cellulysin, Pronase and a fungal carboxyl protease. **a** Cellulysin (ex *Trichodermal*); **b** Pronase (ex *Streptomyces*); **c** Protease (ex *Aspergillus*)

activity of Cellulysin was usually unaffected, although both inhibition and stimulation of activity could occur. There was a different pattern for the effect of dialysis on protease (AAase) activity. With the exception of one batch tested, dialysis consistently inhibited AAase activity by between 27–85% of control values (Table 2). The effects of dialysis and BSA on protease activity in Cellulysin were additive (Table 3), giving overall levels of protease inhibition similar to or greater than PMSF and DIFP in caseinase assays (Lin and Wittenbach 1981). Such complex organic halogenated compounds are specific inhibitors of serine proteases (see e.g. Stryer

Table 3. Inhibition of Cellulysin batch 7 protease activity by dialysis and BSA

Cellulysin treatment	AAase activity	Inhibition (%)	ACase activity	Inhibition (%)
Fresh	0.34	–	0.37	–
Dialyzed	0.20	41	0.25	32
Fresh + BSA	0.17	50	0.30	20
Dialyzed + BSA	0.11	68	0.19	49

1983) but PMSF is relatively insoluble in aqueous media and so its use is limited because of the toxicity of the solvents used. DIFP may prove to be a more useful compound for protoplast isolation medium protease inhibition (Lin 1981), but we have shown here that it is possible to reduce the action of Cellulysin proteases without the need to introduce chemicals which, although they have specific protease inhibition properties, have unknown biological activity.

Class 3 type batches of Cellulysin frequently responded to dialysis by a slight improvement in GCP release over extended incubation periods (see Fig. 1b). In conjunction with the use of Pectolyase Y-23 however, or in the case of Class 1 batches, no evidence could be found for such stimulation of rates of tissue digestion. For one particular batch (4, see Table 2), dialysis reduced GCP viability compared with controls and this enzyme was subsequently used fresh only.

Post-Isolation Properties of GCP's

We have already reported that using Class 1 Cellulysin to isolate GCP's gives populations of cells with very high proportional viability, as determined by neutral red staining and oxygen electrode estimates of metabolic activity (Fitzsimons and Weyers 1983). We have also found that certain treatments can lead to a loss of sphericity in a protoplast population. This can be expressed in terms of the total % of non-spherical cells in a given sample. However, neutral red uptake was always observed in over 99% of these protoplasts, even when non-sphericity was as high as 50%. Clearly, neutral red uptake alone is not a definitive guide to the condition of protoplast membranes. A further test is necessary to describe membrane quality in otherwise apparently healthy, spherical protoplasts because the action of proteases may only be manifest in subtle ways (Vreugdenhil et al. 1980).

GCP's isolated with many batches of Cellulysin have shown osmotic characteristics which fit the Boyle van't Hoff relationship with a correlation coefficient $r \geqslant 0.99$ (see Weyers and Fitzsimons, Chap. 18, this Vol.). In all cases, if supra-optimal conditions of tissue digestion were employed, then protoplast yields were reduced (see Fig. 1a) although the remaining population was viable according to overall sphericity and neutral red staining assessments. Nevertheless, the appearance of such protoplasts was different from that of protoplasts isolated when yields were maximal. Usually the cytoplasm became opaque and chloroplasts and vacuolar spaces became indistinct. The eventual state of the protoplasts could not be attributed to a particular component of Cellulysin. Increasing the overall protease activity of digestion media by ad-

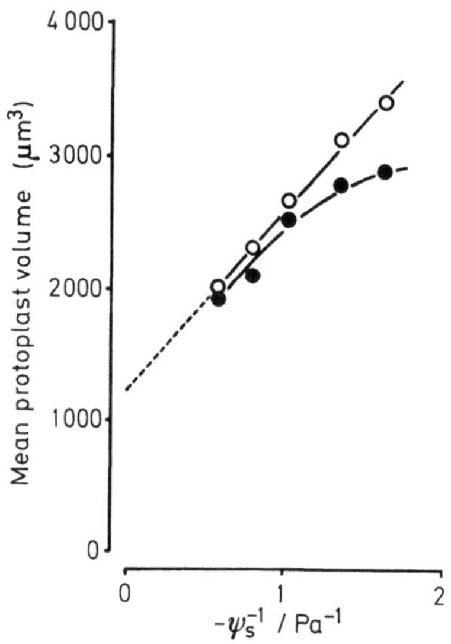

Fig. 3. Boyle van't Hoff analysis of GCPs. Isolation medium: 2% fresh Cellulysin (batch 7) + 0.01% Pectolyase Y-23, 300 mol m^{-3} mannitol, 10 mol m^{-3} MES (pH 5.0, KOH). *Open circles* + 0.5% BSA. *Closed circles* + 2.0% Protease (*ex Aspergillus*). Final levels of AAase activity in each digestion medium were 0.17 and 0.61 respectively (see Table 2)

ding a fungal protease resulted in similar toxic effects. By plotting the data for control and protease-treated protoplast volumes against $-\psi\ s^{-1}$ (Fig. 3), it was clear that the protease-treated cells lost their ability to "fit" the Boyle van't Hoff relationship. This was in part due to their inability to swell beyond a relatively low volume without lysis. These effects can be offset by the addition of BSA (data not shown).

Conclusions

The number of different protoplast isolation media currently in use is probably a reflection of variation in the quality of commercial enzyme products. This is one of the "empirical" elements of protoplast isolation (Evans 1976). The data presented here clarify some aspects of protoplast isolation methods and demonstrate variation in activity between batches of Cellulysin. Many aspects of the effects of protoplast isolation media upon protoplast physiology are not fully understood, but some conclusions can be made.

a) Cellulysin shows a wide range of activity in plant tissue digestion which is not explained by variation in cellulase activity. In our hands inactivity of Cellulysin has been overcome by the addition of a pectinase, Pectolyase Y-23.
b) Protease activity in Cellulysin is comparable with some commercial proteases and its inhibition is desired because these enzymes can effect protoplast membranes. We have shown that treatments of Cellulysin by dialysis and the addition of BSA and DTT can give similar or greater inhibition levels than specific protease inhibitors.

c) Dialysis reduces salt and sugar concentrations in Cellulysin and allows more accurate control of isolation conditions. It may also assist in bringing about protoplast release when Cellulysin is relatively inactive.
d) The "quality" of protoplasts needs to be described in a number of ways to give a composite indication of their relative viability after exposure to the enzymes of tissue digestion. Boyle van't Hoff analysis can be used to assess membrane integrity.

Acknowledgement. We thank the U.K. Science and Engineering Research Council for financial support.

References

Bradford MM (1976) A rapid and sensitive method for the quantitation of microgram quantities of protein utilizing the principle of protein-dye binding. Anal Biochem 72:248–254
Branton D (1971) Freeze-etching studies of membrane structure. Philos Trans R Soc Lond B Biol Sci 261:133–138
Cocking EC (1972) Plant cell protoplasts – isolation and development. Annu Rev Plant Physiol 23:29–50
Dawson RMC, Elliott DC, Elliott WH, Jones KM (1969) Data for biochemical research, 2nd edn. Oxford, London
Evans PK (1976) Higher plant protoplasts: isolation and properties. In: Peberdy et al. (eds) Microbial and plant protoplasts. Academic, New York, pp 51–69
Fitzsimons PJ, Weyers JDB (1983) Separation and purification of protoplast types from *Commelina communis* L leaf epidermis. J Exp Bot 34:55–66
Gotow K, Kondo N, Syōno K (1982) Effect of CO_2 on volume change of guard cell protoplast from *Vicia faba* L. Plant Cell Physiol 23:1063–1070
Gronwald JW, Leonard RT (1982) Isolation and transport properties of protoplasts from cortical cells of corn roots. Plant Physiol (Bethesda) 70:1391–1395
Guy M, Reinhold L, Laties GG (1978) Membrane transport of sugars and amino acids in isolated protoplasts. Plant Physiol (Bethesda) 61:593–596
Jewer PC, Incoll LD, Shaw J (1982) Stomatal responses of Argenteum a mutant of *Pisum sativum* L with readily detachable leaf epidermis. Planta (Berl) 155:146–153
Lin W (1980) Corn root protoplasts. Plant Physiol (Bethesda) 66:550–554
Lin W (1981) Inhibition of anion transport in corn root protoplasts. Plant Physiol (Bethesda) 68:435–438
Lin W, Wittenbach VA (1981) Subcellular localization of proteases in wheat and corn mesophyll protoplasts. Plant Physiol (Bethesda) 67:969–972
Morris P, Thain JF (1980) Comparative studies of leaf tissue and isolated mesophyll protoplasts I. O_2 exchange and CO_2 fixation. J Exp Bot 31:83–95
Morris P, Linstead P, Thain JF (1981) Comparative studies of leaf tissue and isolated protoplasts III. Effects of wall degrading enzymes and osmotic stress. J Exp Bot 32:801–811
Nagata T, Ishii S (1979) A rapid method for isolation of mesophyll protoplasts. Can J Bot 57: 1820–1823
Randall SK, Ruesink AW (1983) Orientation and integrity of plasma membrane vesicles obtained from carrot protoplasts. Plant Physiol (Bethesda) 73:385–391
Stryer L (1981) Biochemistry, 2nd edn. Freeman, San Fransisco
Von Borner (1973) Zur Isolierung von Membranfraktionen aus Protoplasten nach deren präparativer Gewinnung aus Blättern von *Bryophyllum daigremontianum.* Biol Zentralbl 92:583–594
Vreugdenhil D, Harkes PAA, Libbenga KR (1980) Auxin-binding by particulate fractions from tobacco leaf protoplasts. Planta (Berl) 150:9–12

Watts JW, Motoyoshi F, King JM (1974) Problems associated with the production of stable protoplasts of cells of tabacco mesophyll. Ann Bot (Lond) 38:667–671

Weyers JDB, Fitzsimons PJ (1982) The non-osmotic volume of *Commelina* guard cells. Plant Cell Environ 5:417–421

Weyers JDB, Travis AJ (1981) Selection and preparation of leaf epidermis for experiments on stomatal physiology. J Exp Bot 32:837–850

Zeiger E, Hepler PK (1976) Production of guard cell protoplasts from onion and tobacco. Plant Physiol (Bethesda) 58:492–498

Isolation of Maize Protoplasts from the Root Cap and Apex

P.E. PILET, H. HENRY, and C. JOLLÈS

Introduction

Over the last 10 years, several kinds of enzyme treatment for the isolation of maize root protoplasts have been reported. The different methods proposed were generally used for physiological studies on root protoplasts: ion transport (Lin 1980; Lin 1981; Gronwald and Leonard 1982); electrokinetic properties (Pilet 1981; Senn and Pilet 1981, 1982); respiratory and membrane properties (Taylor and Hall 1976; Lin 1982) and herbicide absorption (Darmstadt et al. 1983, 1984). Very few reports on the isolation conditions from several well-delimited root zones have been published (Pilet et al. 1972; Senn and Pilet 1980). The cap, whose importance in root protection is well known, on the penetration of the root through the soil, in graviperception (Wilkins 1976), and in the production and release of different kinds of substances such as growth inhibitors (Pilet 1981) or slime (Paull and Jones 1975), has not yet been used for specific protoplast preparation.

In this report some results on protoplasts obtained from maize cap and apex cells, will be discussed on the capacity of five enzyme digestion media to induce good protoplast release. In addition, two factors influencing protoplast yield have been studied: sorbitol and pectolyase Y-23.

Material and Methods

Preparation of the Plant Material

Caryopses of *Zea mays* L. (cv. LG 11) were germinated in darkness (19 ± 1°C) between moist paper towels on plastic frames as previously described (Pilet 1977). After 48 ± 1 h germination, primary roots (15 ± 3 mm) were selected and divided with a scalpel blade into several parts: cap, apex, elongating and the first part of the differentiating zones (Pilet 1977; Senn and Pilet 1980):

Zones	mm from the tip
Cap	0.0 to 0.5
Apex	0.5 to 1.0
Elongating zone	1.0 to 4.0
Differentiating zone (first part)	4.0 to 6.0

The Physiological Properties of Plant Protoplasts
Edited by P. E. Pilet

Such operations required the use of a macroscope-television system (Pilet et al. 1983) and were done under dim green light (530 ± 20 nm).

Preparation of Protoplasts

For each assay, 40 root segments were treated, just after isolation, with a 3%(w/v) sorbitol solution during 1 h. A cutting in small parts of the elongating (1–4 mm) and differentiating (4–6 mm) zone segments was performed to enhance the penetration of the enzyme mixture.

Enzymatic Isolation of the Protoplasts

Pretreated segments were incubated, after pipetting out the sorbitol solution, with the selectioned enzyme medium (40 segments/2 ml contained in Petri dishes of 4 cm in diameter). Different enzyme media were tested, each one prepared in a buffer solution with 2 mM MES, 1 mM PIPES-Na, 1 mM $CaSO_4$. The final pH was adjusted to 5.7 with NaOH 0.1 N or HCl 0.1 N. With pectolyase Y-23 solutions, BSA (bovine) serum albumine) was also added at a concentration of 1%(w/v).

Incubation was done with gentle agitation on rotating shaker (3 h ± 15 mn, 25°C, 75 r.p.m.) in darkness. The enzyme protoplast mixture was filtered through a nylon sieve (100 μm) and spun down at 75 *g* for 15 mn. Then, the protoplasts pellets were washed twice with a sorbitol solution in the following buffer: 2 mM MES, 1 mM PIPES-Na, 1 mM $CaSO_4$, pH = 5.7; with pectolyase-Y 23 enzyme treatments, 0.1% BSA is further added to the solution. Final protoplast pellets were then suspended in 1 ml of the washing solution. Viable protoplasts were counted by testing their sphericity and their adequate cytoplasm distribution.

Results and Discussion

Effect of Various Enzymes

Several experiments were performed to determine the best mixture for protoplast isolation (Table 1), using a 6%(w/v) sorbitol concentration. Highest yields were obtained when hemicellulase and pectinase are substituted by pectolyase Y-23. Medium E with a high level of pectolyase appears to be necessary for a good protoplast release of the tip cells (cap and apex). For the elongating and differentiating zones, a six times lower concentration of pectolyase induces an efficient release; for higher concentrations, protoplast bursting seems to be responsible for a decrease in the number of protoplasts.

Table 1. Effect of different enzyme mixtures on protoplast release of different maize root zones

Medium	Enzymes[a] (in %, w/v)					Sorbitol (in %, w/v)	Zones (Protoplast number / 10 mg FW)			
	Cine	Hase	Pase	Pol	Pyase		CAP	APEX	1–4 mm	4–6 mm
A	1.0	0.4	0.2	0.0	0.0	6.0	1640	5040	6000	3600
B	2.0	1.0	0.5	0.0	0.0	6.0	410	1940	2320	6500
C	1.0	0.5	0.0	0.5	0.0	6.0	0	2720	2400	3350
D	2.0	0.0	0.0	0.0	0.075	6.0	820	3100	15450	7120
E	2.0	0.0	0.0	0.0	0.5	6.0	3030	5270	10050	2510

FW: Fresh weight

[a] Cine: Cellulysin CALBIOCHEM
Pase: Pectinase SERVA
Pol: Pectionol Fest ROEHM
Pyase: Pectolyase Y-23 SEISHIN

Effect of Sorbitol

It has been already found that the optimal osmotic pressure increases with the differentiation of the cells along the root axis, starting from the elongating and up to the differentiating zone (Pilet and Senn 1980). Osmotic pressure effects on protoplast release limited to the cap and apex parts were analyzed, using the optimal enzyme concentration (Table 1, medium E). Sorbitol was employed for all the experiments as osmoticum in concentrations between 4 and 10% (w/v).

At low sorbitol levels, a significative difference between the two zones was observed (Fig. 1). Cap protoplasts, coming essentially from the peripherical cells of the cap which have a high degree of vacuolisation (Juniper and French 1970; Moore and

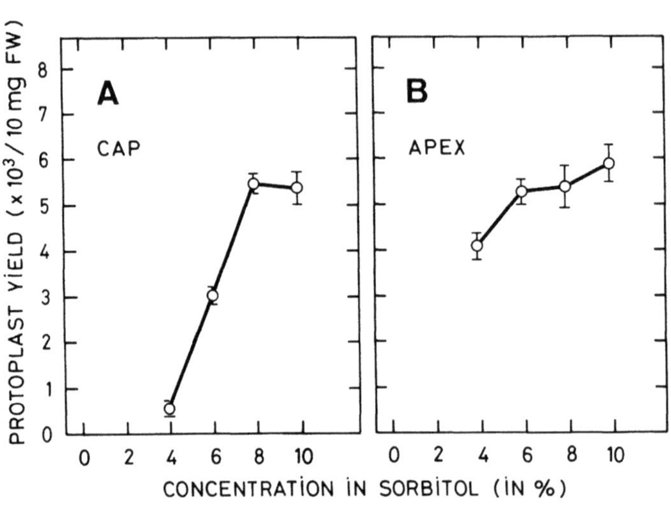

Fig. 1. Effect of sorbitol concentration on protoplast yield (cap and apex)

McClelen 1983a,b), seem to be more sensitive to osmotic effects: an optimal result was reached for a 8% (w/v) sorbitol concentration. Apex protoplasts release does not differ significantly between 6 and 10% (w/v) of sorbitol.

Effect of Pectolyase Y-23

Pectolyase Y-23, a complex of different substances (polygalacturonase, pectin lyase and a factor increasing tissue digestion: Nagata and Ishii 1979; Gronwald and Leonard 1982) was found to be very efficient for the digestion of walls from the root cells. As recommended by Lin (Lin 1981), protective agents have to be added to all pectolyase media for their beneficial effect in the presence of toxic substances in the enzyme preparation or of products released during protoplast isolation (BSA 1% (w/v) in the enzyme medium, 0.1% in the washing medium).

The increase of the pectolyase Y-23 concentration in the medium enhances protoplast yields obtained from the cap and apex, but the stability of the protoplasts seems to be better with a concentration equal to or less than 0.5% (w/v) of pectolyase Y-23 (Fig. 2).

In conclusion, when osmotic conditions and appropriated enzymes have been carefully tested, any kind of cells from roots, including cap cells, may give protoplasts in a number large enough to allow physiological analyses.

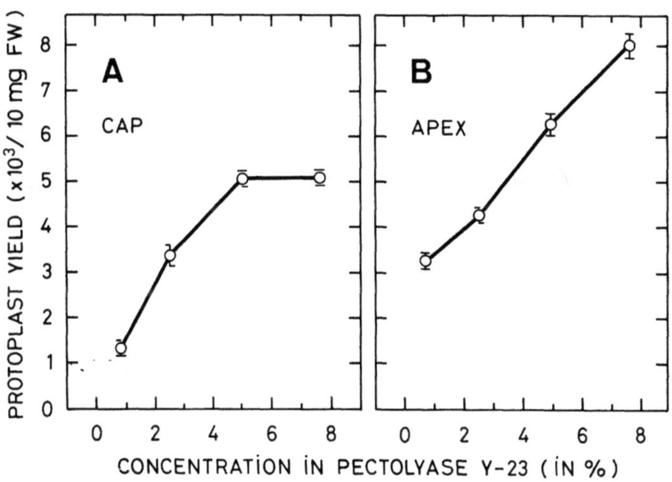

Fig. 2. Effect of pectolyase Y-23 concentration (to be multiplied by 10^{-1}) on protoplasts yield (cap and apex)

References

Darmstadt GL, Balke NE, Schrader LE (1983) Use of corn root protoplasts in herbicide absorption studies. Pectic Biochem Physiol 19:172–183

Darmstadt GL, Balke NE, Price TP (1984) Triazine absorption by excised corn root tissue and isolated corn root protoplasts. Pestic Biochem Physiol 21:10–21

Gronwald JW, Leonard RT (1982) Isolation and transport properties of protoplasts from cortical cells of corn roots. Plant Physiol (Bethesda) 70:1391–1395

Juniper BE, French A (1970) The fine structure of the cells that perceive gravity in the root tip of maize. Planta (Berl) 95:314–329

Lin W (1980) Corn root protoplasts. Isolation and general characterization of ion transport. Plant Physiol (Bethesda) 66:550–554

Lin W (1981) Inhibition of anion transport in corn root protoplasts. Plant Physiol (Bethesda) 68:435–438

Lin W (1982) Responses of corn root protoplasts to exogenous reduced nicotinamide adenine dinucelotide: oxygen consumption, ion uptake and membrane potential. Proc Natl Acad Sci USA 79:3773–3776

Moore R, McClelen CE (1983a) A morphometric analysis of cellular differentiation in the root cap of *Zea mays*. Am J Bot 70:611–617

Moore R, McClelen CE (1983b) Ultrastructural aspects of cellular differentiation in the root cap of *Zea mays*. Can J Bot 61:1566–1572

Nagata T, Ishii S (1979) A rapid method for isolation of mesophyll protoplasts. Can J Bot 57: 1820–1823

Paull RE, Jones RL (1975) Studies on the secretion of maize root cap slime. II. Localization of slime production. Plant Physiol (Bethesda) 56:307–312

Pilet PE (1977) Growth inhibitors in growing and geostimuled maize roots. In: Pilet PE (ed) Proc 9th Int Conf plant growth substances. Springer, Berlin Heidelberg New York, pp 115–128

Pilet PE (1979) Hormonal control of root georeaction, some light effects. In: Skoog F (ed) Plant growth substances. Springer, Berlin Heidelberg New York, pp 450–461

Pilet PE (1981) Root protoplasts: some physiological properties. In: Colloquia Pflanzenphysiologie der Humbold-Univ zu Berlin 4:35–42

Pilet PE, Mayor G, Jaunin F, Nicollier F (1983) Freeze-decapping, a new technique for analysing root growth and gravireaction. Plant Sci Lett 28:121–128

Pilet PE, Prat R, Roland JC (1972) Morphology, RNase and transaminase of root protoplasts. Plant Cell Physiol 13:297–309

Pilet PE, Senn A (1980) Root growth gradients: a critical analysis. Z Pflanzenphysiol 99:121–130

Senn A, Pilet PE (1980) Isolation and some morphological properties of maize root protoplasts. Z Pflanzenphysiol 100:299–310

Senn A, Pilet PE (1981) Electrophoretic mobility, zeta potential and surface charges of maize root protoplasts. Z Pflanzenphysiol 102:19–32

Senn A, Pilet PE (1982) Indolacetic acid and abscisic acid effects on the electrokinetic properties of maize root protoplasts. In: Marme D, Marre E, Hertel R (eds) Plasmalemma and tonoplast, their functions in the plant cell. Elsevier/North Holland Biomedical, Amsterdam, pp 69–76

Taylor ARD, Hall JL (1976) Some physiological properties of protoplasts isolated from maize and tobacco tissues. J Exp Bot 27:383–391

Wilkins MB (1976) Gravity-sensing guidance system in plants. Sci Prog 63:187–217

Plant Protoplast Viability

C.H. BORNMAN and J.F. BORNMAN

Introduction

In its strictest sense a viable protoplast is one capable of regeneration into a plant, a capacity that has been demonstrated for few species and, with a few exceptions, for none in the conifers and monocotyledonous plants. In such cases where whole plant regeneration is potentially possible, there is no single parameter of viability, or collection of parameters, that enable prediction of this potential at the level of the protoplast.

Vital staining of tissues and cells is often used to establish whether or not they are living, and there is extensive literature on vital staining and vital fluorochrome procedures (Drawert 1968; Stadelman and Kinzel 1972), the application and interpretation of which in terms of viability are largely subjective. Apart from exclusion dyes such as Evans blue, methylene blue (Widholm 1972) and trypan blue (Ege et al. 1974), and the sequestration of neutral red into the vacuoles of undamaged cells, two other commonly used dyes to assay cell and protoplast viability are both fluorochromes. The optical brightner calcofluor white was initially used to stain cell walls of algae and fungi (Darken 1961), but later also of cell-wall-regenerating protoplasts (Nagata and Takabe 1970; Herth and Meyer 1977; Hahne et al. 1983). The other fluorochrome, fluorescein diacetate, first shown in 1966 (Rotman and Papermaster 1966) to induce fluorescence in animal cells was subsequently adapted to plant cells (Heslop-Harrison and Heslop-Harrison 1970; Widholm 1972) and protoplasts (Larkin 1976). Estimates of viability, based on the cleavage by esterase activity of fluorescein diacetate to fluorescein, can vary greatly depending on seasonal variation in the plant material, isolation procedures, concentration of stain, duration of staining, choice of fluorescence optics, and bias.

In addition to fluorescein diacetate, a number of other fluorochromes can be used to probe the presence or absence of nuclei, the initiation of the cell cycle, mechanisms of control of mitosis, or to correlate the onset of DNA synthesis with that of protein and cell-wall synthesis (Galbraith and Mauch 1980; Galbraith et al. 1981; Vatsya and Bhaskaran 1981; Bilkey et al. 1982; Bornman and Bornman 1983).

Post-isolation parameters of photosynthetic, respiratory and other physiological activities can provide useful information on the functional status of a protoplast population. Cassells et al. (1980) used the release of ethylene during isolation as a means of predicting the ability of a protoplast population to survive the isolation procedure.

The Physiological Properties of Plant Protoplasts
Edited by P. E. Pilet

In this study we draw attention to some potential indicators of viability and present the argument that it is a compartmentalized phenomenon and one that is difficult to assess objectively.

Materials and Methods

Protoplasts

Protoplasts of winter oilseed rape (*Brassica napus* L., ssp. *oleifera*) were isolated from leaves of 2- to 3-week-old seedlings, axenically-grown under a 16 h photoperiod with a photon fluence rate of 70 μmol m^{-2} s^{-1} (L 40W/77 Fluora, Osram). The isolation procedure was based on that of Shepard (1981). Isolated protoplasts were either held in their rinse or culture medium. The rinse medium consisted of one-quarter strength Murashige and Skoog medium (MSM, 1962) with an osmolality of 385 mOsm kg H_2O^{-1}, and the culture medium of MSM (full-strength macronutrient elements and half-strength iron and micronutrient elements, osmolality 550 mOsm kg H_2O^{-1}) plus 4.5 μM 2,4-dichlorophenoxyacetic acid and 1.8 μM N^6-benzyladenine. For plant regeneration, the protoplasts were plated at 3 x 10^4 ml^{-1} in culture medium containing 0.25% agarose (type VII, Sigma).

Fluorescence Staining of Protoplasts

Acridine orange (AO, Paolillo 1964), 4'6-diamino-2-phenylindole (DAPI, James and Jope 1978), ethidium bromide (EB, McCully 1976) and fluorescein isothiocyanate (FITC, Galbraith and Mauch 1980) were added to the enzyme incubation medium at concentrations of 100, 20, 100 and 12.5 μg ml^{-1}, respectively. Incubation time was from 3 to 4 h. Fluorescein diacetate (FDA, Widholm 1972), from an acetone stock solution, was diluted to 20 μg ml^{-1} (a concentration at which it turns cloudly) with either rinse or culture medium and mixed with the protoplast suspension in a 1:1 (v/v) ratio. Calcofluor white (CF, Hahne et al. 1983) was added to the culture medium at a concentration of 10 μg ml^{-1}. All fluorochromes except calcofluor white (American Cyanamid) were purchased from Sigma.

Secondary fluorescence was observed with a Nikon inverted microscope (Diaphot-TMD) fitted for epifluorescence with filter cassettes incorporating the appropriate combination of dichroic mirror, exciter and barrier filters according to the excitation method of choice: AO, blue; CF, violet; DAPI, violet; EB, violet; FDA, blue-violet; and FITC, blue. All fluorescence observations except the growing cultures were made with the aid of 50 mm rectangular capillaries (microslides, Vitro Dynamics, NJ) with an internal pathlength of 300 μm.

Measurement of Photosynthetic Fluorescence

Variable fluorescence of dark-adapted protoplasts was measured at 0, 3, 12 and 16 h after isolation in a modified Aminco-Bowman spectrophotofluorimeter and calculated according to $F_v = F_{max} - F_O$, where F_v = variable, F_{max} = maximum and F_O = steady state or initial fluorescence (Bornman et al. 1985). The rise time taken for the system to reach $F_O + 0.75\ F_v$ was plotted to guage possible damage to the photosynthetic system with age.

Results and Discussion

Fluorochrome Fluorescence

The main observations are summarized in Table 1 and Fig. 1. Protoplasts were effectively labelled by addition of the fluorochromes AO, DAPI, EB and FITC to the enzyme-containing incubation medium. Since they had been purified by flotation, there was little background fluorescence as a consequence of leaky or damaged protoplasts. Protoplasts stained with FITC displayed the characteristic apple green fluorescence and could be immobilized in an agarose-containing culture medium incorporating CF for monitoring of cell wall resynthesis. Wall reformation appeared to have already commenced after 3 h and was unmistakenly evident after 16 h. Colony and callus formation, organogenesis and also plant regeneration was successfully established, confirming a similar demonstration by Galbraith and Mauch (1980) for FITC and Hahne et al. (1983) for FC and one by Hughes and McCully (1975) that plants can be grown in CF-containing medium. Whereas cell-wall resynthesis and its visualization by CF could be taken as an indication of viability, FITC fluorescence could not, because protoplasts which appeared non-viable due to lack of cyclosis and negative response to CF, still fluoresced.

Earlier unpublished observations on protoplasts isolated from *Brassica napus* and *B. oleracea* had shown that as much as 10 and 25% of the respective populations consisted of enucleated cytoplasts (see also Vatsya and Bhaskaran (1981). Acridine orange and EB confirmed this for *B. napus.* Differentiation of nuclei by EB is particularly clear and the nuclei of non-viable (dead) protoplasts do not appear to stain. Although EB was regarded as not very toxic by O'Brien and McCully (1981), stained protoplasts could not be kept alive (EB is an inhibitor of RNA polymerase). Staining with the metachromatic AO was found difficult to evaluate. Pearse (1972) has cautioned against the common interpretation that living and dead cells give green and red fluorescence respectively. As an example of the ambiguity of response to treatment with AO, Bilkey et al. (1982) detected nuclei by their "bright green fluorescence", whereas Vatsya and Bhaskaran (1981) detected them by their "characteristic orange-green fluorescence." In our experiments protoplasts stained with AO did not survive in culture.

Protoplasts stained with DAPI also did not survive, but this DNA-specific fluorescence probe has value in identifying and distinguishing nucleoplasts and cytoplasts (Fig. 1), as well as providing a method for quantitation of DNA. In contrast to FITC, a non-specific fluorochrome with an intense level of fluorescence, FDA can be used as an indicator of viability, since its hydrolysis to fluorescein does not take place in dead cells. However, the intensity of the fluorescence should be judged carefully (Bornman et al. 1982), particularly in the presence of the red autofluorescence of

Table 1. Observations on response of *Brassica napus* (oilseed rape) protoplasts to staining with various fluorochromes

Fluorochrome	Response
1. Concentration, $\mu g\ ml^{-1}$ 2. Excitation method 3. Staining application	
Acridine orange (AO) 1. 100 2. Blue 3. Pre-isolation	Response affected by pH and concentration of dye: fluorescence metachromasia (living protoplasts: green; dead protoplasts: red) is unreliable as indicator of viability; can discriminate between protoplasts and cytoplasts
Calcofluor white (CF) 1. 10 2. Violet 3. Post-isolation	Indicator of cell wall resynthesis; at low concentrations non-toxic; micro-callus colonies can be established in agarified culture media containing the stain; plant regeneration not affected; useful in combination with FITC
4′6-Diamidino-2-phenylindole (DAPI) 1. 20 2. Violet 3. Pre-isolation	DNA-specific probe that clearly visualizes nuclei and chloroplasts: useful in distinguishing presence or absence of nuclei and chloroplasts
Ethidium bromide (EB) 1. 100 2. Violet 3. Pre-isolation	Differentiation of nuclei very clear; fluorescence is yellowish-orange, but in ageing protoplasts shifts to bright orange (increase in phenolics?); excellent stain for distinguishing non-nucleus-containing subprotoplasts from protoplasts
Fluorescein diacetate (FDA) 1. 10 2. Blue-violet 3. Post-isolation	Degree of esterase activity reflected in degree (intensity) of fluorescein accumulated; intensity of fluorescence used as an indicator of degree of viability; careful interpretation required; concentration and length of staining affect response
Fluorescein isothiocyanate (FITC) 1. 12.5 2. Blue 3. Pre-isolation	Non-specific, but very useful visual fluorescence probe; cannot be used to indicate functional status of protoplasts, but excellent as a marker dye and non-toxic at low concentrations; plant regeneration possible; useful in combination with CF

chlorophyll excited by near ultraviolet and blue light. We have so far not found protoplasts to survive in a culture medium containing FDA.

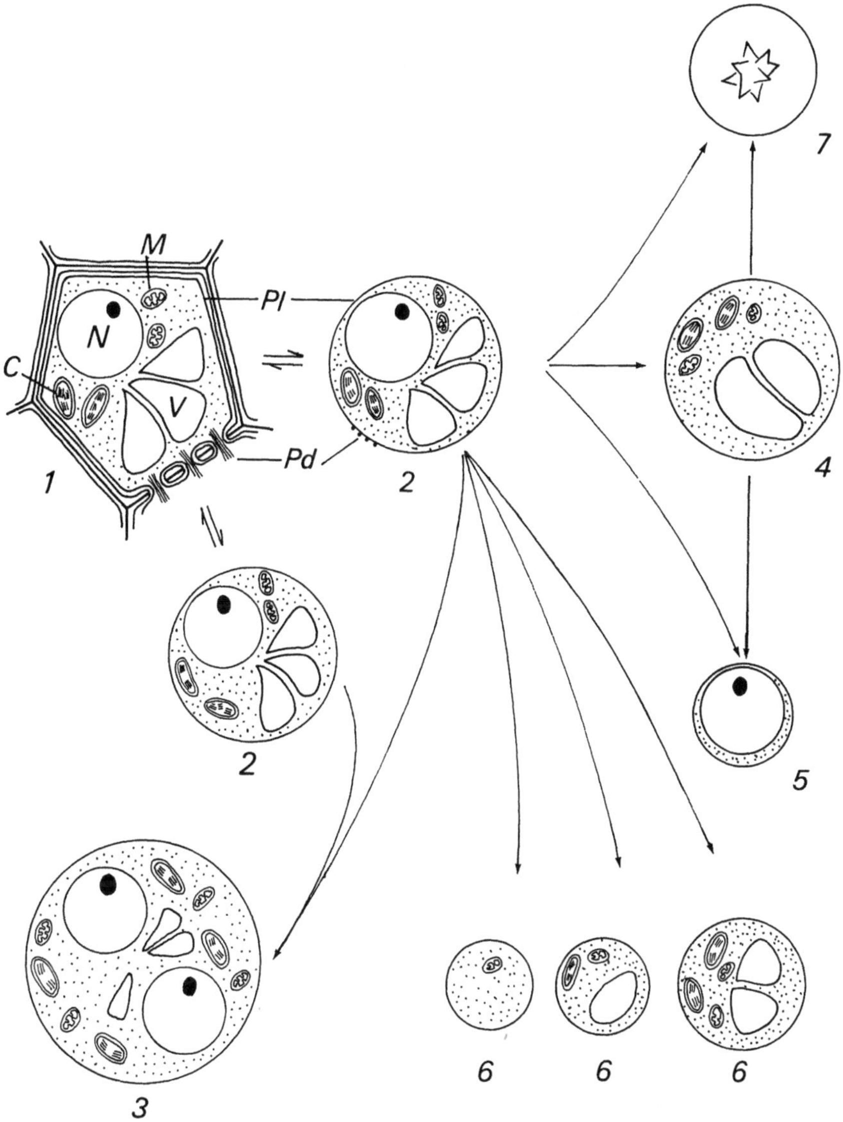

Fig. 1. Diagrammatic representation of a cell (*1*) and the family of protoplasts (*2–7*) obtainable from cells by either enzymatic/osmotic or mechanical manipulation. With the exception of the vacuoplast (*7*), the cell and its derivative proto- and subprotoplasts (*4–6*) all will give positive tests for viability with fluorescein diacetate (FDA), even though cytoplasts (*4*) and microplasts (*6*) are not capable of mitosis and cytokinesis. Cyto- and microplasts can proliferate weakly only by means of limited budding. Fluorochromes which are DNA-specific for example 4′6-diamidino-2-phenylindole (DAPI), may be used to distinguish protoplasts (*2*), multiplasts (*3*) and nucleoplasts (*5*) from cytoplasts (*4*), microplasts (*6*) and vacuoplasts (*7*). For further details see inter alia the following: multiplasts, through spontaneous fusion – Miller et al. (1971); cytoplasts – Bradley (1978), Wallin et al. (1978), Bracha and Sher (1981), Lörz et al. (1981); microplasts – Bilkey et al. (1982); vacuoplasts – Schnabl et al. (1978), Schmidt and Poole (1980); nucleoplasts – Bradley (1978), Wallin et al. (1978), Lörz et al. (1981). *C* chloroplast; *M* mitochondrion; *N* nucleus; *Pd* plasmodesmata; *Pl* plasmalemma

Photosynthetic Fluorescence

As far as photosynthetic fluorescence was concerned (Fig. 2), the time taken for the fluorescence rise (the time taken for this rise reflects the rate of electron transport through photosystem II) did not change much over a 12-h period and had only increased by ca. 17% relative to the control 16 h after isolation. However, although these results may reflect a functional photosynthetic system, they cannot on their own be used to assess total protoplast viability, since the stresses to which the protoplasts are subjected during isolation are likely to cause changes in the level of substrates and products between organelles and the cytoplasm. As shown in Fig. 2, viability as estimated with FDA over the same 16-h period showed virtually no decrease (9%). Prompt fluorescence anf fluorescein fluorescence cannot be directly compared, as the former reflects the photosynthetic ability of the chloroplast and the latter the activity of esterases. These two methods illustrate the differential viability of the chloroplast and the intact protoplast, and emphasize the relativity of the concept.

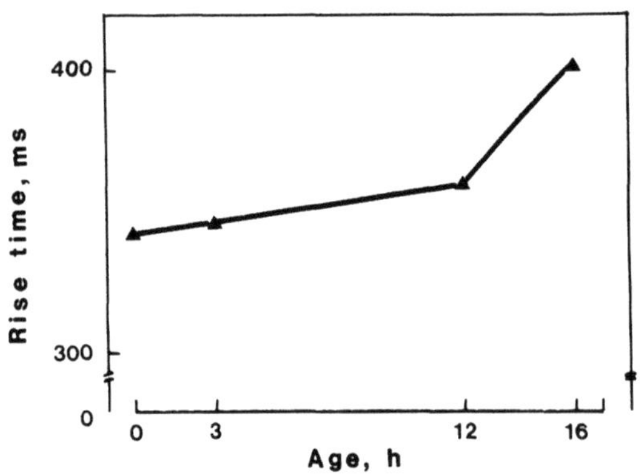

Fig. 2. Viability of *Brassica napus* protoplasts as a function of age, as determined by fluorescence rise time (representing photosynthetic electron transport rate). Assessment of viability (%) by FDA was carried out simultaneously: 95 ± 3, 93 ± 2, 87 ± 7 and 86 ± 5 after 0, 3, 12 and 16 h

Burgess (1978) has questioned the assumption that a protoplast behaves as a cell without a wall, stressing in addition to cell-wall removal, also the traumatic effects of the prior events of dehydration and shearing of the plasmodesmata. These latter effects are not immediately reversed and must influence photosynthetic rates and levels of protein synthesis. Burgess emphasizes that freshly isolated protoplasts represent a stressed system. It is on these highly stressed protoplasts that viability tests are performed.

Many methods have been used to estimate viability and a few examples are the following: cyclosis, usually obvious only after a period of ageing; vacuolar accumulation of neutral red, although active uptake through the tonoplast is questionable (see O'Brien and McCully 1981); esterase hydrolysis of FDA (Larkin 1976); delayed light emission (Bornman et al. 1982); prompt fluorescence (Bornman and Bornman 1983); photosynthetic and respiratory activities; use of exclusion dyes; cell-wall regeneration; plating efficiency; etc., and in the final analysis, whole plant regeneration.

The leaf is the plant organ most used for the isolation of protoplasts, but probably the least suited for this purpose since its tissues are the composites of many different cell types at various levels of differentiation, development, specialization and often ploidy. This is reflected in the variable capacity to regenerate plants from protoplasts of leaves among different species as well as among tissues (of leaf, hypocotyl, shoot tip) within the same species. Magnien et al. (1982) have demonstrated that the cytological heterogeneity of cells of leaf tissues is reflected in the protoplast population in culture. The capacity to regenerate plants from mesophyll protoplasts of oilseed rape and the present incapacity in the case of sugar beet from leaves of similar age is probably a reflection of this variable cytological heterogeneity (cells of sugar beet leaf callus display a high degree of chromosomal aberration). However, based on esterase activity, freshly-isolated protoplasts of both species will be judged as highly viable.

In conclusion, three questions can be asked (1) How valid is the use of protoplast systems for physiological studies? (2) What is viability? (3) How can it best be assessed? The answer to the first is that it will depend on the investigator's objectives. The answer to the second is that viability is a compartmentalized phenomenon, reflecting many different degrees and continuous shifts in physiological activities; as a concept it is diffuse. Finally, with regard to the assessment of viability there is as yet no single absolute parameter that will allow prediction of the potential for survival. This means that in toto viability is best judged retrospectively, that is after whole plant regeneration.

For practical purposes it may therefore be necessary to use more than one test of viability in order to better gauge the functional potential of the system under study.

References

Bilkey PC, Davey MR, Cocking CC (1982) Isolation, origin and properties of enucleate plant microplasts. Protoplasma 110:147–151

Bornman JF, Bornman CH(1983) Protoplast viability – a relative concept. In: Potrykus I et al. (eds). Protoplasts 1983, Poster Proc 6th Int Protoplast Symp Basel. Exper Suppl (Basel) 45:264–265. ISBN 3-7643-1513-X

Bornman JF, Bornman CH, Björn LO (1982) Effects of ultraviolet radiation on viability of isolated *Beta vulgaris* and *Hordeum vulgare* protoplasts. Z Pflanzenphysiol 105:297–306

Bornman JF, Björn LO, Åkerlund H-E (1985) Action spectrum for inhibition by ultraviolet radiation of photosystem II activity in spinach thylakoids. Photobiophys Photobiochem (in press)

Bracha M, Sher N (1981) Fusion of enucleated protoplasts with nucleated mini protoplasts in onion (*Allium cepa* L). Plant Sci Lett 23:95–101

Bradley PM (1978) Production of enucleated plant protoplasts of *Allium cepa*. Plant Sci Lett 13:287–290

Burgess J (1978) Plant cells without walls? Nature (Lond) 275:588–589

Cassells AC, Cocker FM, Austin S (1980) Ethylene release during tobacco protoplast isolation and subsequent protoplast survival. Plant Sci Lett 19:169–173

Darken MA (1961) Natural and induced fluorescence in microscopic organisms. Appl Microbiol 9:354–360

Drawert H (1968) Vitalfärbung und Vitalfluorochromierung pflanzlicher Zellen und Gewebe. Protoplasmatologia II D3. Springer, Berlin Heidelberg New York

Ege T, Hambert H, Krondahl V, Ericsson J, Ringertz NR (1974) Characterization of minicells (nuclei) obtained by cytochalasin enucleation. Exp Cell Res 365–377

Galbraith DW, Mauch TJ (1980) Identification of fusion of plant protoplasts II. Conditions for the reproducible labelling of protoplasts. Z Pflanzenphysiol 98:129–140

Galbraith DW, Mauch TJ, Shields BA (1981) Analysis of the initial stages of plant protoplast development using 33258 Hoechst: reactivation of the cell cycle. Physiol Plant 51:380–386

Hahne G, Herth W, Hoffmann F (1983) Wall formation and cell division in fluorescence-labelled plant protoplasts. Protoplasma 115:217–221

Herth W, Meyer Y (1977) Ultrastructural and chemical analysis of the wall fibrils synthesized by tobacco mesophyll protoplasts. Biol Cell 30:33–40

Heslop-Harrison J, Heslop-Harrison Y (1970) Evaluation of pollen viability by enzymatically induced fluorescence: intracellular hydrolysis of fluorescein diacetate. Stain Technol 47: 115–120

Hughes J, McCully ME (1975) The use of an optical brightner in the study of plant structure. Stain Technol 50:319–329

James TW, Jope C (1978) Visualization by fluorescence of chloroplast DNA in higher plants by means of the DNA-specific probe 4'6-diamidino-2-phenylindole. J Cell Biol 79:623–630

Larkin PJ (1976) Purification and viability determinations of plant protoplasts. Planta (Berl) 128:213–216

Lörz H, Paszkowski J, Dierks-Ventting C, Potrykus I (1981) Isolation and characterization of cytoplasts and miniprotoplasts derived from protoplasts of cultured cells. Physiol Plant 53:385–391

Magnien E, Dalschaert X, Faraoni-Sciamanna P (1982) Transmission of a cytological heterogeneity from the leaf to the protoplasts in culture. Plant Sci Lett 25:291–303

McCully ME (1976) The use of fluorescence microscopy in the study of fresh plant tissues. Proc Microsc Soc Can 3:172–173

Miller RA, Gamborg OL, Keller WA, Kao KN (1971) Fusion and division of nuclei in multinucleated soybean protoplasts. Can J Genet Cytol 13:347–353

Murashige T, Skoog F (1962) A revised medium for rapid growth and bioassays with tobacco tissue cultures. Physiol Plant 15:473–497

Nagata T, Takabe I (1970) Cell wall regeneration and cell division in isolated tobacco mesophyll protoplasts. Planta (Berl) 92:301–308

O'Brien TP, McCully ME (1981) The study of plant structure: principles and methods. Termacarphi, Melbourne

Paolillo DJ Jr (1964) Acridine orange fluorescence in shoot tips of *Ephedra.* Acta Histochem 18:276–282

Pearse AGE (1972) Histochemistry. Theoretical and applied, 3rd edn, vol 2. Whitefriars, London

Rotman B, Papermaster BW (1966) Membrane properties of living mammalian cells as studied by enzymatic hydrolysis of fluorogenic esters. Proc Natl Acad Sci USA 55:134–141

Schmidt R, Poole RJ (1980) Isolation of protoplasts and vacuoles from storage tissue of red beet. Plant Physiol (Bethesda) 66:25–28

Schnabl H, Bornman CH, Ziegler H (1978) Studies on isolated starch-containing (*Vicia faba*) and starch-deficient (*Allium cepa*) guard cell protoplasts. Planta (Berl) 143:33–39

Shepard JF (1981) The isolation, culture, and regeneration of plant protoplasts. Course on plant molecular biology. Cold Spring Harbor Laboratory, USA

Stadelman EJ, Kinzel H (1972) Vital staining of plant cells. In: Prescott DM (ed) Methods in cell physiology, vol 5. Academic, New York

Vatsya B, Bhaskaran S (1981) Production of subprotoplasts in *Brassica oleracea* var. Capitata – a function of osmolarity of the media. Plant Sci Lett 23:277–282

Wallin A, Glimelius K, Eriksson T (1978) Enucleation of plant protoplasts by cytochalasin B. Z Pflanzenphysiol 87:333–340

Widholm JM (1972) The use of fluorescein diacetate and phenosafranine for determining viability of cultured plant cells. Stain Technol 47:189–194

Isolation of Plasma Membrane from Ryegrass (Lolium multiflorum) Endosperm Protoplasts

A. SCHIBECI

Introduction

Biochemical studies on the structure an function of plant plasma membranes have been hampered because satisfactory methods of their isolation have not been available. A major reason is that the high shear forces required to disrupt plant cell walls cause considerable membrane fragmentation and damage to cell organelles. Isolation of plasma membranes from the resulting homogenate was rendered particularly difficult by the lack of unequivocal markers for their identification (Leonard and Hodges 1980) and because they comprise only a small proportion of the total cellular membranes (Morré and Buckhout 1979).

Methods for the isolation of plant plasma membranes have relied on putative chemical and enzymic markers for their identification. These markers include K^+-stimulated ATPase (Hodges and Leonard 1974; Leonard and Vanderwoude 1976) glucan synthase measured with high UDPglucose concentrations (Klaus and Kindl 1979) cellulase (Koehler et al. 1976; Pierce and Hendrix 1979) and staining with phosphotungstic acid/chromic acid (Nagahashi et al. 1978). The validitiy and usefulness of these markers have been reviewed critically by Quail (1979).

The problems imposed by the cell wall can be avoided by using protoplasts as a source of plasma membranes. Unlike intact plant tissues, protoplasts can be disrupted by gentle techniques. The exposed plasma membrane of protoplasts can also be labelled with radioactive probes which provide unequivocal markers during subsequent fractionation procedures (Quail 1979; Hall 1983).

Scarborough (1975) has described a method for the isolation of plasma membranes from cell-wall-less mutants of *Neurospora crassa*. The plasma membrane of intact cells were first stabilized by coating with concanvalin A. After lysis, the plasma membranes were in the form of large sheets which could be sedimented at low centrifugal forces leaving intracellular vesicles in suspension. The method has also been applied successfully to the isolation of plasma membranes from yeast (Dúran et al. 1975). Recently, Jacobson and co-workers introduced the use of cationic silica microbeads to stabilize and coat uniformily the cell surfaces of *Dictyostelium discoideum* (Chaney and Jacobson 1983) and yeast (Schmidt et al. 1983).

In the study described here, plasma membranes were isolated from protoplasts of suspension-cultured ryegrass endosperm cells after radiolabelling and stabilization of the surfaces of intact protoplasts, prior to lysis.

The Physiological Properties of Plant Protoplasts
Edited by P. E. Pilet

Methods

Protoplasts were obtained by treating suspension-cultured ryegrass (*Lolium multiflorum* Lam.) with Driselase (Kyowa Hakka Kogyo Co. Japan) as described previously (Keller and Stone 1978) and fractionated by centrifugation in 0.5M sucrose (10 min, 400 *g*). Protoplasts which sedimented ("heavy") were discarded to avoid contamination of the plasma membrane preparation by unbroken protoplasts. The protoplasts which floated ("light") were collected and used for plasma membrane isolation.

The method for the isolation of plasma membrane is summarized in Fig. 1. The exposed plasma membranes were radiolabelled using a variety of probes (Table 1). Labelled protoplasts were coated with murine myeloma protein J539 (Schibeci et al. 1982) or cationized silica microbeads (Chaney and Jacobson 1983; Schmidt et al. 1983) and disrupted either by osmotic shock (Wagner and Siegelman 1975) or polylysine-induced lysis (Dürr et al. 1975). The resulting lysate was fractionated on a discontinuous sucrose-sorbitol gradient (Schibeci et al. 1982) (Fig. 1).

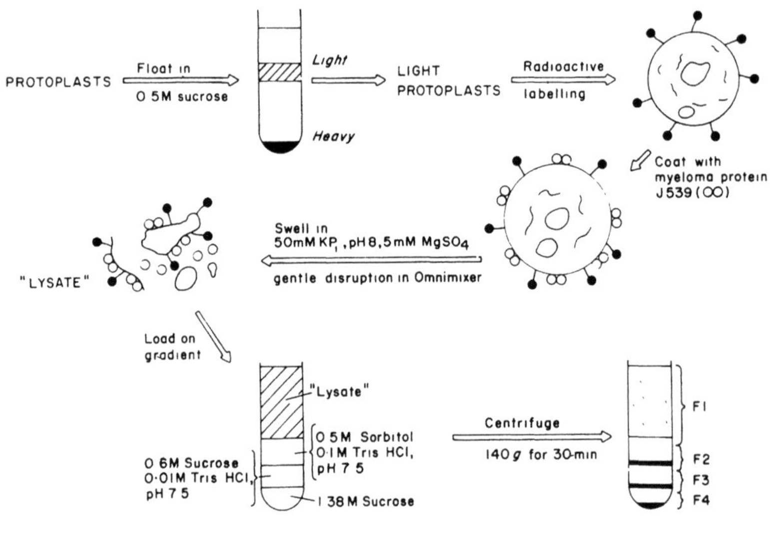

Fig. 1. Procedure for the isolation of plasma membrane from plant protoplasts. The scheme is essentially the same when silica microbeads are used to coat protoplasts, with the exception that $MgSO_4$ is replaced by 1 mM EGTA and the gradient modified as follows: 60% (w/w)sucrose/40% (w/w)sucrose/20%(w/w)sorbitol, 10 mM Tris-HCl, pH 7.5

Results and Discussion

Stabilization of the Plasma Membrane

Protoplasts isolated from ryegrass (*Lolium multiform*) endosperm cells grown in suspension cultures retain their physiological properties including membrane potential (Henry et al. 1984), sugar transport (Tousimis et al. 1985) and the capacity to regenerate cell wall (unpublished observations).

Table 1. Methods used to label the plasma membrane of intact protoplasts

Labelling reagent/ligand	Target groups	Label permeates p.m.	Labelling of i.m.	Reference
^{125}I and Iodgen	Tyrosine, histidine	Yes	No[a]	Fraker and Speck 1978
^{125}I and lactoperoxidase/ glucose oxidase	Tyrosine, histidine	Yes	No[a]	Suzuki et al. 1980
Galactose oxidase and [^{3}H]-borohydride	Terminal galactose N-acetyl galactosamine	Yes	No[a]	Gahmberg and Hakamori 1973
[^{35}S]-sulphanilic acid	Amino groups, tyrosine histidine	Yes	No[a]	Gailbraith and Northcote 1977
[^{125}I]-iodosulphanilic acid	Amino groups, tyrosine histidine	Yes	No[a]	Perlin and Spanswick 1980
[^{125}I]-myeloma protein J539	1,6-β-D-galactopyranosyl	No	No	Schibeci et al. 1982
Fluorescein-labelled axinellin	1,6-β-D-galactopyranosyl	No	No	Schibeci et at. 1982
Fluorescein-labelled myeloma protein J539	1,6-β-D-galactopyranosyl	No	No	Schibeci et at. 1982
Lanthanum	Acidic groups (?)	No	No	Taylor and Hall 1980

[a]Approximately 3% of the label associated with fraction Fl (see Fig. 1) is eluted with the excluded volume (V_0) when gel filtration is carried with Bio Gel A5m

[b]Enzymobeads (Bio-Rad Laboratories Technical Bulletin 1071)
p.m. = plasma membranes, i.m. = internal membranes

In contrast to fungal (Scarborough 1975), yeast (Dúran et al. 1975) and carrot (Boss and Ruesink 1979), ryegrass protoplasts were not agglutinated by concanavalin A. However, they were agglutinated by galactose-binding proteins: murine myeloma protein J539 (Glaudemans 1975) and axinellin obtained from the marine sponge *Axinella polyploides* (Bretting and Kabat 1976).

The observation that galactose-binding proteins agglutinate protoplasts from several other plant species (Larkin 1977) suggests that galactose-containing molecules are common components of plant plasma membranes. Galactose-binding proteins, therefore, may be generally useful for stabilization of plant plasma membranes.

The plasma membrane could also be stabilized by using cationized silica microbeads (Chaney and Jacobson 1983; Schmidt et al. 1983). As observed by Jacobson and co-workers, the protoplast suspension must be free of soluble proteins which are adsorbed onto and cause aggregation of the silica microbeads. Cytosolic proteins released from damaged protoplasts may thus contaminate the plasma membrane fraction. Another observation is that divalent cations cause aggregation of the plasma membrane on lysis of protoplasts. Contaminating cell components are difficult to remove from the aggregated membranes by washing. This membrane aggregation can be reduced by the inclusion of 1 mM EGTA in the lysis buffer. Similar observations were made by Schmidt et al. (1983).

Surface-Labelling of Protoplasts

To enable the plasma membrane to be identified during the fractionation of cell lysates, protoplasts were surface-labelled before lysis. Ideally, only the plasma membrane molecules and no other molecules should be labelled (Carraway 1975, Quail 1979).

The various membrane-specific ligands and labels that were screened for use in the identification of plasma membrane in this study are listed in Table 1. Radioiodination in the presence of the chemical catalyst, Iodogen (Pierce Chemical Co.) and labelling with [^{125}I]-myeloma protein J539 were used routinely. However, with cells containing high levels of endogenous peroxidase activity the use of Na ^{125}I for specific labelling the plasma membrane may be effectively precluded (Gailbraith and Northcote 1977).

In a combined histochemical and fractionation study, Taylor and Hall (1980) showed that when tobacco protoplasts were incubated with $LaCl_3$ or $La(NO_3)_3$, lanthanum, which is electron-dense, binds irreversibly to the surface of protoplasts with no detectable penetration of the plasma membrane and binding to internal organelles. Similar results were obtained with the ryegrass protoplasts in this study.

Disruption of Protoplasts and Fractionation of Lysates

Polylysine-induced lysis was used by Dürr et al. (1975) to lyse yeast protoplasts. However, complete lysis of ryegrass protoplasts could not be obtained using this method. A similar observation was made by Buser and Matile (1977) with protoplasts from *Bryophyllum* leaves. This method showed the highest preservation of intracellular organelles of ryegrass protoplasts as indicated by the survival of vacuoles (up to 40%).

Osmotic swelling of the protoplasts in 50 mM K phosphate, pH 8.0, containing 5 mM $MgSO_4$ (Wagner and Siegelman 1975), followed by blending in a Sorvall Ominmixer for 2 min at low speed, provided a controlled method of disruption. This method was therefore used routinely.

The protoplast lysate containing the labelled plasma membranes coated with myeloma protein J539 was fractionated on a discontinuous sucrose-sorbitol gradient (Fig. 1). As shown in Fig. 2, most of the label is recovered in fraction F3. Similar results are obtained when silica microbeads are used to coat the plasma membrane and most of the label is recovered in fraction FS4 in the modified sucrose gradient (Table 2). Examination of thin sections of fraction F3 by electron microscopy showed the presence of expanses of "sheet-like" membranes.

When snap-frozen in liquid nitrogen and stored at -70°C isolated plasma membranes remained stable for several months. Enzymic activities, measured after stor-

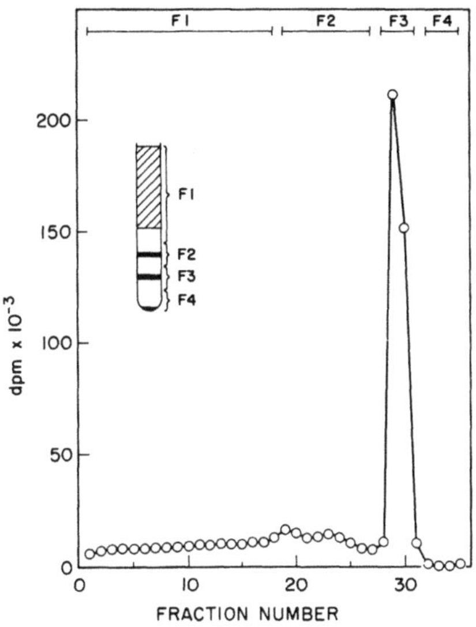

Fig. 2. Distribution of radioactivity after centrifugation of a lysate obtained from protoplasts labelled with $[^{125}I]$-myeloma protein J539. The lysate was fractionated on a discontinuous sucrose-sorbitol gradient as described in Fig. 1

Table 2. Distribution of radioactivity after centrifugation of a lysate obtained from protoplasts labelled with $[^{125}I]$-myeloma protein J539 and coated with silica microbeads. The lysate was fractionated on a discontinuous sucrose-sorbitol gradient as described in legend to Fig. 1. (Data from Briggs et al. 1985)

Fraction	Sucrose/sorbitol concentration	Percent radioactivity
FS1	Top zone	5.6
FS2	20% / 40% interphase	5.6
FS3	40% / 60% interphase	6.9
FS4	Pellet	81.9

age, included β-glucan synthases, assayed at low and high UDPglucose concentrations (Henry et al. 1983), various putative markers for internal membranes (Cohen et al. 1983; Schibeci et al. 1984) and Ca^{2+}-dependent protein phosphorylation (Polya et al. 1984).

Assessment of Purity

Assessment of the purity of the plasma membrane fraction is difficult to determine. Contamination by other membranes can be quantified provided that adequate markers for them are known. This is not always the case (Quail 1979, Parish 1983). Mophometric quantitation (Morré and Buckhout 1979) may be useful in the determination of membrane contamination of isolated plasma membrane "sheets" fractions.

Electron microscopy of thin section of fraction F3 prepared after membrane stabilization with myeloma protein J539 showed two main contaminants: protein bodies and a few starch granules. These protein bodies are frequently observed at the periphery of protoplasts in close proximity to the plasma membrane.

Plasma membranes prepared by stabilization with cationized silica microbeads were frequently contaminated with cytoplasmic proteins, nuclei and ribosomes (Chaney and Jacobson 1983; Briggs et al. 1985) but contained fewer protein bodies.

Possible Mechanisms of Plasma Membrane Stabilization

The mechanism of membrane stabilization by murine myeloma protein J539 or cationic silica microbeads is not well understood. It is possible that this process involves "fixation" of various components in the membrane relative to each other and/or results from the stabilization of the association of cytoskeletal elements (e.g., actin-binding protein, myosin) with the plasma membrane, as postulated by Moore et al. (1978). Plasma membrane "domains" have been isolated from animal cells (Evans et al. 1980; Horst et al. 1980) and microorganisms (Condeelis 1979). While the existence of membrane "domains" is not in question, they could be generated during cell disruption (Brunner et al. 1978; Carraway and Carraway 1982). For example, disruption of plasma membrane-cytoskeletal association during homogenation could lead to reorganization of plasma membrane proteins producing lipid-rich and protein-rich "domains". Thus, procedures for the isolation of plasma membrane not involving surface stabilization prior to cell lysis may exaggerate heterogeneity of plasma membrane preparations.

Acknowledgements. The author wishes to thank Prof. B.A. Stone for continuous interest, Prof. A.B. Wardrop for electron microscopic studies, Dr. B.S. Jacobson for a gift of silica microbeads, Dr. H. Bretting for providing axinellin, Mr. A.B. Briggs for his contribution to the preparation of plasma membrane using silica microbeads and Dr. P.J. Harris for critically reading the manuscript.

References

Boss WF, Ruesink AW (1979) Isolation and characterization of concanavalin A – labelled plasma membranes of carrot protoplasts. Plant Physiol (Bethesda) 64:1005–1011

Bretting H, Kabat EA (1976) Purification and characterization of the agglutinins from the sponge *Axinella polyploides* and a study of their combining sites. Biochemistry 15:3228–3236

Briggs AB, Schibeci A, Stone BA (1985) Characterization of plasma membranes and other organelles from suspension-cultured ryegrass (*Lolium multiflorum*) endosperm cells. Proc Aust Biochem Soc 17:29

Brunner G, Bauer HC, Suter D, Speth V (1978) Artefacts produced during plasma-membrane isolation I. Cell disruption causes alterations in the structure of the plasma membrane of thymocytes. Biochim Biophys Acta 507:419–424

Buser C, Matile P (1977) Malic acid in vacuoles isolated from *Bryophyllum* leaf cells. Z Pflanzenphysiol 82:462–466

Carraway KL (1975) Covalent labelling of membranes. Biochim Biophys Acta 415:379–410

Carraway KL, Carraway CAC (1982) Isolation and characterization of plasma membranes. In: Marchalonis JJ, Warr GW (eds) Antibody as a Tool. Wiley, Chichester, pp 509–559

Chaney LK, Jacobson BS (1983) Coating cells with colloidal silica for high yield isolation of plasma membrane sheets and identification of transmembrane proteins. J Biol Chem 258: 10062–10072

Cohen PB, Schibeci A, Fincher GB (1983) Biosynthesis of arabinogalactanprotein in *Lolium multiflorum* (ryegrass) endosperm cells. Plant Physiol (Bethesda) 72:754–758

Condeelis J (1979) Isolation of concanavalin A caps during various stages of formation and their association with actin and myosin. J Cell Biol 80:751–758
Dúran A, Bowers B, Cabib E (1975) Chitin synthetase zymogen is attached to the yeast plasma membrane. Proc Natl Acad Sci USA 72:3952–3955
Dürr M, Boller T, Wiemken A (1975) Polybase induced lysis of yeast spheroplasts. A new gentle method for preparation of vacuoles. Arch Microbiol 105:319–327
Evans WH, Flint NA, Vischer P (1980) Biogenesis of hepatocytes plasmamembrane domains. Biochem J 192:903–910
Fraker PJ, Speck JC (1978) Protein and cell membrane iodination with a sparingly soluble chloramide 1,3,4,6 tetra-chloro 3α,6α diphenyl glycoluril. Biochem Biophys Res Commun 80:849–857
Gahmberg CG, Hakamori S-I (1973) External labelling of cell surface galactose and galactosamine in glycolipid and glycoprotein of human erythrocytes. J Biol Chem 248:4311–4317
Gailbraith DW, Northcote DH (1977) The isolation of plasma membrane from protoplasts of soybean suspension cultures. J Cell Sci 24:295–310
Glaudemans CPJ (1975) The interaction of homogeneous, murine myeloma immunoglobins with polysaccharide antigens. Adv Carbohydr Chem Biochem 31:313–346
Hall JL (1983) Plasma membranes. In: Hall JL, Moore AL (eds) Isolation of membranes and organelles from plant cells. Academic, London, pp 55–81
Henry RJ, Schibeci A, Stone BA (1983) Localization of β-glucan synthases on the membranes of cultured *Lolium multiflorum* (ryegrass) endosperm cells. Biochem J 209:627–633
Henry RJ, Schibeci A, Stone BA (1984) Proton motive potential difference across the plasma membrane of plant cells: A comparison of cultured cells and protoplasts from italian ryegrass (*Lolium multiflorum*) endosperm. Aust J Plant Physiol 11:119–128
Hodges TK, Leonard RT (1974) Purification of a plasma membrane – bound adenosine triphosphate from plant roots. Methods Enzymol 32:392–406
Horst MN, Baumbach GA, Olympic MA, Roberts RM (1980) Isolation of a domain of the plasma membrane in CHO cells which contain iodinatable, acidic glycoproteins of high molecular weight. Biochim Biophys Acta 600:48–61
Keller F, Stone BA (1978) Preparation of *Lolium* protoplasts and their purification using an anti-galactan-Sepharose conjugate. Z Pflanzenphysiol 87:167–172
Klaus UF, Kindl H (1979) Plant plasma membrane: Correlation between glucan synthase II activity and mannosyl transferase activity. H-S Z Physiol Chem 360:1517–1523
Koehler DE, Leonard RT, Vanderwoude WJ, Linkins AE, Lewis LN (1976) Association of latent cellulase activity with plasma membranes from kidney bean abscission zones. Plant Physiol (Bethesda) 58:324–330
Larkin PJ (1977) Plant protoplast agglutination and membrane bound β-lectins. J Cell Sci 26: 31–46
Leonard RT, Vanderwoude WJ (1976) Isolation of plasma membranes from corn roots by sucrose density gradient centrifugation. Plant Physiol (Bethesda) 57:105–114
Leonard RT, Hodges TK (1980) The plasma membrane. In: Stumpf PK, Conn EE (eds) The biochemistry of plants, vol 1. Academic, New York, pp 163–182
Moore PB, Ownby CL, Carraway KL (1978) Interactions of cytoskeletal elements with the plasma membrane of Sarcoma 180 ascites tumor cells. Exp Cell Res 115:331–342
Morré DJ, Buckhout TJ (1979) Isolation of Golgi apparatus. In: Reid E (ed) Plant organelles. Horwood, Chichester, pp 117–134
Nagahashi G, Leonard RT, Thomson WM (1978) Purification of plasma membranes from roots of barley. Plant Physiol (Bethesda) 61:993–999
Parish RW (1983) Plasma membrane in *Dictyostelium*. Mol Cell Biochem 50:75–95
Perlin DS, Spanswick RM (1980) Labelling and isolation of plasma membranes from corn leaf protoplasts. Plant Physiol (Bethesda) 65:1053–1057
Pierce WS, Hendrix DL (1979) Utilization of enzyme markers to determine the location of plasma membrane from *Pisum* epicotyls on sucrose gradients. Planta (Berl) 146:161–169
Polya GM, Schibeci A, Micucci V (1984) Phosphorylation of membrane proteins from cultured *Lolium multiflorum* (ryegrass) endosperm cells. Plant Sci Lett 36:51–57

Quail PH (1979) Plant cell fractionation. Annu Rev Plant Physiol 30:425–484

Scarborough GA (1975) Isolation and characterization of *Neurospora crassa* plasma membranes. J Biol Chem 250:1106–1111

Schibeci A, Fincher GB, Stone BA, Wardrop AB (1982) Isolation of plasma membrane from protoplasts of *Lolium multiflorum* (ryegrass) endosperm cells. Biochem J 205:511–519

Schibeci A, Pnjak A, Fincher GB (1984) Biosynthesis of arabinogalactanprotein in *Lolium multiflorum* (Italian ryegrass) endosperm cells. Subcellular distribution of galactosyl transferases. Biochem J 218:633–636

Schmidt R, Ackermann R, Kratky Z, Wasserman B, Jacobson B (1983) Fast and efficient purification of yeast plasma membranes using cationic silica microbeads. Biochim Biophys Acta 732:421–427

Suzuki T, Sadasivan R, Saito-Taki T, Stechschulte DJ, Balentine L, Helmkamp GM (1980) Studies of $F_{c\gamma}$ receptors of human B lymphocyotes: Phospholipase A_2 activity of $F_{c\gamma}$ receptors. Biochemistry 19:6037–6044

Taylor ARD, Hall JL(1980) Labelling and isolation of plasma membrane from higher plant protoplasts. In: Ferenczy L, Farkas GL (eds) Advances in protoplast research. Pergamon, Oxford, pp 463–468

Tousimis T, Schibeci A, Stone BA (1985) Uptake of sugars by suspension cultured ryegrass endosperm cells. Aust J Plant Physiol (submitted for publication)

Wagner GJ, Siegelman HW (1975) Large scale isolation of intact vacuoles and isolation of chloroplasts from protoplasts of mature plant tissue. Science (Wash DC) 190:1298–1299

The Use of Protoplasts in the Study of Coated Vesicles

B.G. MERSEY, L.R. GRIFFING and L.C. FOWKE

Introduction

In plant cells the plasma membrane functions as a selectively permeable boundary around the cytoplasm, acting in concert with the cell wall as an interface with the external environment. Organelles associated with the inner surface of the plasma membrane of walled cells include endoplasmic reticulum associated with the plasmodesmata, cytoskeletal elements such as cortical microtubules and microfilaments, and smooth and coated vesicles. As disruption of the plasmodesmata due to enzymatic degradation of the cell wall during the production of protoplasts also disrupts the association of the endoplasmic reticulum with the plasma membrane, in protoplasts the organelles associated with the plasma membrane are primarily cytoskeletal elements and smooth and coated vesicles of the cortical cytoplasm.

Some of the smooth single-membrane vesicles in the cortical cytoplasm appear to originate at the dictyosomes and may carry material such as cell wall matrix polysaccharides to the surface of the cell (review: Robinson and Kristen 1982). Coated vesicles consist of a single membrane surrounded by a distinctive protein coat composed primarily of a class of polypeptides called "clathrin". The overall diameter of coated vesicles in most plant cells is approximately 100 nm (Newcomb 1980). The clathrin coat appears to be a dynamic structure which can readily associate with or dissociate from membranes. Some of the smaller smooth vesicles in the cytoplasm may thus have been derived from coated vesicles. "Coated pits" are localized coated regions of larger membranes, and occur on the plasma membrane and on some cytoplasmic organelles such as the Golgi apparatus, and a recently defined membrane system, "partially coated reticulum" (Pesacreta and Lucas 1984; Tanchak et al. 1984). Coated vesicles are generally considered to be derived from coated pits.

There is a transient quality to microtubules and coated vesicles in the cortical region of the cytoplasm. The type and abundance of these organelles vary depending on the source of the cells or protoplasts (e.g. species) and their metabolic states (e.g. degree of differentiation). Fowke et al. (1983b) found that tobacco protoplasts derived from suspension cultured cells contained substantially greater numbers of coated vesicles, coated pits, and microtubules associated with the plasma membrane than did protoplasts derived from tobacco mesophyll cells. The increased number of cortical organelles appeared to correlate with the increased capacity of the suspension culture-derived protoplasts for cell wall regeneration.

NRCC # 23851

The Physiological Properties of Plant Protoplasts
Edited by P. E. Pilet

The functional interrelationships of the cortical organelles with the plasma membrane and with each other are generally uncharacterized in plants. The cytoskeleton may act to maintain the organization and stability of plasma membrane components such as cellulose synthases during wall synthesis. Cortical microtubules appear to be involved in the orientation of cellulose microfibrilis (Robinson and Quader 1982; Heath and Seagull 1982). Smooth, Golgi-derived secretory vesicles may act as a source of several plasma membrane components (Robinson and Kristen 1982), while coated membranes may function to take up plasma membrane components. An exocytotic role for plant coated vesicles is also envisaged (Newcomb 1980; Fowke et al. 1983a).

Protoplasts may be used as a model system to study the function of cortical organelles. For examination of the organization and dynamic properties of plasma membrane components, the plasma membrane must be accessible to experimentation. The external surface of the plasma membrane of protoplasts can be examined and manipulated, and many researchers have used this feature of protoplasts to study cell-wall regeneration, cell-surface glycoproteins and glycolipids, and the interaction of the plasma membrane with a variety of materials (reviews: Fowke and Gamborg 1980; Fowke et al. 1983a, 1985). The inner surface of the plasma membrane can also be studied using a technique originally described by Marchant (1978) for algal protoplasts. Protoplasts are attached to the surface of coverslips or coated electron microscope grids using polycations such as poly-L-lysine and then are osmotically ruptured, leaving a "patch" of plasma membrane on the substrate. This technique has been employed in examining, by immunofluorescence cytochemistry and by electron microscopy, the orientation and distribution of cortical organelles on the inner surface of the plasma membrane from a variety of plants (reviewed in Fowke et al. 1983a, 1985).

As the use of protoplasts in the study of cortical microtubules, microfilaments, and the cell wall has been discussed in previous reviews (e.g. Fowke and Gamborg 1980, Fowke et al. 1983a, 1985), we will concentrate on the usefulness of protoplasts in the study of coated vesicles. In the following section, a method is described to recover coated vesicles from plant protoplasts. Protoplasts are superior to walled cells for the isolation of coated vesicles because lysis can be accomplished by very low shear forces. Damage to other organelles and membranes is thus minimized and contamination of the coated vesicle fractions by these membranes is reduced. Entrapment of the plasma membrane and cortical cytoplasm by cell-wall fragments during homogenization is also eliminated. The application of recently developed procedures for isolating coated vesicles from plant cells (Mersey et al. 1982) and protoplasts (Mersey et al. 1983, 1985) has provided biochemical information about this organelle, which had previously been described only ultrastructurally (Newcomb 1980; Fowke et al. 1983a). Applications of this method in studies of the function of coated vesicles in plants will be discussed.

The Isolation of Coated Vesicles from Protoplasts

Materials and Methods

Coated vesicles have recently been prepared from protoplasts obtained from soybean cells cultured in suspension in liquid 1-B5 medium at 25°C and constant light of 1500 lux (Mersey et al. 1983, 1985). A flow diagram of this method is shown in Fig. 1.

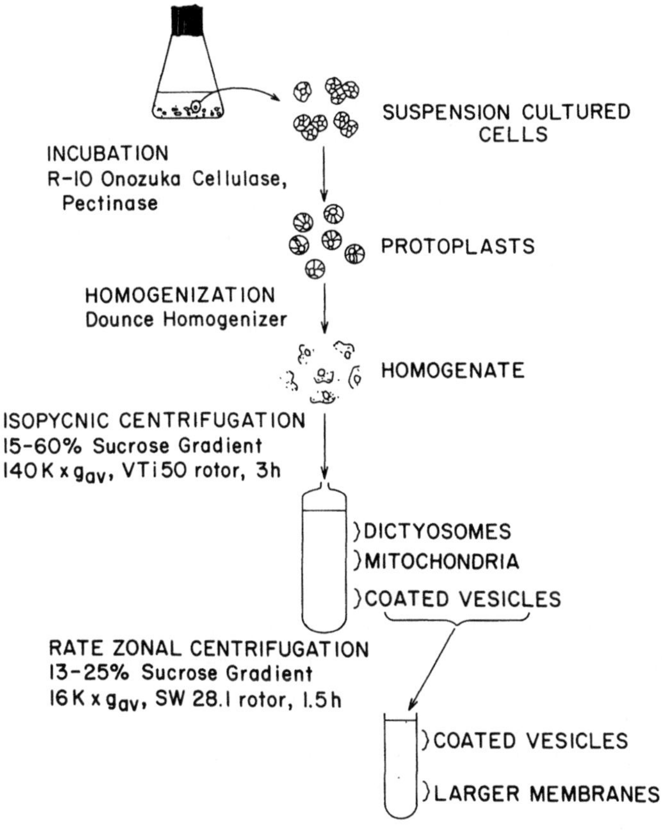

Fig. 1. Flow diagram for the preparation of coated vesicle-enriched fractions from protoplasts

Cells are harvested 3 days after transfer to fresh medium, combined in a volume ratio of 3:2 with an enzyme mixture (1.25% (w/v) R-10 Onozuka cellulase, 0.5% (w/v) hemicellulase, 0.5% (w/v) pectinase in 0.7 M sorbitol, 3 mM MES (2-N-morpholinoethanesulphonic acid), 6 mM $CaCl_2.2H_2O$, 0.7 mM $NaH_2PO_4.H_2O$, pH 5.8[1],

1 Sources of enzyme: R-10 Onozuka cellulase: Kanematsu-Gosha (USA) Inc., Los Angeles, CA, USA; Hemicellulase rhozyme, HP-150: Genencor Inc., Corning, NY,USA; Pectinase (purified, fungal): Sigma Chemical Co., St. Louis, MO, USA

and incubated overnight with gentle agitation at room temperature. Over 90% of the protoplasts are viable according to the fluorescein diacetate assay and they begin to regenerate cell wall within 6 h (unpublished data).

The protoplasts are filtered through 85 μm nylon mesh, purified by flotation centrifugation through a dextran cushion, and homogenized using a chilled Dounce homogenizer in a medium similar to that used for stabilizing microtubules (100 mM Pipes [piperazine-N,N'-bis (2-ethanesulphonic acid)], pH 6.9, 2 mM DTE, 2 mM EDTA, 1 mM $MgCl_2$, 0.3 M sorbitol). The homogenate is centrifuged at 1000 g_{max} for 10 min to pellet larger organelles. The supernatant is then layered onto a linear 15 to 60% (w/w) sucrose gradient in gradient buffer (45 mM MES, pH 6.5, 1 mM DTE, 1 mM EDTA, 0.1 mM $MgCl_2$). The gradients are centrifuged in a vertical rotor at 140,000 g_{av} for 3 h at 4°C. The 48 to 52% sucrose fractions from this isopycnic centrifugation step are most enriched in coated vesicles. These fractions are pooled and the coated vesicles pelleted by centrifugation (75,000 gmax, 70 min), resuspended in gradient buffer to 10% sucrose, layered onto linear 13 to 25% (w/w) sucrose gradients and centrifuged in a swinging bucket rotor at 16,000 g_{av} for 90 min. Purified coated vesicles from this rate zonal centrifugation step are found in the 10 to 18% sucrose region of the gradient.

Fractions from the gradients were examined by electron microscopy, either by negative staining or by thin sectioning techniques. SDS polyacrylamide gel electrophoresis was used to determine the polypeptide composition of the fractions (Mersey et al. 1985), and marker enzymes for various organelles and membranes in the isopycnic gradient were assayed.

Results

Coated vesicles (Figs. 2, 3) from soybean protoplasts have ultrastructural characteristics similar to those isolated from other plant sources (Mersey et al. 1982), animal tissues (e.g. Pearse 1975), and yeast (Mueller and Branton 1984). Protoplast coated vesicles isolated using the procedure described consist of a single membrane-bounded vesicle (arrows, Figs. 2, 3) surrounded by a characteristic basket-like coat. Their equilibrium density in sucrose gradients is approximately 1.23 g ml^{-1}. The isopycnic centrifugation procedure effectively separates coated vesicles from most other cellular membranes and organelles, including ER, Golgi membranes and mitochondria. Some coated vesicles may be found in the Golgi fraction (density = 1.13 g ml^{-1}) and these probably represent coated pit regions of those membranes (Griffing et al. 1984). Residual contamination of the isolated coated vesicle fraction by large membrane structures is significantly reduced by the subsequent rate zonal centrifugation (Mersey et al. 1983, 1985).

Analysis of the fractions by SDS polyacrylamide gel electrophoresis (Mersey et al. 1983, 1985) has shown that the major polypeptide of the coated vesicle fractions is a 190,000 M.W. polypeptide. This polypeptide is electrophoretically distinct from bovine brain clathrin, which is a 180,000 polypeptide. Other polypeptides in the protoplast coated vesicle fractions are approximately 105, 100, 96, 64, 50, 38 and 32,000 M.W. (Mersey et al. 1985). Many of these polypeptides have electrophoretic

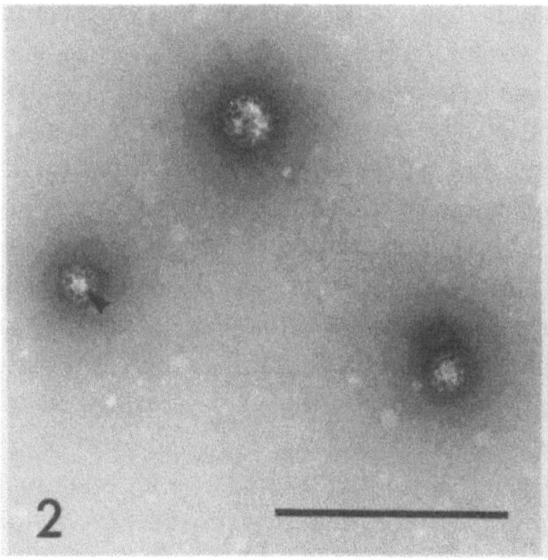

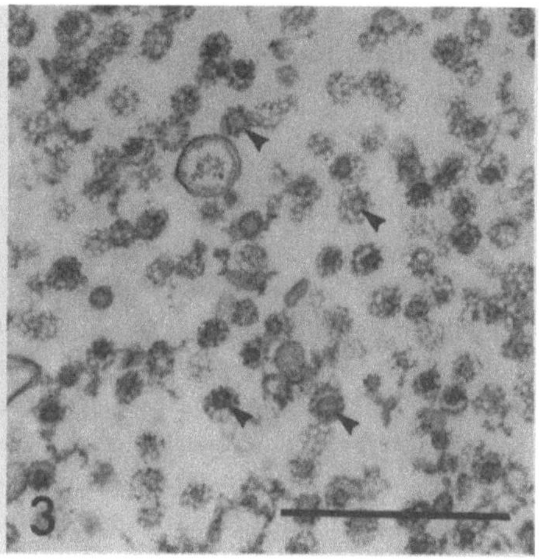

Fig. 2. Negatively stained, glutaraldehyde-fixed coated vesicles from soybean protoplasts (phosphotungstic acid). Bar = 0.5 μm

Fig. 3. Ultrathin section of pelleted, glutaraldehyde-fixed, osmium tetroxide-post-fixed coated vesicles from soybean protoplasts (uranyl acetate-lead critrate stain). Bar = 0.5 μm

mobilities that are strikingly similar to those found in coated vesicle preparations from animal and yeast cells (discussed in Mersey et al. 1985).

Discussion

This section will discuss four applications of the coated vesicle isolation method which could aid in understanding functions of plant coated vesicles. First, the method may allow insight into the physical relationship between coated membranes and the cytoskeleton. Second, it may be used in studying the transient nature of coated vesicles. Third, it may help in understanding the involvement of coated vesicles in the movement and sorting of plasma membrane components. Fourth, it may aid in defining the role of pinocytosis in higher plant cells.

1. The relationship between coated vesicles and the cytoskeleton: The physical relationship between coated vesicles and cortical microtubules is unclear. Coated pits in the plasma membrane can occur in regions of the cell periphery adjacent to microtubules, and some workers have noted small, filamentous "stalks" attaching coated vesicles to microtubules (Palevitz 1982). Bonnett (1969) shows cytoplasmic coated vesicles apparently attached to a microtubule which curves through the cytoplasm and terminates adjacent to the plasma membrane. In animal cells, isolated coated vesicles have been shown to bind actin (Schook et al. 1979) and to contain both the α and β subunits of tubulin (see Kelly et al. 1983; Pfeffer et al. 1983). It would be interesting to assay plant coated vesicles for actin and tubulin. This could be done by probing isolated coated vesicles or separated polypeptides with appropriate antisera.

2. The transient nature of coated vesicles: Cortical organelles, the cytoskeleton and the coated and smooth membranes, are in a state of flux. All these organelles also exist in other areas of the cytoplasm and the relationships among these spatially separate populations are unclear. It is difficult to say whether a cytoplasmic microtubule can change position and become a cortical microtubule without depolymerizing and reforming at nucleating sites (Wick and Duniec 1983). Likewise, it is not known whether Golgi-associated coated membranes give rise to the coated vesicles and coated pits associated with the plasma membrane, or vice versa, or whether there is any relationship at all between these two populations of coated membranes (Fowke et al. 1983a). With regard to the latter point, it should be noted that Golgi-associated coated vesicles are generally smaller (average diameter 69 nm) than cortical coated vesicles (average diameter 100 nm) (van der Valk and Fowke 1981).

A related problem is whether coated regions form by assembly from a common pool of coat subunits (triskelions) or whether the coat subunits pre-exist on the parent membrane. In animal cells, the problem of coat dynamics has been approached in three general ways, each requiring the isolation of a relatively pure fraction of coated vesicles.

First, intact coated vesicles have been used to study the components of the coat and cytoplasm required for the assembly and linkage of the coat to the membrane. The polymerization/depolymerization of clathrin can be driven by a self-assembly process (Keen et al. 1979), but linkage to the membrane and controlled assembly of triskelions appear to require accessory proteins and ATP (Patzer et al. 1982; Zaremba and Keen 1983). Hence, coats could conceivably form from a common cytoplasmic pool of subunits, given the right primer conditions.

Second, isolated coated vesicles have been used as antigens to produce antibody which can be visualized by immunofluorescence techniques. In fibroblast cells, clathrin labelled with fluorescent antibody was seen to move within the cell cortex, changing from a diffuse fluorescence to form discreet pits (Willingham et al. 1981; Wehland et al. 1981). It was suggested that since coated vesicle antibody microinjected into the cell did not influence the function of coated membranes (endocytosis of α_2-macroglobulin), therefore the coat must never leave the membrane, not even in the form of free coated vesicles. This suggestion was based on the assumption that the injected antibody binds to unattached coat subunits as well as to coated pits and vesicles. These experiments have served as a basis for questioning the existence of free, cytoplasmic coated vesicles. However, other workers remain convinced of the existence of free coated vesicles (see van Deurs and Nilausen 1982 and Willingham and Pastan 1983 for ultrastructural arguments addressing this debate).

A third approach using isolated coated vesicles has indirectly addressed the hypothesis that the vesicles may act as vehicles to transport substances from one organelle or membrane to another. Altsteil and Branton (1983) found that while the uncoated membranes readily fuse with other membranes, the presence of a coat inhibits fusion. Consequently, if coated vesicles do act as shuttles between membranes, they would have to uncoat prior to fusing with the target membrane.

3. Sorting and arrangement of plasma membrane components: From animal studies, it is evident that coated pits can influence the flow of molecules within the plane of the membrane. Although it is unlikely that coated structures act as physical constraints on the membrane surrounding them, they could act to restrict movement within the coated domain. As an alternative to constraining membrane flow, it has been hypothesized that coated structures could help to generate membrane flow by endocytosis of plasma membrane previously added via smooth vesicle exocytosis (Bretscher et al. 1980; Pearse and Bretscher 1981). Coated membranes would act as "molecular filters," i.e., they would exclude some components of the parent membrane and select for other components. The selective nature of coated vesicles in animal cells is now well established. They are not only enriched for certain receptors which specifically bind to extracellular ligands of low abundance during receptor-mediated endocytosis (review: Brown et al. 1983), but they also exclude some surface glycoproteins (Bretscher et al. 1980).

The selective association of specific plant cell membrane components with coated regions can be examined directly by assaying the contents of isolated coated vesicles. Since isolated coated vesicle fractions from soybean protoplasts probably contain a mixture of both Golgi-derived and plasma membrane-derived coated vesicles, Griffing et al. (1984) assayed these fractions for Golgi and plasma membrane marker enzymes. A glucan synthase, putatively associated with the plasma membrane (see Griffing et al. 1984) was not found in the coated vesicle fraction. They also found that while isolated coated vesicles did not contain the Golgi marker, latent IDPase, they were enriched for glucan synthase I, an enzyme also found in Golgi membranes. These data indicate that coated vesicles can selectively exclude certain components while accumulating other components of the plasma membrane and Golgi membranes. Based on histochemical evidence, coated vesicles in soybean protoplasts also appear to carry peroxidase, an enzyme which shows a discontinuous distribution across the dictyosome stack and is not apparently associated with the plasma membrane (Griffing and Fowke 1985).

4. Pinocytosis in plant cells: Two recent reports provide evidence that coated regions of the protoplast plasma membrane can act to take up cationized ferritin, a membrane-impermeant, electron-dense, surface label (Tanchak et al. 1984; Joachim and Robinson 1984). Shortly after the protoplasts were exposed to the protein, label could be found in coated pits and coated vesicles. Labelled material later appeared in uncoated vesicles and finally in other organelles. The ultrastructure and kinetics of the uptake was analogous to that found during receptor-mediated endocytosis (e.g. Willingham and Pastan 1983) and endocytosis of cationized ferritin by animal cells (e.g. van Deurs and Nilausen 1982).

Cram (1980) expressed several arguments against pinocytosis in plants. He discounted the possibility of receptor-mediated endocytosis of ions, based on the calculation that the surface area of the membrane could not accommodate the required number of receptors. Since coated vesicle-mediated endocytosis has been demonstrated in plant protoplasts and, since pinocytosis is not apparently involved in large-scale uptake of simple solutes, what are the protoplasts taking up via this pathway? Perhaps plant cells take up macromolecules of low abundance via specific receptors which are associated with coated pits and vesicles.

Conclusion

The special properties of protoplasts have been widely exploited in research dealing with the plasma membrane and associated organelles. Protoplasts from suspension-cultured soybean cells are an excellent system for studying plant coated pits and vesicles. Enriched fractions of coated vesicles are readily prepared from protoplasts, and can be used to examine the biochemistry, structural dynamics, and function of these ubiquitous plant cell organelles.

Acknowledgements. We wish to acknowledge the helpful comments and discussions of our colleagues, P.J. Rennie, F. Constabel and A.J. Cutler.

References

Altsteil L, Branton D (1983) Fusion of coated vesicles with lysosomes: measurement with a fluorescence assay. Cell 32:921–929

Bonnett HT Jr (1969) Cortical cell death during lateral root formation. J Cell Biol 40:144–159

Bretscher MS, Thomson JN, Pearse BMF (1980) Coated pits act as molecular filters. Proc Natl Acad Sci USA 77:4156–4159

Brown MS, Anderson RGW, Goldstein JL (1983) Recycling receptors: the round-trip itinerary of migrant membrane proteins. Cell 32:663–667

Cram WJ (1980) Pinocytosis in plants. New Phytol 84:1–17

Fowke LC, Gamborg OL (1980) Applications of protoplasts to the study of plant cells. Int Rev Cytol 68:9–51

Fowke LC, Griffing LR, Mersey BG, van der Valk P (1983a) Protoplasts for studies of the plasma membrane and associated cell organelles. Protoplasts 1983, Lecture Proc 6th Int Protoplast Symp Basel. Exper Suppl (Basel) 46:101–110

Fowke LC, Rennie PJ, Constabel F (1983b) Organelles associated with the plasma membrane of tobacco leaf protoplasts. Plant Cell Rep 2:292–295

Fowke LC, Griffing LR, Mersey BG, Tanchak M (to be published 1985) Protoplasts for studies of cell organelles. In: Fowke LC, Constabel F (eds) Plant protoplasts. CRC, Boca Raton

Griffing LR, Fowke LC (1985) Localization of peroxidase in soybean suspension culture cells and protoplasts: intracellular vacuole differentiation and presence of peroxidase in coated vesicles and multivesicular bodies (manuscript submitted for publication)

Griffing LR, Mersey BG, Fowke LB (1984) A coated vesicle-enriched fraction from soybean protoplasts contains glucan synthetase I activity. Plant Physiol Suppl 75:2

Heath IB, Seagull RW (1982) Oriented cellulose fibrils and the cytoskeleton: a critical comparison of models. In: Lloyd CW (ed) The cytoskeleton in plant growth and development. Academic, London, pp 163–182

Joachim S, Robinson DG (1984) Endocytosis of cationic ferritin by bean leaf protoplasts. Eur J Cell Biol 34:212–216

Keen JH, Willingham MC, Pastan IH (1979) Clathrin-coated vesicles: isolation, dissociation and factor-dependent reassociation of clathrin baskets. Cell 16:303–312

Kelly WG, Passaniti A, Woods JW, Daiss JL, Roth TF (1983) Tubulin as a molecular component of coated vesicles. J Cell Biol 97:1191–1199

Marchant HJ (1978) Microtubules associated with the plasma membrane isolated from protoplasts of the green alga *Mougeotia*. Exp Cell Res 115:25–30

Mersey BG, Fowke LC, Constabel F, Newcomb EH (1982) Preparation of a coated vesicle-enriched fraction from plant cells. Exp Cell Res 141:459–463

Mersey BG, Griffing LR, Rennie PJ, Fowke LC (1983) Coated vesicles from plant protoplasts. Protoplasts 1983, Poster Proc 6th Int Protoplast Symp Basel. Exper Suppl (Basel) 45:216–217

Mersey BG, Griffing LR, Rennie PJ, Fowke LC (to be published 1985) The isolation of coated vesicles from protoplasts of soybean. Planta (Berl) 163

Mueller SC, Branton D (1984) Identification of coated vesicles in *Saccharomyces cerevisiae.* J Cell Biol 98:341–346

Newcomb EH (1980) Coated vesicles: their occurrence in different plant cell types. In: Ockleford CD, Whyte A (eds) Coated vesicles. Cambridge Univ Press, New York, pp 55–68

Palevitz BA (1982) The stomatal complex as a model of cytoskeletal participation in cell differentiation. In: Lloyd CW (ed) The cytoskeleton in plant growth and development. Academic, New York, pp 345–376

Patzer EJ, Schlossman DM, Rothman JE (1982) Release of clathrin from coated vesicles dependent upon a nucleoside triphosphate and a cytosol fraction. J Cell Biol 93:230–236

Pearse BMF (1975) Coated vesicles from pig brain: purification and biochemical characterization. J Mol Biol 97:93–98

Pearse BMF, Bretscher MS (1981) Membrane recycling by coated vesicles. Annu Rev Biochem 50:85–101

Pesacreta TC, Lucas WJ (1984) The plasma membrane coat and a coated vesicle-associated reticulum of membranes: their structure and possible interrelationship in *Chara corallina.* J Cell Biol 98:1537–1545

Pfeffer SR, Drubin DG, Kelly RB (1983) Identification of three coated vesicle components as α- and β-tubulin linked to a phosphorylated 50,000-dalton polypeptide. J Cell Biol 97: 40–47

Robinson DG, Kristen U (1982) Membrane flow via the Golgi apparatus of higher plant cells. Int Rev Cytol 77:89–127

Robinson DG, Quader H (1982) The microtubule-microfibril syndrome. In: Lloyd CW (ed) The cytoskeleton in plant growth and development. Academic, London, pp 109–126

Schook W, Puszkin S, Bloom W, Ores C, Kochwa S (1979) Mechanochemical properties of brain clathrin: interactions with actin and α-actinin and polymerization into basketlike structures or filaments. Proc Natl Acad Sci USA 76:116–120

Tanchak MA, Griffing LR, Mersey BG, Fowke LC (1984) Endocytosis of cationized ferritin by coated vesicles of soybean protoplasts. Planta (Berl) 162

van der Valk P, Fowke LC (1981) Ultrastructural aspects of coated vesicles in tobacco protoplasts. Can J Bot 59:1307–1313

van Deurs B, Nilausen K (1982) Pinocytosis in Mouse L-fibroblasts: ultrastructural evidence for a direct membrane shuttle between the plasma membrane and the lysosomal compartment. J Cell Biol 94:279–286

Wehland J, Willingham MC, Dickson R, Pastan I (1981) Microinjection of anticlathrin antibodies into fibroblasts does not interfere with receptor-mediated endocytosis of α_2-macroglobulin. Cell 25:105–119

Wick SM, Duniec J (1983) Immunofluorescence microscopy of tubulin and microtubule arrays in plant cells I. Preprophase band development and concomitant appearance of nuclear envelope-associated tubulin. J Cell Biol 97:235–243

Willingham MC, Pastan I (1983) Formation of receptosomes from plasma membrane coated pits during endocytosis: analysis by serial sections with improved membrane labelling and preservation techniques. Proc Natl Acad Sci USA 80:5617–5621

Willingham MC, Rutherford AV, Gallo MG, Wehland J, Dickson RB, Schlegel R, Pastan IH (1981) Receptor-mediated endocytosis in cultured fibroblasts: cryptic coated pits and the formation of receptosomes. J Histochem Cytochem 29:1003–1013

Zaremba S, Keen JH (1983) Assembly polypeptides from coated vesicles mediate reassembly of unique clathrin coats. J Cell Biol 97:1339–1347

Membrane Transport in Protoplasts

P. MORRIS

Introduction

Studies of the mechanisms of solute transport through the plasmalemma of plant cells is often hindered by the presence of the cell wall and other problems, such as cuticular, stomatal and free space resistances, encountered when working with complex tissues. Isolated protoplasts in theory represent a far simpler system than intact tissue which should allow studies of transport at the cellular rather than at the tissue level. Although isolated protoplasts from higher plant tissues and plant cell cultures have been used as experimental systems for many years, their general physiological state, especially in regard to their membrane transport properties, has only recently been investigated in any detail. Since membrane transport in protoplasts has not previously been reviewed, the aim of this report is to briefly assess our current knowledge of transport systems in protoplasts and to outline some of the problems which may be encountered in such studies. The existing methods used to study transport will also be discussed and a new transport assay for protoplasts described. Some preliminary results on the uptake of alkaloid drugs by protoplasts will also be presented.

Macromolecular and particle uptake will not be considered here (see Power 1977), nor will the increasing use of isolated vacuoles in studies of solute transport, as this subject has been excellently reviewed recently (Leigh 1983). It has been about a decade since isolated protoplasts were first used for studies of leucine uptake (Reusink 1973) and measurements of membrane potentials (Rona 1973; Heller et al. 1974) with a gradual increase in publications on ion transport (Taylor and Hall 1976; Mettler and Leonard 1979a,b), amino acid uptake (Robinson and Mayo 1977; Guy et al. 1978; Ruesink 1978), sugar uptake (Guy et al. 1978) and studies of membrane potentials (Racusen et al. 1977; Bridskin and Leonard 1979) during the 1970's. Within the past several years, however, the use of protoplasts has increased significantly both for ion transport studies (Lin 1980, 1981, 1982, 1984; Morris and Thain 1980; Schmidt and Poole 1980; Gronwald and Leonard 1982; Rahat and Reinhold 1983; Akerman et al. 1983) and for studies of the transport or organic solutes (Rubinstein and Tatter 1980; Guy et al. 1980, 1981; Suzuki 1981, 1982; Singer and McDaniel 1982). The primary aim of the early studies was to establish protoplasts as a useful system for studying membrane transport. This has now been done and in most respects the basic parameters of transport have been defined. Protoplast membranes exhibit specificity and can distinguish between K^+ and Na^+ (Mettler and Leonard 1979b; Morris and Thain 1980) and L & D isomers of sugars (Guy et al. 1978, 1980).

The Physiological Properties of Plant Protoplasts
Edited by P. E. Pilet

A light-dependent modulation of the sugar transport system which acts synergistically with protonation has been found in pea mesophyll protoplasts which show both light-enhanced sugar and K^+ uptake (Guy et al. 1980, 1981; Rahat and Reinhold 1983). Although light stimulated K^+ uptake does not involve proton co-transport in these cells, an ATP fuelled K^+ porter or K^+-H^+ antiporter may operate in the light. Whereas in the dark K^+ may be passively distributed (Rahat and Reinhold 1983). Recent results also suggest that an ATP-linked Ca^{2+} efflux pump is present in the plasmalemma of wheat mesophyll protoplasts which is not influenced by plant hormones or under phytochrome control (Akterman et al. 1983). The most notable development recently in membrane transport studies with protoplasts is the demonstration of an NADH-O_2 electron transfer system associated with the plasmalemma of corn root protoplasts (Lin 1982). Evidence for H^+-ATPase involvement in energizing ion uptake in plant cells is extensive, but there is now evidence, at least in corn root protoplasts, that a redox reaction driven by NADH oxidase can also drive H^+ efflux and K^+ influx and hyperpolarize the protoplast membrane potential (Lin 1984). The protein can be removed from the plasmalemma with trypsin with a concomitant reduction is NADH stimulated O_2 uptake and ^{86}Rb influx and gives rise to an extra peptide band of 42,000 molecular weight (Lin 1982). A detailed account of this is presented by Lin in his report on ion transport in root protoplasts (Chap. 9, this Vol.). An important parameter when considering membrane transport is the membrane potential ($\Delta\psi$, which is a result of and tends to regulate the transport of charged and uncharged species. Some controversy surrounds the magnitudes of $\Delta\psi$ in isolated protoplasts. Using microelectrodes membrane potentials of +10 to +20 mV (inside positive) have been reported in several protoplast systems (Rona 1973; Heller et al. 1974; Racusen et al. 1977), while others using microelectrodes and a suction micropippette have reported values of $\Delta\psi$ of -40 mV (inside negative) in protoplasts from cultured tobacco cells (Briskin and Leonard 1979).

The distribution of lipophyllic cations gave a Nernst potential of -62 mV (inside negative) for isolated oat mesophyll protoplasts (Rubinstein 1978) and -59 mV to -120 mV in the dark and light in pea mesophyll protoplasts (Rahat and Reinhold 1983).

Materials and Methods

Protoplast Isolation and Purification

Mesophyll Protoplasts

Plants of *Catharanthus roseus.* G. Don var ocellata, were grown in a glasshouse and were propagated from cuttings every 6 to 9 months. Fully expanded leaves were taken for protoplast isolation by the mixed enzyme method of Power and Cocking (1970). The leaf lamina were cut into 1 mm wide slices but were not detached from the mid-rib (feathered leaves). The sliced leaves were floated on a plamolysing medium of 0.7 M mannitol for 1 h in 15 cm Petri dishes. Protoplasts were isolated by incubating the leaves statically for 18 h at 30°C under 3000 lx white light in a medium containing Gamborgs B5 salts, 0.7 M mannitol and the enzymes Cellulysin (0.5 w/v,

Calbiochem) and Maceraze (0.1% w/v Calbiochem), at pH 5.5. After filtration through 150 μm nylon mesh, protoplasts were purified by initial centrifugation at 60 *g* for 5 min, followed by three successive floatations (200 *g* for 5 min) in B5 salts with 23% w/v sucrose. Purified protoplasts were resuspended to a concentration of 2 x 10^6 protoplasts ml^{-1} in B5 salts containing 0.7 M mannitol and were used immediately for uptake experiments.

The experimental methods used for the ion transport studies with tobacco protoplasts, including surface area determinations have been described previously (Morris and Thain 1980, 1983).

Transport Assay

Protoplast suspensions (2 x 10^6 protoplasts ml^{-1}) were incubated either in 10-ml flasks containing 3-ml suspension for long-term uptake or in individual wells of a culture plate (Nuclon multidish 24) with 250 μl suspension well, for short-term uptake. Flasks or plates were incubated at 25°C on an orbital shaker at a speed just sufficient to keep the protoplasts in suspension (50–60 rpm). At the start of the uptake period 0.2 μCi ml^{-1} of [^{3}H]-morphine (65 C $mmol^{-1}$) were added.

At intervals 4 x 50 μl of suspension containing about 8 x 10^4 protoplasts were pipetted into 1.5 ml plastic eppendorf tubes containing 1.4 ml of ice cold PEG (polyethylene glycol)/Dextran mixture made up with 0.7 M mannitol. The PEG/Dextran mixture consisted of 7% w/w (final concentration) Dextran T500 (Pharmacia Batch FD16027) and 0.5% w/w (final concentration) PEG 6000 (Union Carbide) which resulted in a phase system with a volume ratio of about 1:20 (bottom dextran:top PEG). To achieve phase-partition the tubes were centrifuged in an Eppendorf 5412 micro-centrifuge for 15 s after thorough mixing by inversion. The upper PEG-rich phase was aspirated away and 1.4 ml of icecold PEG/Dextran mixture added to each tube. The tubes were then mixed by inversion and recentrifuged for 15 s. After aspiration of the upper phase, the interface containing the intact protoplasts was removed (30–50 μl) and diluted into 1.4 ml of ice cold 0.7 M mannitol. One ml of this was used for scintillation counting after addition of 15 ml of Scintran Cocktail T (BDH Ltd). The remaining suspension was used to determine protoplast numbers, viability and size distribution.

Results and Discussion

Since the solutes in the external medium are in direct contact with the exposed plasmalemma, many investigations have utilized protoplasts for membrane transport studies. However, several important changes during protoplast isolation need to be considered. Isolated protoplasts are under severe osmotic stress and in most cases are in a new solute environment both externally and internally (due to volume reduction). Removal of the cell wall disrupts plasmodesmata and removes proteins located between the membrane and the cell wall as well as destroying the plasmalemma/cell wall interface. Moreover enzyme preparations used to isolate protoplasts are rarely if ever pure and protoplast transport systems may be modified by contaminants in these enzymes (Reusink 1978; Morris et al. 1981). It would not, therefore, be surprising to find differences between cells and protoplasts in the functioning of their transport systems. However protoplasts isolated from a wide variety of tissues appear to show similar quantitative transport properties to intact cells, although detailed comparisons of cells and protoplasts have yet to be undertaken.

Solute Concentrations and Membrane Potentials

Determining the nature and concentration of solutes in protoplasts and the membrane potential must be regarded as the first steps in any approach to studies of transport in protoplasts. Unfortunately solute concentrations have not been determined in the majority of cases while the magnitude of $\Delta\psi$ appears to depend upon the method of measurement (Racuson et al. 1977; Rubinstein 1978). Briskin and Leonard (1979) concluded that plasmolysis and enzymatic release of protoplasts from cultured tobacco cells had no significant effect on $\Delta\psi$. The results obtained with lipophilic cation distribution would tend to support the conclusion that isolated protoplasts have a negative membrane potential of about -50 to -60 mV (inside negative) in 1 mM KCl, the positive values observed previously possible being due to surface electrical phenomenon due to electrode insersion into agar. These potentials are considerably less negative, however, than in intact cells and the finding that some component in the enzyme preparations used for cell-wall digestion causes membrane depolarization of tobacco leaf cells (Morris et al. 1981) may be related to the low potentials found in protoplasts. The effects of these enzymes on $\Delta\psi$ in protoplasts requires further study.

Comparison of Fluxes in Cells and Protoplasts

In transport experiments the measured solute fluxes should be related to unit area of membrane. As none of the standardization parameters commonly used to relate the quantity of material used is a direct measure of membrane surface area, and the variations in some are often not related to changes in surface area, care has to be taken in the choice of standardization parameters. This is particularly important when comparing fluxes in protoplasts and cells of intact tissue since large reductions in surface area occur on protoplast isolation (Morris and Thain 1980, Ruesink 1978).

Protoplast surface areas can be calculated from measurements of protoplast diameters and their frequency distribution but in the cells of intact tissue and in cultured cells this is more difficult (Morris and Thain 1983; Thain 1983).

Despite an increased amount of research, few studies have compared solute transport in protoplasts with the parent cells and no comparisons have yet been made on the basis of the surface area of the membrane of cells and protoplasts. Figure 1 shows a comparison of the uptake kinetics of K^+ ($^{42}K^+$), K^+ ($^{86}Rb^+$), Na^+ ($^{22}Na^+$) and

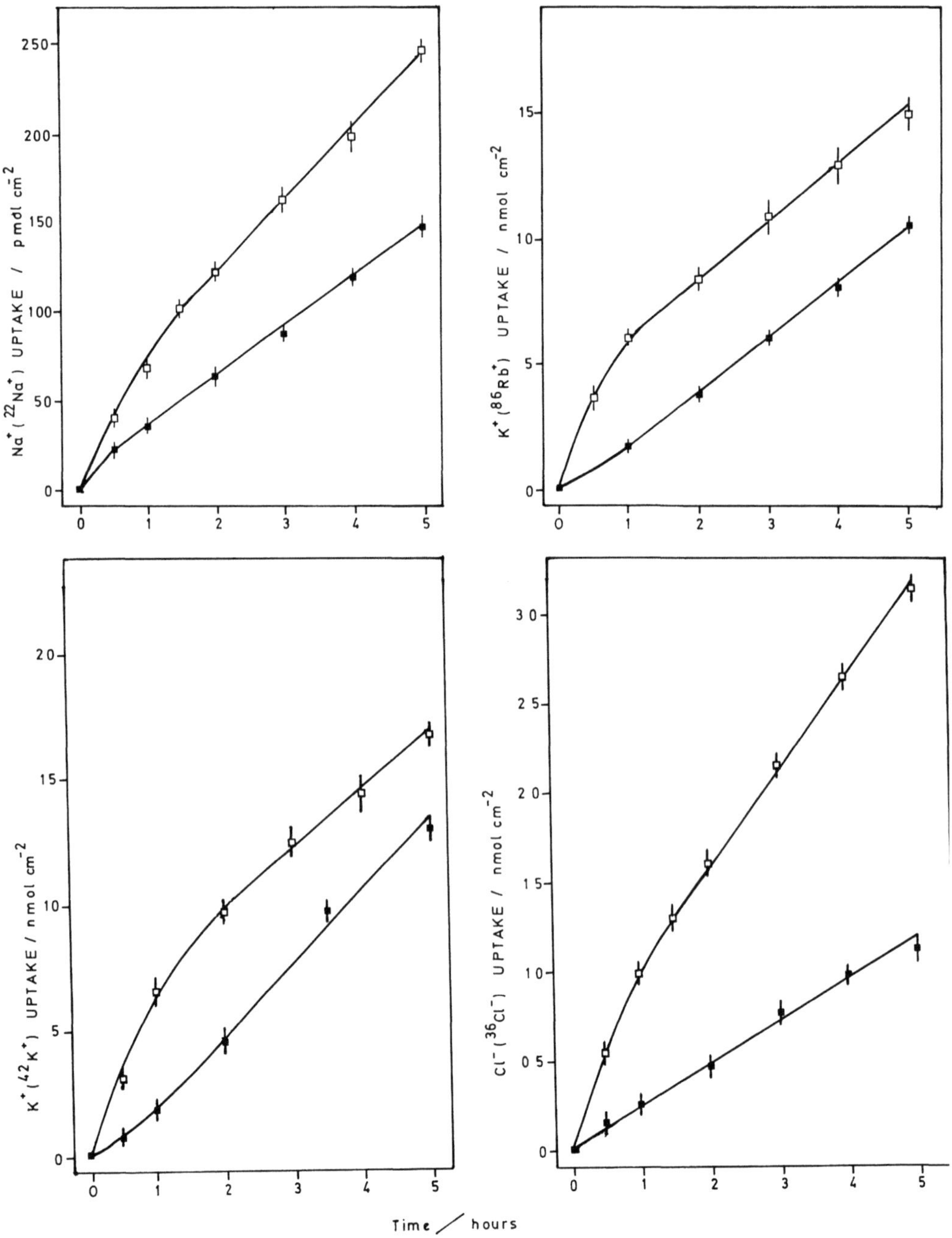

Fig 1. Time courses of $^{42}K^+$, $^{86}Rb^+$, $^{36}Cl^-$ and $^{22}Na^+$ uptake by tobacco leaf tissue (epidermis free discs) and isolated mesophyll protoplasts. Each point is the mean of 3 replicates ± SEM □ – leaf tissue, ■ – isolated protoplasts

Cl^- ($^{36}Cl^-$) in tobacco protoplasts and epidermis free leaf discs on this basis. While the net tracer uptake rate of K^+ (either as $^{42}K^+$ or $^{86}Rb^+$) did not differ between protoplasts and mesophyll cells, Na^+ uptake was slightly inhibited and Cl^- uptake was severly depressed in protoplasts. This may reflect partial inhibition of transport systems since active transport of Na^+ and Cl^- is implied. While in these cells K^+ may be passively distributed according to its electrochemical potential gradient (Morris 1977), active H^+-transport as the driving force for passive K^+-transport or a H^+/K^+ exchange system may operate as in pea mesophyll cells (Rahat and Reinhold 1983). It is also possible that the increased internal salt concentrations caused by converting cells to protoplasts with the subsequent reduction in volume, or a change in the membrane potential may contribute to these discrepancies, although it is difficult to see how this would occur. Further comparisons of membrane fluxes in cells and protoplasts on a surface area basis are required both for inorganic and organic solutes.

Comparisons of solute uptake in tissues exposed to high osmolarities with uptake into protoplasts under similar conditions (Taylor and Hall 1976; Ruesink 1978; Guy et al. 1978; Rubinstein and Tatter 1980; Morris and Thain 1980; Schmidt and Poole 1980) also require careful interpretation in view of the detremental effects plasmolysis may have on the membrane properties of only a proportion of the cells in the tissue (Morris et al. 1981).

Determination of Membrane Fluxes

Evaluation of unidirectional fluxes of solutes at individual membranes requires accurate measurements of influx and efflux kinetics.

The initial rate of influx can be determined in protoplasts provided that the assay allows rapid separation of the protoplasts from the labelled medium. Efflux kinetics in protoplasts are more difficult to measure and have not been widely studied, although the efflux of 3-O-methyl glucose in pea mesophyll protoplasts has been reported (Guy et al. 1981). Influx and efflux can be compared by calculating the influx and efflux rate constants, which being concentration-independent, can be directly compared. Furthermore as the general flux equations only apply under steady-state conditions, it is often advisable to use rate constants for flux comparisons (Thain 1984). Realization of the full potential of isolated protoplasts for use in quantitative studies of membrane transport depends therefore upon the development of methods for obtaining sufficiently accurate measurements of these fluxes. Currently a variety of assays are being used to measure solute transport in protoplasts but there is no standard method, each laboratory having developed its own procedures (Table 1). There is a need for a general method for separating protoplasts from the influx or efflux medium. Such a method should be generally applicable to different protoplast types, be capable of separating intact protoplasts from damaged protoplasts, organelles and cell debris rapidly, with little contamination from the extracellular medium. McDaniels (1983) has recently discussed the development of transport assays for cell suspension cultures in some detail and these observations are also relevant to transport assays with isolated protoplasts. The increasing use of silicons oils to separate

Table 1. Assay methods used to determine membrane fluxes in isolated protoplasts

Assay method	Reference	Separation time and conditions	Protoplast systems	Solutes studied
Filtration by suction onto glass fibre discs	Taylor and Hall 1976	Not stated	Tobacco mesophyll and maize root protoplasts	$K^+(^{86}Rb^+)$, $^{32}PO_4{}^{2-}$
Filtration through 0.45 μ millipore filters (not to dryness)	Schmidt and Poole 1980	Not stated	Red beet root protoplasts	$^{86}Rb^+$
	Rubinstein and Tatter 1980	Not stated	Oat mesophyll protoplasts	$[^3H]$-α aminoisobutyric acid
Dilution of external medium by repeated centrifugation to form pellet	Rubinstein 1978	$<$ 15 min 45 x *g*	Oat mesophyll protoplasts	$[^3H]$-triphenylmethyl phosphonium bromide
	Rubinstein and Tatter 1980	$<$ 15 min 45 x *g*	Oat mesophyll protoplasts	$[^3H]$-α aminoisobutyric acid
	Robinson and Mayo 1977	$<$ 15 min 80 x *g*	Cultured tobacco mesophyll protoplasts	$[^3H]$-UTP, $[^3H]$-uracil $[^3H]$-leucine $[^{14}C]$ glucose, $^{32}PO_4{}^{=}$
	Singer and McDaniel 1982	Not given 800 x *g*	Tobacco mesophyll protoplasts	$[^{14}C]$-leucine $[^{14}C]$ 3 amino 1,2,4 triazole (Amitrole)
	Suzuki 1981, 1982	15 min 120 x *g*	Cultured cells of Catharanthus roseus	$[^3H]$-leucine $[H^3]$-D-glucose
	McDaniel 1983	15 min 45 x *g*	Cultured cells of Nicotiana tabacum	$[^3H]$-leucine
Dilution of external medium by repeated floatation in 21% sucrose	Morris and Thain 1980	20 min 200 x *g*	Tobacco and Antirrhinum mesophyll protoplasts	$^{42}K^+$, $^{86}Rb^+$, $^{36}Cl^-$, $^{22}Na^+$

Table 1 (cont.)

Centrifugation through two step gradient of 5%	Mettler and Leonard 1979a,b	10 min 100 x *g*	Cultured cells of tobacco	K($^{86}Rb^+$), $^{36}Cl^-$, $^{32}PO_4^=$, $^{45}Ca^{++}$
Ficol and 0.7 M mannitol and 4°C	Gronwald and Leonard 1982	10 min 100 x *g*	Cortical corn root protoplasts	$^{36}Rb^+$, $^{32}PO_4^=$, $^{36}Cl^-$
Centrifugation through 2 step gradient of KCl:	Guy et al. 1978, 1980, 1981	Not given	Pisum sativum mesophyll protoplasts	[^3H]-3-0-methylglucose, [^3H]-D-glucose
$CaCl_2$ into dibutyl phalate	Rahat and Reinhold 1983	Not given	Pisum sativum mesophyll protoplasts	$^{86}Rb^+$
Centrifugation through ice cold 10% Ficol	Morris and Thain 1980	5 min 200 x *g*	Tobacco and Antirrhinum mesophyll protoplasts	$^{86}Rb^+$
Floatation through two phase (21 and 25%) sucrose solutions	Morris and Thain 1980	5 min 200 x *g*	Tobacco and Antirrhium mesophyll protoplasts	$^{86}Rb^+$
Centrifugation through silicone oils into mannitol or phosphoric acid	Volakita et al. 1981	< 1 min	Pea mesophyll protoplasts	$^{14}CO_2$, $H^{14}CO_3^-$, [^{14}C]-5, 5 dimethyl oxaoldine 2,4 dione
	Rahat and Reinhold 1983	< 1 min	Pea mesophyll protoplasts	[^3H]-tetraphenyl phosphonium bromide
	Akerman et al. 1983	< 1 min	Wheat mesophyll protoplasts	$^{45}Ca^{++}$ + [$^3H]_2$0
	Lin 1980, 1984	30 s	Corn root protoplasts	$^{86}Rb^+$, $^{32}PO_4^=$

protoplasts from labelled media (Table 1) may well meet many of the requirements for a transport assay, except perhaps the ability to selectively separate protoplasts damaged during the incubation period, from intact protoplasts. This method may also have other limitations in its general applicability as mesophyll protoplasts isolated from mature tobacco leaves in 0.7 M mannitol were found not to penetrate silicone oils (Morris 1977). The assay appears suitable however for separation of the denser, less vacuolated and more robust protoplasts from immature leaves, roots and some cell cultures which are generally isolated in less dense media.

The transport assay described here (see materials and methods) is based on the observation that isolated protoplasts partition in one phase of a two-phase PEG/Dextran mixture at phase equilibrium (Kanai and Edwards 1973). This method was originally used to purify protoplasts from cell debris, but has been modified here as a transport assay. Conditions were determined which, while maintaining partition of the protoplasts in one phase only, increased the partition coefficient of the compound under study such that it preferentially partitioned in the opposite phase. Table 1 illustrates the variations in partition coefficient which were obtained for different alkaloids under different conditions in two-phase PEG/Dextran systems. This differential phase partitioning and the volume ratio of the phases provides an effective method of separating protoplasts from labelled solutions. The optimum phase compositions to maximize the value of k have to be determined for each solute however. The major disadvantage of such an assay is that as PEG is used to fuse protoplast, exposure to PEG may alter the properties of the membranes. These possible effects can be minimized by short exposure (< 1 min) to PEG concentrations well below that used to fuse protoplasts. Our results show that over 98% of the proto-

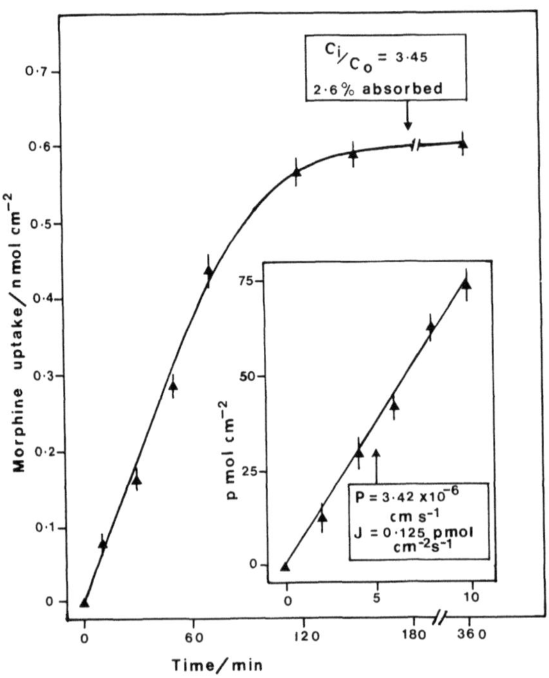

Fig. 2. Uptake of [^{3}H]-morphine by isolated mesophyll protoplasts of *Catharanthus roseus* [^{3}H]-morphine was added t = O to give a final concentration of 35 μM. Ci/Co = accumulation ratio, P = apparent permeability coefficient. J = unidirectional influx. Each *point* is the mean of 4 replicates ± SEM

plasts recovered were viable [by the FDA (fluorescence diacetate) method], and were free from organelles and cell debris which pelleted at the bottom of the tube, thus only intact viable protoplasts were assayed. Furthermore, no indication of protoplast fusion or membrane irregularities typical of PEG treatment were observed. Figure 2 illustrates the results obtained using this method in some studies of the uptake of morphine by mesophyll protoplasts of *C. roseus*. Here uptake periods have been reduced to 1–2 min, which allows measurements of initial tracer fluxes and hence calculation of membrane permeabilites. In this respect there is a notable lack of information in the literature regarding the membrane permeability of different solutes in protoplasts.

Identification of Active Transport Mechanisms

The increased use of inhibitors to study active transport mechanisms has added much to our general knowledge of transport in protoplasts (Akerman 1983; Lin 1984; Mettler and Leonhard 1979b; Rahat and Reinhold 1983; Suzuki 1982). Much of this work, which tends to avoid thermodynamic characterization of solute fluxes and driving forces, has been concerned with the energy supply for solute transport and has provided valuable information about the coupling of solute transport to energy producing reactions in the cells. However, the effects of these inhibitors on membrane permeability and electrochemical potentials has not always been studied, with the result that it is often difficult to determine if a reduced influx in the presence of an inhibitor is due to direct inhibition of an active transport system or indirect inhibition of active co-transport or exchange reactions. Furthermore, these reduced uptake rates may be a consequence of alterations in passive membrane permeability directly.

Whether these difficulties can be resolved and active transport of particular solutes established without prior thermodynamic characterization of solute movements still remains an area of some controversy.

Alkaloid Uptake by Protoplasts

It is apparent from Fig.2 that even though morphine is not produced by this species it is accumulated by *Catharanthus* mesophyll protoplasts. Recent evidence suggests that plant cells may have species specific transport mechanisms for secondary metabolites which are located at the tonoplast (Deus and Zenk 1982). We are currently studying the transport of alkaloid drugs in protoplasts to determine if the plasmalemma also shows specificity for species – specific compounds and if active transport mechanisms are involved.

Conclusions

Used with due caution, protoplasts may be favourable tools to study membrane transport, as well as other properties of the plasmalemma. Provided that there is reasonable assurance that the membranes have not altered with respect to the investigated phenomenon during protoplast isolation and that the specific properties of the protoplasts, especially in regard to their fragility are taken into account during experimentation.

Our understanding of the mechanisms of solute transport across the plasmalemma of isolated protoplasts has progressed rapidly since they were first used for transport studies some 10 years ago. Much remains to be done, however, especially with regard to the measurements of unidirectional fluxes at each membrane and in quantification of the electro-chemical gradients, since to prove active transport is occuring it is necessary to show that energy-linked transport is against an electro-chemical potential gradient. Further studies are also required in order to characterize transport mechanisms at each membrane, (the use of isolated vacuoles may be particularly useful here) and to understand the molecular basis of active transport systems. Furthermore, it is still not known to what extent the proton gradient generated by active H^+-transport creates conditions for the passive distribution of other solutes and to what extent such systems operate in different cell types. It will be interesting to determine if

Table 2. Distribution of alkaloids in two phase PEG/Dextran systems (Sharp 1984)

a)

Alkaloid	*Partition coefficient (k)*
Morphine	1.10
Narceine	1.75
Serpentine	1.30
Quinine	1.40
Quinidine	1.90

Conditions: pH7, phase ratio (7:7) positive charged top phase

b)

Alkaloid	*Phase ratio*	*k*
Morphine	7:5	1.15
Morphine	7:1	1.12
Morphine	7:0.5	1.10
Morphine	9:1	1.05

Conditions: pH7, no charge distribution

c)

Alkaloid	*pH*	*k* Top phase (+ve)	*k* Top phase (−ve)
Serpentine	7.0	1.23	1.37
Serpentine	3.0	1.25	2.0

Conditions: phase ratio 7:7 charge distribution achieved with Cl^- (−ve), $PO_4^=$ (+ve pH 7.0) or $SO_4^=$ (+ve pH 3.0).

For composition of positive and negative charged phase systems see Sharpe and Warren (1984).

$$k = \frac{\text{Con}^n \text{ of alkaloid in top phase (PEG)}}{\text{Con}^n \text{ of alkaloid in bottom phase (Dextran)}}$$

an NADH-oxidizing system can be found to operate in root protoplasts in general; in mesophyll protoplasts (especially those showing light stimulated cation and sugar uptake) and in protoplasts from cultured cells. This, and the relationship of this system to the ATP-driven H^+-sugar and H^+-amino acid co-transport system seems to be an area for intensive future study.

Although the mechanisms of solute transport in protoplasts have been studied primarily to increase our understanding of the functioning of plant cell membranes, the characterization of the transport of secondary plant metabolites may have practical benefits. These compounds are usually accumulated in plant cells but are often translocated in the whole plant from sites of synthesis to sites of storage. Little is known about the factors which control the accumulation and transport of these compounds in plant cells. Using isolated protoplasts the mechanisms of transport of these plant derived drugs is being studied with a view to their eventual modification. By modifying the transport system higher concentrations of products may be accumulated, or released by cultured plant cells which synthesize these drugs.

Acknowledgements. I would like to thank Dr. P. Sharpe for his help and advice in developing the aqueous two-phase systems and for allowing me to use the data in Table 2, and Professor M.W. Fowler in whose laboratory this work was conducted. I would also wish to thank the Wolfson Foundation for financial support.

References

Akerman KEO, Proudlove MO, Moore AL (1983) Evidence for a Ca^{2+} gradient across the plasma membrane of wheat protoplasts. Biochem Biophys Res Commun 113:171–177

Briskin DP, Leonard RT (1979) Ion transport in isolated protoplasts from tobacco suspension cells III. Membrane potential. Plant Physiol (Bethesda) 64:959–962

Deus B, Zenk MH (1982) Exploitation of plant cells for the production of natural compounds. Biotechnol Bioeng 24:1965

Gronwald JW, Leonard RT (1982) Isolation and transport properties of protoplasts from cortical cells of corn root. Plant Physiol (Bethesda) 70:1391–1395

Guy M, Reinhold L, Laties GG (1978) Membrane transport of sugars and amino acids in isolated protoplasts. Plant Physiol (Bethesda) 61:593–596

Guy M, Reinhold L, Rahat M (1980) Energisation of the sugar transport mechanism in the plasmalemma of isolated mesophyll protoplasts. Plant Physiol (Bethesda) 65:550–553

Guy M, Reinhold L, Rahat M, Seiden A (1981) Protonation and light synergistically convent plasmalemma sugar carrier system in mesophyll protoplasts to its fully activated form. Plant Physiol (Bethesda) 67:1146–1150

Heller R, Grignon C, Rona JP (1974) Importance of the cell wall in the thermodynamic equilibrium of ions in free cells of *Acer pseudoplantanus* L. In: Zimmerman U, Dainty J (eds) Membrane transport in plants. Springer, Berlin Heidelberg New York, pp 239–243

Kanai R, Edwards GE (1973) Purification of enzymatically isolated mesophyll protoplasts from C_3, C_4 and CAM plants using an aqueous Dextran-PEG two phase system. Plant Physiol (Bethesda) 52:484–490

Leigh RA (1983) Methods, progress and potential for the use of isolated vacuoles in studies of solute transport in higher plant cells. Physiol Plant 57:370–396

Lin W (1980) Corn root protoplasts isolation and general characterization of ion transport. Plant Physiol (Bethesda) 66:550–554

Lin W (1981) Inhibition of anion transport in corn root protoplasts. Plant Physiol (Bethesda) 68: 435–438

Lin W (1982) Responses of corn root protoplasts to exogenous NADH: oxygen consumption, ion uptake and membrane potential. Proc Natl Acad Sci USA 79:3773–3776

Lin W (1984) Further characterisation of the transport property of plasmalemma NADH oxidation system in isolated corn root protoplasts. Plant Physiol (Bethesda) 74:219–222

McDaniels C (1983) Transport of ions and organic molecules. In: Evans DA, Sharp WR, Ammirato PV, Yamada Y (eds) Handbook of plant cell culture, vol 1. MacMillian, New York, pp696–714

Mettler IJ, Leonard RT (1979a) Ion transport in isolated protoplasts from tobacco suspension cells I. General characteristics. Plant Physiol (Bethesda) 63:183–190

Mettler IJ, Leonard RT (1979b) Ion transport in isolated protoplasts from tobacco suspension cells II. Selectivity and kinetics. Plant Physiol (Bethesda) 63:191–194

Morris P (1977) Ion transport and metabolism in leaf tissue and isolated protoplasts. Ph D thesis, Univ of East Anglia, Norwich, England

Morris P, Thain JF (1980) Comparative studies of leaf tissue and isolated mesophyll protoplasts II. Ion relations. J Exp Bot 31:97–104

Morris P, Thain JF (1983) Improved methods for the measurement of total cell surface area in leaf mesophyll tissue. J Exp Bot 34:95–98

Morris P, Linstead P, Thain JF (1981) Comparative studies of leaf tissue and isolated protoplasts III. Effects of wall degrading enzymes and osmotic stress. J Exp Bot 32:801–811

Power JB (1977) The physiology of isolated plant protoplasts. In: Smith H (ed) The molecular biology of plant cells. Blackwells botanical monographs, vol 14. Blackwells, Oxford, pp 418–428

Power JB, Cocking EC (1970) Isolation of leaf protoplasts: macromolecular uptake and growth response. J Exp Bot 21:64–70

Racusen RH, Kinnersley AM, Galston AW (1977) Osmatically induced changes in electrical properties of plant protoplast membranes. Science (Wash DC) 198:405–407

Rahat M, Reinhold L (1983) Rb^+ uptake by isolated pea mesophyll protoplasts in light and dark. Physiol Plant 59:83–90

Robinson DJ, Mayo MA (1977) Changing rates of uptake of [^{3}H] leucine and other compounds during culture of tobacco mesophyll protoplasts. Plant Sci Lett 8:197–204

Rona MJP (1973) Premieres mesures du potential électrique sur des protoplastes et des vacuoles isolees d'*Acer pseudoplantanus* L. Comptes rendus 227:185–188

Rubinstein B (1978) Use of lipophilic cations to measure the membrane potential of oat leaf protoplasts. Plant Physiol (Bethesda) 62:927–292

Rubinstein B, Tatter TA (1980) Regulation of amino acid uptake into oat mesophyll cells: a comparison between protoplasts and leaf segments. J Exp Bot 31:269–279

Ruesink AW (1978) Leucine uptake and incorporation by convolvalus tissue culture cells and protoplasts under severe osmotic stress. Physiol Plant 44:48–56

Schmidt R, Poole RJ (1980) Isolation of protoplasts and vacuoles from storage tissue of red beet. Plant Physiol (Bethesda) 66:25–28

Sharpe PT (to be published 1984) Partition of alkaloids in aqueous two-phase systems

Sharpe PT, Warren Gs (1984) The incorporation of glycolipids with defined carbohydrate sequence into liposomes and the effects on partition in aqueous two-phase system. Biochim Biophys Acta 772:176–182

Singer SR, McDaniel CN (1982) Transport of the herbicide 3-amino, 1, 2, 4-triazole by cultured tobacco cells and leaf protoplasts. Plant Physiol (Bethesda) 69:927–929

Suzuki M (1981) L-Leucine transport in isolated protoplasts of vinca suspension culture: characteristics of uptake. Plant Cell Physiol 22:1209–1278

Suzuki M (1982) Sugar co-transport with H^+ in vinca protoplasts: inhibition of the uptake by ATPase inhibitors. Plant Sci Lett 27:173–180

Taylor ARD, Hall JL (1976) Some physiological properties of protoplasts isolated from maize and tobacco tissues. J Exp Bot 27:383–391

Thain JF (1983) Curvative correction factors in the measurement of cell surface areas in plant tissues. J Exp Bot 34:87–94

Thain JF (1984) The analysis of radioisitopic tracer flux experiments in plant tissues. J Exp Bot 35:444–453

Volokita M, Kaplan A, Reinhold L (1981) Evidence for mediated HCO_3- transport in isolated pea mesophyll protoplasts. Plant Physiol (Bethesda) 67:1119–1123

The Binding of Anion Transport Inhibitors on the Plasmalemma Isolated from Corn Root Protoplasts

W. LIN

Introduction

DIDS[1] and SITS[2], two amino-reactive disulfonic stilbene derivatives, have been shown to inhibit anion transport processes in animal cell membrane irreversibly [8, 18]. These chemical modifiers have been shown to covalently bind to the plasmalemma and also have been used successfully to identify and characterize the band 3 membrane protein as the carrier for Cl^- and $SO_4^=$ transport in red blood cells [2,3,4,8].

Recently, in corn root tissues, we have shown that DIDS and SITS are effective inhibitors of Cl^- and $SO_4^=$ transport in isolated protoplasts and root segments [10]. The inhibitory effect of DIDS on Cl^- transport has also been found in *Chara coralline* cells [7]. Isolated plasmalemma ATPase from corn roots [10], oat roots [5] and tobacco callus [19] was found to be inhibited by DIDS. The detail mechanism of DIDS and SITS inhibition in plant cells is not known. In corn roots, it was suggested that DIDS blocks the plasmalemma Cl^- channel while the Cl^- activated H^+-ATPase is not affected [1]. However, in oat roots, Churchill et al. [5] showed that DIDS inbited a Cl^- stimulated-vanadate-insensitive microsomal ATPase activity.

In the present paper, we report on the attempt of using DIDS and SITS to identify the anion transport carrier in isolated corn root protoplasts. We found that the binding of DIDS and SITS to the plasmalemma can be demonstrated by the fluorescence property of these chemical modifiers and radiolabeled $^3H_2$DIDS.

Materials and Methods

Root protoplasts were isolated from 3-day-old etiolated corn root (*Zea mays* L. Pioneer Hybrid 3320) seedlings as previously described [9, 10] and suspended for all experiments in a basic medium of 0.65 M mannitol, 0.2 mM $CaCl_2$ and 1 mM Hepes buffer, pH 6.0. The previously described rapid separation technique [9] was used to measure Cl^- influx into protoplasts. About 0.5 million protoplasts were used in each measurement. Carrier-free $^{36}Cl^-$ (New England Nuclear) in 1 mM KCl was added in the basic medium to measure Cl^- influx. A 20-min radioactive exposure period was employed.

1 DIDS, 4,4'-diisothiocyano-2,2'-stilbenedisulfonic acid
2 SITS 4-acetamido-4'-isothiocyano-2,2'-stilbene disulfonic acid

The Physiological Properties of Plant Protoplasts
Edited by P. E. Pilet

Plasmalemma was isolated from protoplasts by resuspending protoplast pellet with 15 ml of 0.6 M sorbitol, 50 mM phosphate buffer (pH 7.9). Fifteen ml of 50 mM Tris-Mes (pH 7.9), 2 mM EDTA and 6 mM DTT were then added. After 5 min at room temperature, the mixture was forced through a 20 nm nylon mesh 5 times and layered on a 10 ml solution of 35% sucrose and centrifuged at 10,000 *g* for 10 min. The upper layer of gradient was then layered on a sucrose gradient consisting of 10 ml each of 31 and 38% sucrose solution. After 45 min at 80,000 *g* centrifugation, plasmalemma was collected at the 31/38% sucrose interface and washed with 20 ml of 0.25 M sucrose, 10 mM Tris-Mes pH 6.0 buffer. A 80,000 *g* 30 min centrifugation was used to pellet the plasmalemma.

Binding of DIDS or SITS was performed by preincubating protoplasts with different concentration of DIDS (0.05 and 0.01 mM, see legend to Fig.1 for the concentration in each experiment) in basic medium for 10 min. The excess amount of DIDS or SITS was removed by 3 washes of treated protoplasts with mannitol basic medium.

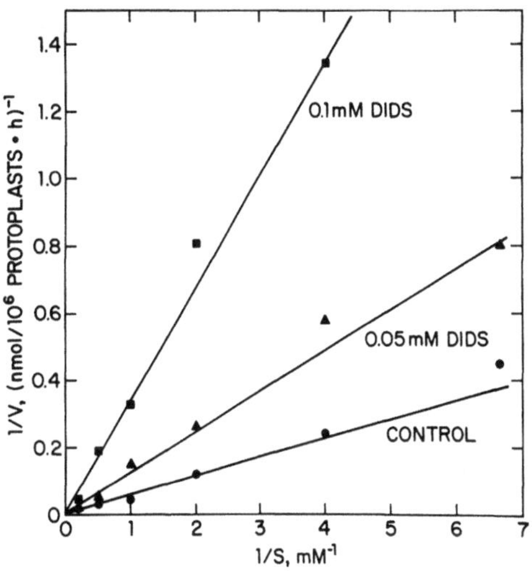

Fig. 1. Double reciprocal plots of the effect of DIDS on the Cl^- uptake kinetics in isolated corn root protoplasts

Fluorescence spectra were taken with a Hitachi Perkin Elmer MPF-2A Fluorescence Spectrophotometer with a Hitachi QPD33 recorder.

$^3H_2$DIDS was synthesized according to the method of Cabantchik and Rothstein [3] by Amersham.

Membrane ATPase activity was assayed as previously described [10].

Results and Discussion

The inhibition of anion uptake in red blood cells by SITS and DIDS was proposed to be the result of a covalent binding of these chemicals to the outer peptide chain of the band 3 protein, the major anion transport carrier [4, 8, 18]. DIDS has also been reported as an effective inhibitor of anion transport in plant cells [7, 10], and the information in the mechanism of its inhibition has yet to be emerged. In *Chara coralline* cells, the DIDS inhibition on Cl^- influx was found to be reversible and has been suggested that DIDS is acting as a competitor for the Cl^- binding site [7]. However, in isolated corn root membrane vesicles, the inhibition by DIDS and SITS on Cl^- stimulated H^+-ATPase was shown to be noncompetitive [1]. If DIDS molecule is covalently reacting with ϵ-amino groups of two lysine residues on the transport protein [4, 18], and if these two lysine residues are the active site of Cl^- transport protein, no competitive inhibition kinetic for the DIDS inhibition on Cl^- transport can be expected unless the geometry for DIDS binding on the transporter protein is low. Kinetic analyses of DIDS inhibition on Cl^- influx in isolated corn root protoplasts (Fig. 1) showed that as that reported in *Chara coralline* cells [7], DIDS is inhibiting Cl^- influx competitively [17]. The result further substantiates that DIDS is a competitive for the Cl^- binding site. Detail binding studies of DIDS and anion transport protein in plant cell as that being performed in red blood cells [4, 8, 18] have become critical tests in the understanding of anion transport protein in plant cells.

Fluorescence technique and radioisotope labeling of membrane surface were used to investigate the binding properties of DIDS and membrane protein in isolated corn root protoplasts. Preliminary results showed that the conjugated, double-bond arrangement of DIDS and SITS molecules gave a strong and distinct fluorescence signal. The reaction between these chemicals with free amino groups of protein does not alter this arrangement. Therefore, one should be able to use this fluorescence properties to investigate the binding of these chemicals with plasmalemma surface protein. In red blood cells, $^3H_2$DIDS has been successfully used to identify and characterize the anion transport protein in the plasmalemma. Combination of the DIDS fluorescence measurement and radiolabeling of the plasmalemma had enabled us to show that DIDS is bound to the plasmalemma (may be on the anion transport protein) of isolated corn root protoplasts.

Fluorescence spectra of SITS and DIDS solution and suspension of plasmalemma isolated from protoplasts which have been treated with these chemicals were shown in Fig. 2. A greater than 10 nm peak shift in both the emission and excitation spectra was found in the plasmalemma suspension. Since membrane was isolated after the incubation of protoplasts with DIDS or SITS, the fluorescence signal must come from the chemical modifier which was tightly bound to the membrane. The peak shift in the fluorescence spectra reflect the formation of phospholipid (membrane) micelles with DIDS molecules enclosed in the center. SDS polyacrylamide gel electrophoresis of these membranes showed that DIDS is binding to 3 to 4 out of about 20 polypeptides of the membrane (Lin, unpublished result).

Time course of the DIDS binding with plasmalemma both from the fluorescence measurement (Fig. 3) and radioisotope labeling (Fig. 4) showed a faster saturation of DIDS-binding sites on the plasmalemma. When protoplasts were incubated with 0.01 mM DIDS solution, a maximal binding (or equilibrium) was reached within 30 min. This is in consistence with the notion that binding takes place on the membrane surface as have been shown in several phytohormone-binding studies in plant tissues [14, 15].

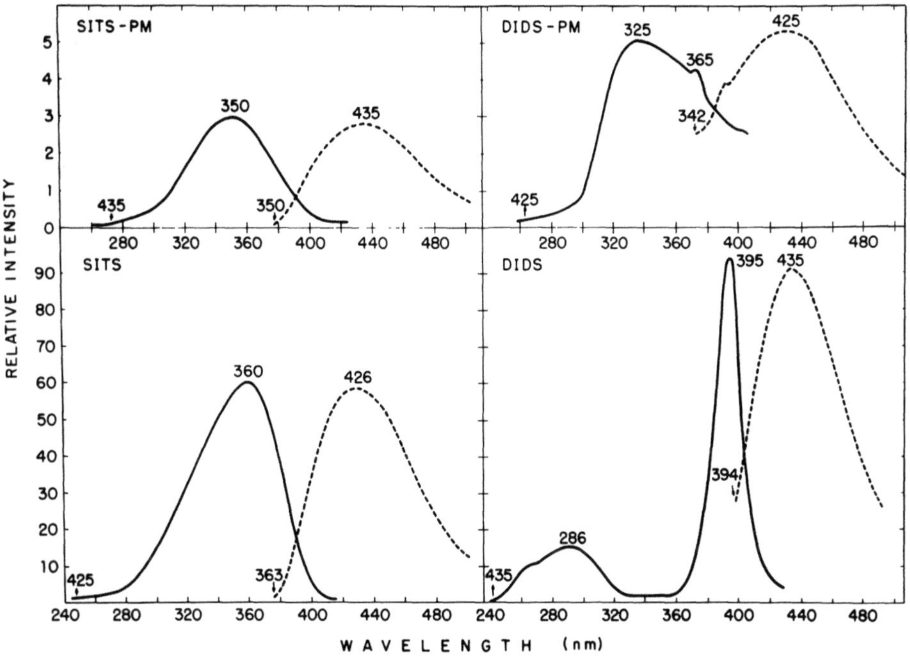

Fig. 2. Fluorescence spectra of 0.1 mM SITS and DIDS solution and the plasmalemma suspension (SITS-PM and DIDS-PM) isolated from protoplasts which have been treated with 0.1 mM DIDS and SITS (see Methods and Materials). *Downward arrows* represent the excitation wavelength used to take emission spectra and *upward arrows* represent the emission wavelength used to take excitation spectra

Fig. 3

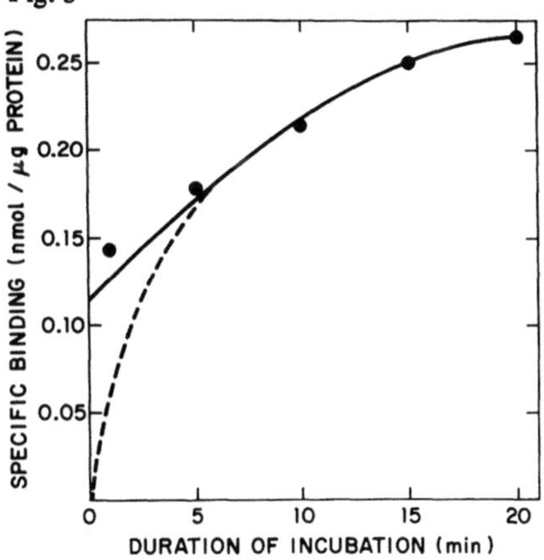

Fig. 4

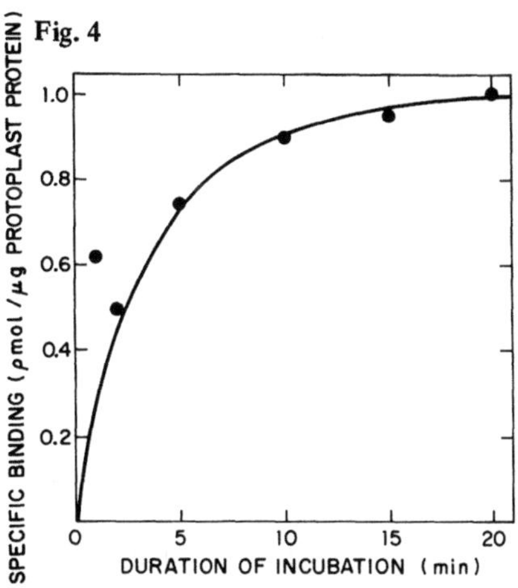

Fig. 3. Time course of the DIDS binding to the plasmalemma by the measurement of fluorescence intensity. At different time intervals, plasmalemma was isolated from the protoplasts which have incubated in a 0.01 mM DIDS solution. Excitation wavelength was 342 nm and emission was measured at 425 nm

Fig. 4. Time course of the $^3H_2$DIDS binding on the isolated corn root protoplasts. 0.01 mM $^3H_2$DIDS was used in the protoplasts incubation medium. At each time interval, protoplasts were washed 3 times and the radioactivity was measured

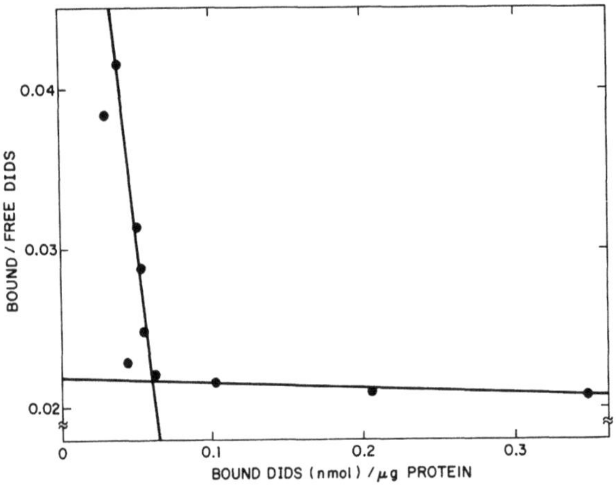

Fig. 5. Scatchard plot of DIDS binding to the plasmalemma by the measurement of fluorescence intensity. Excitation wavelength was 342 nm and emission was measured at 425 nm. Protein was measured by Bio-rad reagent with BSA as standard

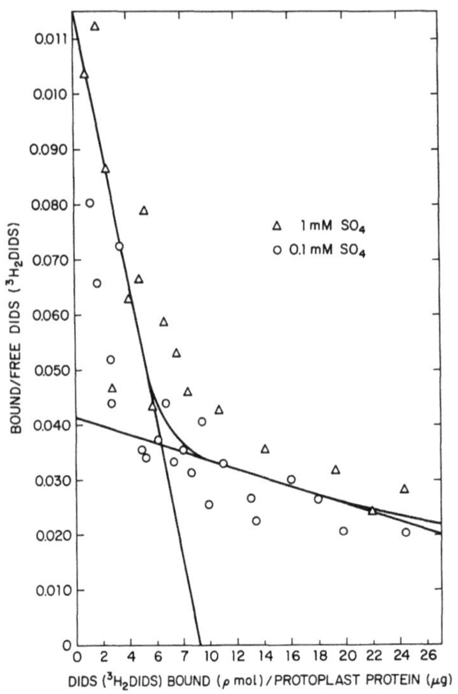

Fig. 6. Scatchard plot of $^{3}H_2$DIDS binding to the protoplasts isolated from corn root tissues. $SO_4^{=}$ was added in the DIDS incubation media

Scatchard analyses [16] of DIDS binding with protoplast plasmalemma (Figs. 5 and 6) showed that there are two DIDS-binding sites on the membrane; one with high binding affinity, probably the anion-binding sites, and a low affinity nonspecific binding sites [6, 15, 16]. Figure 6 also shows that the binding of DIDS with the membrane cannot be protected by high concentrations of SO_4^{2-} in the medium. This is in contrast with the result from the kinetic study (Fig. 1). If the inhibition of DIDS is a competitive kinetic, one would expect a protection of binding site by substrate as result from the competition of DIDS with substrate (Cl^- or SO_4^{2-}). However, since the molecular sizes of DIDS and Cl^- or SO_4^{2-} are so different, it is hard to visualize that they are binding competitively at the exact same site of the carrier. Also, if DIDS is covalently bound to the active site of the carrier protein, one should not see a competitive inhibition of DIDS or Cl^- uptake. The only possible explanation we can offer at the present time is that DIDS or SITS reacts with free amino groups of the anion carrier protein which is very close to the Cl^- or SO_4^{2-} binding or entering site. The binding of DIDS can then block the anion entrance. However, this blockage can be forced to reopen with a higher amount of substrate and resulting a competitive inhibition on the Cl^- uptake but not on the binding kinetic. A more detailed study on the direct binding of DIDS to the membrane protein is needed to provide a definitive answer.

Previously, we reported that DIDS inhibited ATPase activity of isolated corn root plasmalemma [10]. A similar result was observed by Churchill et al. in oat roots [5]. However, the inhibition by DIDS on membrane ATPase activity depends on the experimental condition as noticed by Keifer et al. [7]. When the membrane ATPase activity was assayed from the plasmalemma which was isolated from the protoplasts previously treated with DIDS, no loss of ATPase activity could be detected in the treated plasmalemma (Table 1). The lack of DIDS sensitivity was not due to the absence of DIDS as shown in Fig. 1, but rather that DIDS has never been in direct contact with ATPase when the chemical was added in the intact protoplasts. This result is in consistance with that reported in *Chara coralline* [7]. The inhibiting effect on ATPase activity we observed previously [10] (and Table 1) may have resulted from the fact that after tissue homogenation, the orientation of some membrane vesicles was changed and ATPase molecule was exposed to the external solution which was then become directly in contact with DIDS.

Table 1. Effect of DIDS on plasmalemma ATPase activity isolated from corn root tissue

	ATPase Activity μmol Pi mg^{-1} protein h^{-1}			
	+KCl	−KCl	Δ	%
I. Tissue Homogenate[a]				
Control	34.78	19.93	14.85	100
DIDS	15.98	11.96	4.02	27
II. Protoplasts[b]				
Control	38.36	27.24	11.12	100
DIDS	38.02	27.13	10.89	98

[a] 20–25 μg protein of plasmalemma fraction was used in the assay

[b] 5-10 μg protein of plasmalemma fraction was used in the assay

Assay medium consisted of 1 mM ATP, 1.5 mM $MgSO_4$, 50 mM KCl, and 33 mM Tris-Mes buffer, pH 6.5. Incubation was 20 mm at 38°C

Conclusion

Exposure of membrane surface of plant cells through the removal of the cell wall, and isolation of plant protoplasts, has facilitated the study of membrane surface properties which is important in the understanding of transport mechanism as evidenced by the present paper and that reported previously [10,–12]. Caution must be exercised during the removal of the cell wall. Commonly used cell-wall digestion enzymes in the protoplast isolation media possess high activity of protease which could have some adverse effect on cell surface protein [13]. Nevertheless, the use of isolated protoplasts has enabled us to use chemical probes and other special techniques which have been widely used in the red blood cell membrane transport research [4, 8, 18] to characterize and identify transport proteins in plant cells. Isolated root protoplasts and tightly sealed membrane vesicles [19] have provided two ideal complementary materials for the in vivo and in vitro study of ion transport mechanism in plant cells.

Acknowledgements. The technical assistance of R.M. Schreiner and J.M. Barker and the manuscript preparation by T. Sparre are greatly appreciated.

References

1. Bennett A, Spanswick RM (1983) Optical measurements of ΔpH and $\Delta\psi$ in corn root membrane vesicles. Kinetic analysis of Cl^- effects on a proton-translocating ATPase. J Membr Biol 71:95–107
2. Cabantchik ZI, Rothstein A (1972) The nature of the membrane sites controlling anion permeability of human red blood cells as determined by studies with disulfonic stilbene derivatives. J Membr Biol 10:311–330
3. Cabantchik ZI, Rothstein A (1974) Membrane proteins related to anion permeability of human red blood cells I. Localization of disulfonic stilbene binding sites in proteins involved in permeation. J Membr Biol 15:207–226
4. Cabantchik ZI, Kraaf PA, Rothstein A (1978) The anion transport system of the red blood cells. The role of membrane protein evaluated by the use of probes. Biochim Biophys Acta 515:239–302
5. Churchill KA, Holaway B, Sze H (1983) Separation of two types of electrogenic H^+-pumping ATPases from oat rots. Plant Physiol (Bethesda) 73:921–928
6. Kalbitzer HR, Stehlik D (1979) On the analysis of competitive binding of various ligands to cooperative and independent binding sites of macromolecules. Z Naturforsch Sect C Biosci 34:757–769
7. Keifer DW, Franceschi VR, Lucas WJ (1982) Plasmalemma chlorite transport in *Chara coralline*. Inhibition by DIDS. Plant Physiol (Bethesda) 70:1327–1334
8. Knauf PA (1979) Erythrocyte anion exchange and the band 3 protein: transport kinetics and molecular structure. Curr Top Membr Transp 12:251–363
9. Lin W (1980) Corn root protoplasts: isolation and general characterization of ion uptake properties. Plant Physiol (Bethesda) 66:550–554
10. Lin W (1981) Inhibition of anion transport in corn root protoplasts. Plant Physiol (Bethesda) 68:435–438
11. Lin W (1982) Isolation of NADH oxidation system from the plasmalemma of corn root protoplasts. Plant Physiol (Bethesda) 70:326–328

12. Lin W (1982) Responses of corn root protoplasts to exogenous NADH. Oxygen consumption, ion uptake and membrane potential. Proc Natl Acad Sci USA 79:3773–3776
13. Lin W, Wittenbach VA (1981) Subcellular localization of proteases in wheat and corn mesophyll cells. Plant Physiol (Bethesda) 67:969–972
14. Murphy GJP (1979) Plant hormone receptors: comparison of naphthaleneacetic acid binding by maize extracts and by a nonplant protein. Plant Sci Lett 15:183–191
15. Rubery PH (1981) Auxin receptors. Annu Rev Plant Physiol 32:569–596
16. Scatchard G (1949) The attractions of proteins for small molecules and ions. Ann NY Acad Sci 51:660–672
17. Segel IH(1975) Enzyme kinetics. Wiley, New York, pp 210–212
18. Shami Y, Rothstein A, Knauf PA (1978) Identification of the Cl^- transport site of human red blood cells by a kinetic analysis of the inhibitory effects of a chemical probe. Biochim Biophy Acta 508:357–363
19. Sze H (1982) Characterization of nigericin-stimulated ATPase from sealed microsomal vesicles of tobacco callus. Plant Physiol (Bethesda) 70:498–505

Intracellular Transport of Metabolites in Protoplasts: Transport Between Cytosol and Vacuole

Th. BOLLER

Introduction

One of the intriguing characteristics of the eukaryotic cell is the compartmentation of its metabolism. The biochemist's charts of metabolic pathways, derived initially from prokaryotes, show a very complex network of enzymatic reactions that can be regulated allosterically. However, these charts are inadequate to indicate the additional complexity of intracellular compartmentation in eukaryotes, where metabolites can be accumulated in one compartment, excluded from another, and exchanged between them in a regulated manner. In plant cells in particular, the transport of metabolites between the cytosol and many intracellular compartments, including plastids, mitochondria, peroxisomes, and vacuoles, is a centerpiece of many metabolic pathways.

In intact tissue, information on intracellular transport can be collected only by indirect means in most cases. A well-known technique is the compartmental analysis which is based on measurements of the efflux of radioactively labelled metabolites. However, it is difficult to perform and interpret this indirect approach correctly (Zierler 1981).

In protoplasts, in contrast, a direct approach to intracellular transport is possible because intact organelles can be isolated quickly from protoplasts (Galun 1981). Therefore, protoplasts are an important tool to understand the flow of metabolites between different intracellular compartments. Here, I want to discuss the use of protoplasts to investigate transport between the cytosol and the vacuole. Part of this subject has recently been reviewed by Leigh (1983). Transport processes between the cytosol and chloroplasts and mitochondria are considered in Chap. 11, this Vol. (Hampp).

Significance of Metabolite Transport Between Cytosol and Vacuole

The large central vacuole fills ca. 90% of the volume of a mature plant cell. It has only few biosynthetic capacities, but it is extremely important for cellular metabolism: It is the main storage site for metabolites that are transiently or permanently withdrawn from the biosynthetically active cytoplasm (Matile 1978; Leigh 1983). Examples include the permanent deposition of secondary metabolites, the seasonal

The Physiological Properties of Plant Protoplasts
Edited by P. E. Pilet

deposition and remobilization of oligosaccharides in storage organs, and the diurnal accumulation and release of organic acids in plants exhibiting crassulacean acid metabolism. In these and in many more cases, knowledge about the intracellular transport between cytosol and vacuole is a key to the understanding of plant metabolism.

The use of isolated vacuoles for transport studies in higher plants has recently been reviewed by Leigh (1983). Here, two case studies are discussed in order to illustrate the possibilities and problems of the different experimental approaches to study intracellular transport with protoplasts. The first deals with accumulation of arginine in yeast and other fungi, the second with intracellular transport of sugars in higher plants.

Intracellular Transport of Arginine in Yeast and Other Fungi

Vacuolar Localization of Arginine in Yeast and Other Fungi

Yeast cells are used as model eukaryotic microorganisms, much like bacteria, by many biochemists. However, with regard to compartmentation, they can also be considered model plants (Matile 1978). Like plant cells, they have a large central vacuole, at least in certain stages of growth, and they deposit metabolites in the vacuole. It has long been known that yeast cells accumulate large amounts of amino acids during normal growth; particularly important among them is arginine. Arginine can be accumulated in very large quantities when present in the growth medium, and the stored arginine can later serve as a nitrogen source for one full growth cycle when the exogenous nitrogen supply is suddenly withdrawn.

Compartmental analysis in yeast and other fungi indicated that most of the free amino acids, and most of the arginine in particular, were metabolically inactive, i.e., were not used for protein synthesis. Indirectly, this indicated that these amino acids were located, to a large extent, in a compartment separated from the metabolically active cytosol (Davis 1975).

When techniques were developed for the preparation of protoplasts from yeast cells, and of vacuoles from protoplasts, it became possible, for the first time, to study arginine compartmentation directly. Experiments with isotonically prepared vacuoles showed that the isolated, purified vacuoles contained a large part of the intracellular amino acid pool and most of the intracellular arginine (Wiemken and Dürr 1974). The vacuolar localization of most of the free arginine was corroborated by differential extraction of the soluble pools from the cytosol and the vacuoles (Huber-Wälchli and Wiemken 1979).

Arginine Transport in Isolated Vacuoles in Vitro

Isolated vacuoles could be used to study vacuolar transport directly. They were incubated with radioactive amino acids for short periods of time, and then separated from the incubation medium rapidly by centrifugation through an isotonic density gradient (Boller et al. 1975). Using this technique, arginine transport of yeast vacuoles was studied in some detail (Boller et al. 1975; Boller 1977).

It was found that a specific, saturable arginine permease was present in vacuolar membranes. (A permease, often also called carrier or translocator, is a membrane enzyme catalyzing a physical reaction – the diffusion through a membrane – instead of a chemical reaction.) After less than 5 min, radioactive arginine was tenfold more concentrated inside the vacuoles than outside. However, the transport of arginine was not active; it proceeded in the absence of any energy source and did not lead to net arginine accumulation. The radioactive arginine taken up from the incubation medium was exchanged against unlabelled arginine already present in isolated vacuoles. ATP stimulated transport of ^{14}C-labelled arginine into the vacuoles about twofold; $NADPH_2$ and Na^+ also stimulated transport (Fig. 1). Still, this was not an active process; efflux of [^3H]arginine from the vacuoles was similarly stimulated by ATP, $NADPH_2$ and Na^+ (Fig. 1).

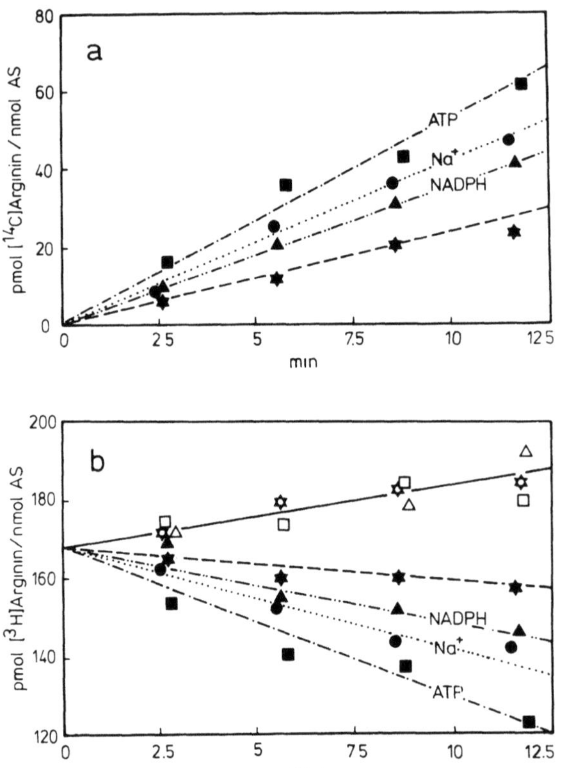

Fig. 1A,B. Stimulation of arginine transport in isolated yeast vacuoles by ATP, Na^+, and NADPH. Isolated vacuoles were incubated for 18 h at $0^{\circ}C$ with [^3H]arginine to label the internal arginine pool, and then purified on an isotonic density gradient. The [^3H]arginine-loaded vacuoles were incubated in the presence (*closed symbols*) or in the absence (*open symbols*) of 0.15 mM [^{14}C]arginine without further additions (★, ☆) or with 1 mM Na_2ATP + 2 mM Na^+ (■, □), 4 mM Na^+ (●), or 1 mM Na_4 NADPH (▲,△) for the time indicated and then centrifuged and analyzed. Incubation conditions: $25^{\circ}C$, Tris-PIPES-buffer, 5 mM, pH 6.8. All solutions added were adjusted to pH 6.8 with Tris or PIPES. **A** Influx of [^{14}C] arginine, referred to the total amino acid (AS) content of the vacuoles. **B** Efflux of [^3H]arginine, referred to the total amino acid (AS) content of the vacuoles. Note that in the absence of exogenous [^{14}C]arginine, the [^3H]arginine content of the vacuoles (open symbols) appears to increase. This is due to a decrease in total amino acid content of the vacuoles during incubation. (Boller 1977)

How was it possible, then, that vacuoles retained the accumulated arginine? Vacuoles were found to accumulate polyphosphate, a polyanion that cannot pass membranes (Urech et al. 1978). Phosphate residues in polyphosphate and arginine were present in nearly equimolar amounts (Dürr et al. 1979). This finding suggested that arginine, a cation, associated with polyphosphate like with a cation exchanger and thereby accumulated in the vacuole (Boller 1977; Matile 1978).

An arginine transport system similar to the one discovered in yeast was found in vacuoles purified from *Neurospora crassa* (Martinoia et al. 1979).

Energy Coupling of Arginine Transport in Isolated Vesicles of the Vacuolar Membrane

The vacuolar membrane (tonoplast) can be prepared from isolated vacuoles by osmotic shock (Ohsumi and Anraku 1981). Such vesicles do not contain stored arginine. However, the vesicles were found to take up arginine from the medium. Arginine transport could be detected only when ATP was given as an energy source. The accumulated arginine was lost rapidly when the vesicles were treated with an uncoupler. These results strongly suggest that a proton-translocating ATPase which builds up an electrochemical potential difference for protons (proton-motive force) functions as an energy-donating system for active arginine uptake in isolated vesicles. These findings called for a study of ATPase on the vacuolar membranes. Kakinuma et al. (1981) indeed found such an ATPase and could distinguish it from other membrane ATPases by its inhibitor sensitivities. A similar ATPase was found in vacuolar membranes of *Neurospora crassa*, and it was hypothesized that this enzyme functioned similarly in transport and sequestration of arginine in the vacuoles of this organism (Bowman and Bowman 1982). The vacuolar ATPase was inhibited by 50 mM nitrate while the ATPases of plasma membrane and of mitochondria were not; the vacuolar ATPase was insensitive to the inhibitor of mitochondrial ATPase, azide, and to the inhibitor of plasma membrane ATPase, vanadate (Bowman 1983).

While the studies with isolated tonoplast vesicles emphasize the role of ATPase in the vacuolar accumulation of arginine, the studies with intact vacuoles suggest that arginine compartmentation can also be brought about in the absence of active transport, by an ion trap mechanism (Matile 1978; Dürr et al. 1979).

Compartmental Analyses in Intact Fungal Cells

Arginine and phosphate residues in polyphosphate were present in yeast vacuoles in approximately equimolar concentrations (Urech et al. 1978). However, when cells were starved for phosphate, they no longer accumulated polyphosphate, but still stored considerable quantities of arginine (Dürr et al. 1979). Since vacuoles could not be isolated from phosphate-starved cells, it remained an open question whether such arginine was still present in the vacuoles. Cramer et al. (1980) studied arginine compartmentation in *Neurospora crassa.* They found that *Neurospora*, under normal growth conditions, contained equal amounts of polyphosphate-phosphate and arginine. The intracellular arginine level was not changed when the cells were starved for phosphate, although 90% of the polyphosphate was mobilized. Compartmental analysis indicated that almost all the arginine was still sequestered and not available for protein synthesis in the cytoplasm. Thus, arginine was accumulated in the *Neurospora* vacuoles not only together with polyphosphate, but also in its absence. The anions accompanying vacuolar arginine in the latter case were not identified.

Drainas and Weiss (1982) studied the energy requirement for arginine compartmentation in *Neurospora* by compartmental analysis in vivo. Interestingly, they found that arginine remained vacuolar in the presence of uncouplers which strongly lowered ATP levels and energy charge in the cells. Thus, arginine did not leave the vacuoles in vivo when the energy supply was stopped. This result resembles the findings obtained with isolated vacuoles but is in sharp contrast to the situation described for tonoplast vesicles where uncouplers caused an immediate loss of arginine previously accumulated (Oshumi and Anraku 1981).

Intracellular Transport of Sugars in Higher Plants

Accumulation of Sugars in Fruits and Storage Organs in Vivo

Sucrose plays a primary role in long-distance transport of photosynthetic products from the source leaves via the phloem to the sinks. In addition, sucrose may accumulate in large quantities in the fruits and storage organs of many plants. Sometimes, glucose and fructose are stored instead of sucrose; in various plant species, other oligosaccharides may fulfill storage functions (see Keller et al. , Chap. 14, this Vol.). The sheer quantity of the sugars accumulated in storage organs or fruits indicates that a large part of it must be present in the vacuole.

Analysis of isolated vacuoles from sugar beet (Doll et al. 1979) and from sugar cane cells (Thom et al. 1982) corroborated this. Interestingly, in gentian roots, gentianose was stored in the vacuoles while sucrose was much less concentrated in vacuoles than in the cytosol (Keller and Wiemken 1982).

The presence of large amounts of sugars does not necessarily mean that the sugars are accumulated in the vacuole as compared to the cytosol; they might be present in equal concentrations in both compartments. However, compartmental analysis of sucrose in sugar beet taproot tissue indicated that the cytosolic sucrose concentration was 76 mM, while the vacuolar concentration was 514 mM (Saftner et al. 1983). This result suggests that sucrose is in fact accumulated in the vacuole.

Sugar Transport in Isolated Vacuoles in Vitro

Vacuoles isolated from various plant tissues have been used to study sugar transport in vitro. Guy et al. (1979) studied uptake of a glucose analog, 3-0-methylglucose, in vacuoles isolated from pea mesophyll protoplasts. They found time-dependent uptake of this compound but not of the unnatural enantiomer of glucose, L-glucose, indicating the presence of a stereospecific permease in the tonoplast. ATP stimulated transport by about 50%, while a proton ionophore inhibited it. While the data were in agreement with a possible contribution of ATP and of a proton gradient to sugar transport, the authors noted that they did not actually find "uphill" transport. Moreover, a high transport rate was found in control vacuoles, in the absence of any energy source.

Doll et al. (1979) and Willenbrink and Doll (1979) studied sucrose transport in red beet vacuoles isolated directly from the tissue. These vacuoles contained large amounts of sucrose. They found a net increase in vacuolar sucrose concentration when the vacuoles were incubated with 10 mM unlabelled sucrose, even in the absence of an energy source. However, it is not clear, from the published information, how these experiments could be performed with the accuracy needed: The increase in sucrose is very small compared to the large sucrose content present initially in the vacuoles. The results obtained with [^{3}H]sucrose seem more reliable; these results indicated a slight stimulation of sucrose transport by ATP (Doll et al. 1979). The transport of sucrose could be inhibited by raffinose and other sugar analogs (Willenbrink and Doll 1979).

More recently, Thom et al. (1982) performed an analysis of sugar uptake by vacuoles isolated from sugarcane suspension cultures. These authors found transport systems for 3-0-methylglucose and sucrose. The pH optimum of both was around pH 5.5; the K_m values for 3-0-methylglucose and sucrose transport were 170 and 70 μM, respectively. At the end of a transport experiment, 3-0-methylglucose was more concentrated in the vacuoles than outside. However, the vacuoles contained large amounts of glucose at the beginning; since 3-0-methylglucose is a glucose analogue, it may have entered the vacuoles in exchange against glucose (Thom et al. 1982). Therefore, as noted by the authors, the transport processes found in isolated sugar cane vacuoles were not necessarily active.

In summary, sugar transport in isolated vacuoles resembled arginine transport in yeast vacuoles: Transport could be measured in the absence of an energy source. The sugars found in vacuoles after isolation did not flow out in the course of experimental manipulation. It will be interesting to find out how the sugars are held back in the isolated vacuoles.

Energization of the Tonoplast of Higher Plants

Vacuolar sap is generally more acidic than the cytoplasm. Therefore, it has long been postulated that a proton-pumping ATPase operates at the tonoplast and supplies energy, in form of a "proton-motive force", for various secondary active transport systems at the tonoplast. The search for this tonoplast ATPase proved to be difficult. Initially, ATPase was measured in preparations of intact vacuoles. This imposed two problems, as reviewed by Briskin and Leonard (1980) and Boller (1982). First, enzymes hydrolyzing ATP unspecifically are present in isolated vacuoles, most notably an unspecific acid phosphatase, and much of the initial work did not distinguish between specific and unspecific activities. Second, ATPases are present in several other membranes, including the plasma membrane, and ATPase activities in preparations of isolated vacuoles might come from membrane contaminants.

A more direct way to establish the contribution of ATP to tonoplast energization is to measure its influence on the membrane potential in isolated vacuoles. The membrane potential can be determined indirectly from the distribution of lipophilic cations or weak acids. Using such techniques, Doll and Hauer (1981) found, in isolated red beet vacuoles, a hyperpolarization from -55 mV to more negative values upon incubation with ATP. Wagner and Lin (1982), in contrast, measured an ATP-dependent acidification of the sap of isolated, intact tulip vacuoles; these results were indicative of a depolarization of the tonoplast. Thom and Komor (1984), working with vacuoles from cultured sugar cane cells, similarly found an ATP-induced increase of the membrane potential from 0 to +20 mV.

Similar techniques were introduced to measure ATP-dependent ion-pumping activity in sealed plant microsomes (Sze 1980). Microsomes can be obtained directly from tissue homogenates. If tonoplast vesicles could be identified from a microsomal preparation, this would obviate the need for preparation of protoplasts and vacuoles in studies of tonoplast properties. Separation of microsomes in density gradients yielded two membrane fractions with ATP-dependent ion pumping (Mandala et al. 1982; Churchill et al. 1983). One membrane fraction was identified, on the basis of marker enzymes, as the plasma membrane. Its ATPase was vanadate-sensitive. The second membrane fraction, which did not migrate with any known marker, was hypothesized to consist of tonoplast vesicles since it had the same density as a tonoplast preparation from [^{14}C]choline-labelled tobacco protoplasts (Boller and Kende 1979). Its ATPase was nitrate-sensitive (Churchill et al. 1983). Recently, the identity of the tonoplast-like microsomal fraction was experimentally verified. Tonoplasts obtained from red beet vacuoles (isolated mechanically) had a nitrate-sensitive proton-pumping ATPase which migrated like the nitrate-sensitive ATPase of microsomes obtained from tissue homogenates in sucrose density gradients (Bennett et al. 1984).

Intracellular Sugar Transport in Vivo

While work with isolated vacuoles demonstrated the presence of sugar transport systems and of ATP-dependent proton pumps on the tonoplast, it remained to be investigated how these transport processes functioned in vivo. Protoplasts are a key to study intracellular transport in vivo: The intact protoplasts are pulse-labelled with radioactive metabolites for a short period of time, and intracellular organelles are then rapidly separated and their contents of label determined.

In protoplasts from sugar cane cell cultures, such a technique enabled Thom et al. (1982) to compare the transport of sugars into the vacuole in vivo and in vitro. They found that 3-0-methylglucose was transported in a similar way in vivo and in vitro. Sucrose, however, formed in the protoplasts from exogenously supplied [^{14}C]glucose, accumulated much more rapidly in the vacuoles in vivo than expected from the transport data in vitro. Thus transport processes in vivo differed from those in vitro. The difference could be based on the existence of a proton motive force across the tonoplast in vivo as opposed to its absence in vitro (Komor et al. 1982). An alternative possibility was proposed by Brown and Coombe (1982) for sucrose transport in grape vacuoles. They hypothesized that sucrose might be transported in an entirely different way in vivo, namely, by group translocation, i.e., by transport coupled to the synthesis of sucrose.

The intracellular compartmentation of sucrose was examined also in photosynthesizing protoplasts (Kaiser et al. 1982; Boller and Alibert 1983). Protoplasts were exposed to $H^{14}CO_3^-$ for a short period of time and then used for the rapid isolation of vacuoles. In barley, it was found that sucrose, along with other photosynthetic products, entered the vacuoles rapidly after being synthesized in the cytosol (Kaiser et al. 1982). In *Melilotus*, however, the vacuoles appeared to keep out the newly made sucrose from the vacuoles, rather than to accumulate it (Boller and Alibert 1983); it was estimated that the cytoplasmic sucrose concentration of newly synthesized sucrose reached ca. 100 mM, while the vacuolar concentration did not exceed 10 mM even after 2 h. Taken together, the results of Kaiser et al. (1982) and of Boller and Alibert (1983) indicate that photosynthesizing cells regulate the compartmentation between cytosol and vacuole. In some cells, like in barley protoplasts, the assimilation products may be deposited transiently in the vacuoles (and, in the form of starch, in the chloroplasts); in other cells, like in the *Melilotus* protoplasts, the assimilation products may be excluded from the vacuoles and may be transferred, as rapidly as possible, to the phloem. Compartmental analysis of newly formed sucrose in intact leaves of *Vicia faba* indicated, in fact, that sometimes all of the newly formed sucrose was transferred to the phloem, while in other cases, a part of the sucrose remained accumulated in the leaf (Outlaw and Fisher 1975).

It should be noted, at this point, that protoplasts differ from intact mesophyll cells in one important point. Intact cells transfer a large part of the newly formed photosynthetic products to the phloem (Giaquinta 1983); protoplasts, however, do not lose the photosynthetic products (Kaiser et al. 1979; Kaiser et al. 1982; Boller and Alibert 1983).

Conclusions

Protoplasts can be used in two ways to study intracellular transport between cytosol and vacuole. Firstly, intracellular transport processes can be studied in vitro: Intact vacuoles can be purified from protoplasts and used for transport studies. Furthermore, membrane vesicles can be isolated from the vacuoles and used to investigate tonoplast properties without the possible complications by the vacuolar contents. Since techniques for the direct mechanical isolation of vacuoles (see Leigh 1983) and for the separation and identification of tonoplast vesicles from tissue homogenates (Boller and Kende 1979; Bennett et al. 1984) are available, protoplasts are not always necessary for these studies. Protoplasts are indispensable, however, for a second approach, namely, for the direct study of metabolite transport in vivo. Here, the intracellular metabolites are labelled in the protoplasts for a short period of time, and the vacuoles are rapidly isolated thereafter. The two approaches complement each other and might be used more often in combination.

Sometimes, it is difficult to use the results of studies with isolated vacuoles or tonoplast vesicles for the interpretation of transport processes operating in vivo. In addition to the examples discussed above, this may well be seen in the case of malate transport in plants exhibiting crassulacean acid metabolism. In vivo, malic acid is very efficiently accumulated in the vacuole during the night, and theoretical considerations indicate that this is due to an active malate transport energized by a proton pump (Lüttge et al. 1981). During the day, malic acid is mobilized again. In vitro, a malate permease (Buser-Suter et al. 1982) and a very active, specific ATPase (Aoki and Nishida 1984) have been characterized; however, malate transport was found to be a passive exchange flux in vitro and could not be stimulated with ATP (Buser-Suter et al. 1982). Thus, the in vitro studies give no indications, so far, how the rhythmic accumulation and mobilization of malic acid operates in vivo.

While making use of the advantages of protoplasts for intracellular transport studies, it should be kept in mind that a protoplast is an artificially altered plant cell. For example, the isoosmotic condition required to prepare and to store protoplasts is a serious interference with normal metabolism. It would cause plasmolysis in intact tissue; plasmolysis has been shown to perturb intracellular transport processes in certain tissues (Saftner et al. 1983). Therefore, compartmental analysis of intact tissue should complement transport studies with protoplasts or vacuoles isolated from them. In fungi, the elegant work on arginine sequestration by means of indirect compartmental analysis, in combination with the results on isolated vacuoles, provide a rather complete picture of intracellular arginine transport. In higher plants, studies of intracellular transport by indirect compartmental analysis and by direct analysis of protoplasts or isolated vacuoles are rarely brought into connection. It would seem advantageous to combine the different approaches as much as possible in the future.

References

Aoki K. Nishida K (1984) ATPase activity associated with vacuoles and tonoplast vesicles from the CAM plant, *Kalanchoë daigremontiana*. Physiol Plant 60:21–25

Bennett AB, O'Neill SD, Spanswick RM (1984) H^+-ATPase activity from storage tissue of *Beta vulgaris*. Plant Physiol (Bethesda) 74:538–544

Boller T (1977) Die Arginin-Permease der Hefevakuole. Dissertation ETH Nr 5928. Buchdruckerei Wattwil

Boller T (1982) Enzymatic equipment of plant vacuoles. Physiol Veg 20:247–257

Boller T, Alibert G (1983) Photosynthesis in protoplasts from *Melilotus alba*: distribution of products between vacuole and cytosol. Z Pflanzenphysiol 110:231–238

Boller T, Kende H (1979) Hydrolytic enzymes in the central vacuole of plant cells. Plant Physiol (Bethesda) 63:1123–1132

Boller T, Dürr M, Wiemken A (1975) Characterization of a specific transport system for arginine in isolated yeast vacuoles. Eur J Biochem 54:81–91

Bowman EJ (1983) Comparison of the vacuolar membrane ATPase of *Neurospora crassa* with the mitochondrial and plasma membrane ATPases. J Biol Chem 258:15238–15244

Bowman EJ, Bowman BJ (1982) Identification and properties of an ATPase in vacuolar membranes of *Neurospora crassa*. J Bacteriol 151:1326–1337

Briskin DP, Leonard RT (1980) Isolation of tonoplast vesicles from tobacco protoplasts. Plant Physiol (Bethesda) 66:684–687

Brown SC, Coombe BG (1982) Sugar transport by an enzyme complex at the tonoplast of grape pericarp cells. Naturwissenschaften 69:43–44

Buser-Suter C, Wiemken A, Matile P (1982) A malic acid permease in isolated vacuoles of a crassulacean acid metabolism plant. Plant Physiol (Bethesda) 69:456–459

Churchill KA, Holaway B, Sze H (1983) Separation of two types of electrogenic H^+-pumping ATPases from oat roots. Plant Physiol (Bethesda) 73:921–928

Cramer CL, Vaughn LE, Davis RH (1980) Basic amino acids and inorganic polyphosphates in *Neurospora crassa*: independent regulation of vacuolar pools. J. Bacteriol 142:945–952

Davis RH (1975) Compartmentation and regulation of fungal metabolism: genetic approaches. Annu Rev Genet 9:39–65

Doll S, Hauer R (1981) Determination of the membrane potential of vacuoles isolated from red beet storage tissue. Planta (Berl) 152:153–158

Doll S, Rodier F, Willenbrink J (1979) Accumulation of sucrose in vacuoles isolated from red beet tissue. Planta (Berl) 144:407–411

Drainas C, Weiss RL (1982) Energy requirement for the mobilization of vacuolar arginine in *Neurospora crassa.* J Bacteriol 150:779–784

Dürr M, Urech K, Boller T, Wiemken A, Schwencke J, Nagy M (1979) Sequestration of arginine by polyphosphate in vacuoles of yeast (*Saccharomyces cerevisiae*). Arch Microbiol 121: 169–175

Galun E (1981) Plant protoplasts as physiological tools. Annu Rev Plant Physiol 32:237–266

Giaquinta RT (1983) Phloem loading of sucrose. Annu Rev Plant Physiol 34:347–387

Guy M, Reinhold L, Michaeli D (1979) Direct evidence for a sugar transport mechanism in isolated vacuoles. Plant Physiol (Bethesda) 64:61–64

Huber-Wälchli V, Wiemken A (1979) Differential extraction of soluble pools from the cytosol and the vacuoles of yeast (*Candida utilis*) using DEAE-dextran. Arch Microbiol 120:141–149

Kaiser WM, Paul JS, Bassham JA (1979) Release of photosynthates from mesophyll cells in vitro and in vivo. Z Pflanzenphysiol 94:377–385

Kaiser G, Martinoia E, Wiemken A (1982) Rapid appearance of photosynthetic products in the vacuoles isolated from barley mesophyll protoplasts by a new fast method. Z Pflanzenphysiol 107:103–113

Kakinuma Y, Ohsumi Y, Anraku Y (1981) Properties of H^+-translocating adenosine triphosphatase in vacuolar membranes of *Saccharomyces cerevisiae.* J Biol Chem 256:10859–10863

Keller F, Wiemken A (1982) Differential compartmentation of sucrose and gentianose in the cytosol and vacuoles of storage root protoplasts from *Gentiana lutea* L. Plant Cell Rep 1: 274–277

Komor E, Thom M, Maretzki A (1982) Vacuoles from sugarcane suspension cultures III. Protonmotive potential difference. Plant Physiol (Bethesda) 69:1326–1330

Leigh RA (1983) Methods, progress and potential for the use of isolated vacuoles in studies of solute transport in higher plant cells. Physiol Plant 57:390–396

Lüttge U, Smith JAC, Marigo G, Osmond CB (1981) Energetics of malate accumulation in the vacuoles of *Kalanchoë tubiflora* cells. FEBS Lett 126:81–84

Mandala S, Mettler IJ, Taiz L (1982) Localization of the proton pump of corn coleoptile microsomal membranes by density gradient centrifugation. Plant Physiol (Bethesda) 70:1743–1747

Martinoia E, Heck U, Boller T, Wiemken A, Matile P (1979) Some properties of vacuoles isolated from *Neurospora crassa* slime variant. Arch Microbiol 120:31–34

Matile P (1978) Biochemistry and function of vacuoles. Annu Rev Plant Physiol 29:193–213

Ohsumi Y, Anraku Y (1981) Active transport of basic amino acids driven by a proton motive force in vacuolar membrane vesicles of *Saccharomyces cerevisiae.* J Biol Chem 256:2079–2082

Outlaw WH, Fisher DB (1975) Compartmentation in *Vicia faba* leaves I. Kinetics of ^{14}C in the tissues following pulse labeling. Plant Physiol (Bethesda) 55:699–703

Saftner RA, Daie J, Wyse RE (1983) Sucrose uptake and compartmentation in sugar beet taproot tissue. Plant Physiol (Bethesda) 72:1–6

Sze H (1980) Nigericin-stimulated ATPase in microsomal vesicles of tobacco callus. Proc Natl Acad Sci USA 77:5904–5908

Thom M, Komor E (1984) Role of the ATPase of sugar-cane vacuoles in energization of the tonoplast. Eur J Biochem 138:93–99

Thom M, Komor E, Maretzki A (1982) Vacuoles from sugarcane suspension cultures II. Characterization of sugar uptake. Plant Physiol (Bethesda) 69:1320–1325

Urech K, Dürr M, Boller T, Wiemken A, Schwencke J (1978) Localization of polyphosphate in vacuoles of *Saccharomyces cerevisiae.* Arch Microbiol 116:275–278

Wagner GJ, Lin W (1982) An active proton pump of intact vacuoles isolated from *Tulipa* petals. Biochim Biopyhs Acta 689:261–266

Wiemken A, Dürr M (1974) Characterization of amino acid pools in the vacuolar compartment of *Saccharomyces cerevisiae*. Arch Microbiol 101:45–57
Willenbrink J, Doll S (1979) Characteristics of the sucrose uptake system of vacuoles isolated from red beet tissue. Kinetics and specificity of the sucrose uptake system. Planta (Berl) 147:159–162
Zierler K (1981) A critique of compartmental analysis. Annu Rev Biopyhs Bioeng 10:531–562

Compartmentation of Metabolite Pools in Protoplasts: Chloroplasts, Mitochondria, Cytosol/Vacuole

R. HAMPP

Introduction

Isolated plant organelles such as plastids and mitochondria are an important tool to investigate steps of biochemical reactions, characteristic for these compartments like, e.g., photosynthetic CO_2 fixation, photorespiration, respiration, or provision of biosynthetic precursors. Usually the tissue is homogenized by mechanical grinding and the organelles are separated from the resulting homogenate by differential or density gradient centrifugation. Due to considerable amounts of phenolics or thick cell walls, the separation of active plastids and mitochondria has been largely restricted to a few plant species. In addition, such a system offers only limited access to questions regarding the interaction and compartmentation of metabolic processes, such as the regulation of the cellular energy status or protein, lipid and carbohydrate metabolism, as they depend on a coordinated interaction of subcellular compartments.

Recent work indicates that some of the problems involved in the isolation of organelles can be reduced by the preparation and fractionation of protoplasts, thereby increasing the range of suitable tissue considerably. However, fractionation of subcellular compartments from protoplast homogenates via sucrose or silica sol gradients only is advisable when, e.g., the compartmentation of enzymes is to be assessed. It is too slow to measure subcellular pool sizes of metabolites, especially if they exhibit high rates of turnover.

Considerable methodological progress in the determination of the subcellular compartmentation of metabolites was achieved by Robinson and Walker (1979) and somewhat later by Wirtz et al. (1980), by introducing a system for integrated homogenation and fractionation of protoplasts over silicon oil. Using this technique, chloroplasts could be separated from the rest of the protoplast within seconds. The principle of this method has been developed further by including a second oil layer which, in addition to plastids, permitted the rapid separation of mitochondria from a cytosolic/vacuolar supernatant in one centrifugation step (Hampp 1980; Goller et al. 1982). Recently a different approach was made by McLilley et al. (1982), using a membrane filtration technique. Both methods (Hampp 1980; McLilley et al. 1982) for the first time offered the possibility of assessing mitochondrial pool sizes of metabolites under nearly in vivo conditions, and in dependency on changes in environmental conditions.

Preferentially based on silicon oil filtration (Goller et al. 1982), but to some extent also on membrane filtration of oat mesophyll protoplasts, this paper reports on

The Physiological Properties of Plant Protoplasts
Edited by P. E. Pilet

changes during dark/light transition in compartmented pool sizes of a wider range of metabolites. The results, compared to the in vitro alteration of mitochondrial respiration by products of photosynthesis, offer a new idea for the regulation of plant mitochondrial activities in the light.

Materials and Methods

Plant Material and Culture Conditions

Seedlings of *Avena sativa* L. (cv. Arnold) were grown in hydroponic culture for 7 d. Illumination (about 9 W m^{-2}; Osram HQLS, 400 W, plus two Osram concentra PAR spots, 75 W each) was started after 4 to 5 d of germination in the dark at 26°C and 80% rel. humidity.

Isolation of Protoplasts

Enzymatic isolation and purification of protoplasts from 0.5- to 1-mm-wide leaf segments was carried out as reported earlier (Hampp and Ziegler 1980). Protoplast numbers were counted using a Fuchs-Rosenthal haemocytometer.

Incubation of Protoplasts

Purified protoplasts were suspended in 0.5 M sorbitol, 5 mM $KHCO_3$, 7.5 mM $CaCl_2$, 25 mM 2-(N-morpholino) ethane-sulfonic acid (MES), pH 6.0 (about 0.12 mg chlorophyll ml^{-1}; Arnon 1949) and incubated in a Clark-type oxygen electrode at 20°C in a final volume of 1.5 ml. Samples were taken at the end of a 2-min dark period and after different times of illumination (two opposing 250 W slide projectors, about 600 W m^{-2} at the surface of the electrode vessel). Care was taken to pertain the same conditions as during incubation (darkness or illumination) until the end of the fractionation procedure.

Fractionation of Protoplasts

Oil Filtration

Fifty-μl aliquots of the protoplast suspension were pipetted separately into 400-μl centrifuge tubes that contained a 15-μm nylon mesh and several hydrophobic and hydrophilic layers. The lowermost layer comprised 0.6 M sucrose (determination of marker enzymes) and either 0.1 M NaOH (for reduced pyridine nucleotides) or 0.1 M HCl (other metabolites), in 0.6 M sucrose each. For preparation and handling of these microgradients see Goller et al. (1982). As shown in Fig. 1, this integrated system of protoplast homogenation and fractionation delivers three fractions (P; M; S: chloroplasts; middle layer, enriched with mitochondria; remainder of the cell) in addi-

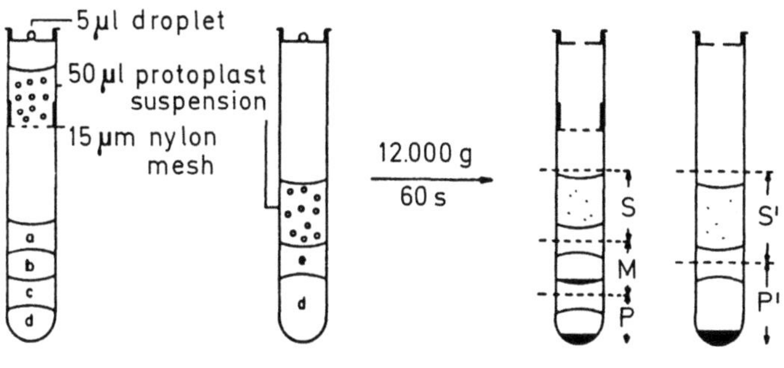

a

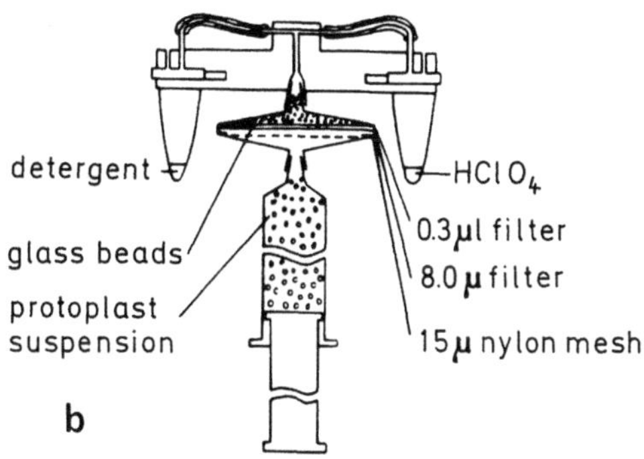

b

Fig. 1a,b. Methods for the fractionation of protoplasts. **a** Oil filtration. In the gradient tube with the inserted nylon net, protoplasts are ruptured during centrifugation and the resulting homogenate forms three aqueous fractions: *P* pellet (chloroplasts); *M* middle fraction (mainly mitochondria); *S* supernatant (cytosolic and vacuolar constituents); these are separated by layers of silicon oil. The second tube (no net) filtrates intact protoplasts through silicon oil and delivers two fractions: *P'* pellet of intact protoplasts; *S'* supernatant; material released from broken protoplasts; *a, c, e* silicon oil layers; *b, d* aqueous layers. **b** Membrane filtration. The suspension of protoplasts contained in a syringe which is attached to a membrane filter holder, is forced through a nylon net and membrane filters. If only the net is inserted, a protoplast homogenate results. From this homogenate chloroplasts can be withdrawn by an additional 8-µm membrane filter ("homogenate – chloroplasts"). If a second membrane filter (0.3 µm) is inserted, a protoplast filtrate results which has a reduced level of mitochondrial markers ("cytosol + vacuole"). By calculation, material associated with chloroplasts and mitochondria can be assayed from quenched ($HClO_4$) or solubilized filtrates (detergent). (After McLilley et al. 1982; see also for details)

tion to the separate filtration of intact protoplasts (P', S'; see also legend to Fig. 1). Procedures for the correction of cross-contamination have been given in detail (Goller et al. 1982). In preceding papers we also assessed possible alterations of the metabolite pools during this fractionation procedure, using adenylates (Hampp 1980; Hampp et al. 1982) and pyridine nucleotides (Hampp et al. 1984), which exhibit high rates of turnover, as an example. There we showed that metabolic reactions are terminated within less than 2 s in the pellet (chloroplasts) and the supernatant fractions (mainly cytosol), and obtained the amount of metabolites located in mitochondria by subtracting metabolite pools associated with chloroplasts and cytosol from those found in intact protoplasts (Goller et al. 1982).

Membrane Filtration

For reasons of comparison we also used the technique of membrane filtration exactly as described by McLilley et al. (1982) (see also Fig. 1 and Results and Discussion).

Isolation of Mitochondria

Mitochondria from oat mesophyll were prepared and purified as described by Neuburger et al. (1982).

Determination of Metabolites

Adenine nucleotides were assayed in neutralized $HClO_4$ extracts directly (ATP) or after enzymatic conversion to ATP (ADP) by bioluminescence (Lumit PM, Abimed, Düsseldorf, F.R.G.; Biolumat, Berthold, Wildbad, F.R.G.; see Hampp et al. 1982; Hampp 1984).

Oxidized and reduced pyridine nucleotides were determined from neutralized HCl (0.1 M) or NaOH (0.1 M) extracts, brought to pH 6.5 and 7.8 respectively, by enzymatic cycling (Lowry and Passonneau 1972; Kato et al. 1973). For the determination of all other metabolites (TP, PGA, Pi, malate, oxaloacetate, aspartate, glutamate, 2-oxoglutarate and citrate) reactions were used, which were selectively coupled to pyridine nucleotides (Lowry and Passonneau 1972; "specific step"). These were followed by enzymatic cycling of the pyridine nucleotides formed in each specific reaction. Detailed informations on all these reactions are given elsewhere (Hampp et al. 1984).

Results and Discussion

Critical Assessment of the Fractionation Methods

1. Silicon Oil Filtration of Protoplast Homogenates. In detailed studies (Hampp 1980; Goller et al. 1982; Hampp et al. 1984) it has been demonstrated that the technique of oil filtration allows for a rapid quenching of metabolic reactions in bulk quantities of chloroplasts, reaching the acid layer within a few s, and of the supernatant, mainly containing cytosolic and vacuolar constituents. Although the internal standardization of the gradient fractions with the metabolites under investigation resulted in good rates of recovery (Hampp 1980; Hampp et al. 1984), there is some uncertainty about the data given for the fraction enriched with mitochondria (M in Fig. 1). As the effectors used for halting mitochondrial metabolism require about 30 s for full inhibition (Hampp et al. 1982), artifactual changes of mitochondrial levels of metabolites are to be expected. In order to reduce this problem, the levels of metabolites located in mitochondria were also calculated by subtracting those assayed in the rapidly quenched chloroplast and cytosol fractions from the total contents of denatured protoplasts (P' in Fig. 1) as described elsewhere (Goller et al. 1982).

2. Membrane Filtration of Protoplast Homogenates. The membrane filtration technique, developed by McLilley et al. (1982) is considerably faster than oil filtration as described above (the authors state 0.1 to 0.2 s between the formation of a protoplast homogenate and $HClO_4$ quench of metabolic activity). However, in contrast to the silicon method, membrane filtration does not yield a pure chloroplast fraction, separated from an only slightly contaminated mitochondrial layer (see Fig. 1b). In our hands, and with *Avena* mesophyll protoplasts, the 0.3 μm filtrate ("cytosol + vacuole") contained significant amounts of mitochondrial and chloroplast markers (10 to 20% of the respective total level of activity), the 8 μm filtrate (cytosol, vacuole, mitochondria) also exhibited a considerable contamination by chloroplast markers. Thus, due to the difficult correction for cross contamination of the fractions by each other, [chloroplast pool = (total homogenate) – (8 μm filtrate); mitochondrial pool = (8 μm filtrate) – (0.3 μm filtrate); for exact description see McLilley et al. 1982] this method can only give an approximation of the metabolite pools associated with the respective compartments.

We recently compared both techniques with oat mesophyll protoplasts and obtained similar results for the compartmentation of adenylates, although, at least in our hands, results obtained with membrane filtration showed considerably higher variations.

Compartmented Metabolite Pools During Dark/Light Transition of *Avena* Mesophyll Protoplasts

In Table 1 results on the distribution of a range of metabolites between chloroplasts, mitochondria and the extraorganellar space, obtained by oil filtration, are summarized. Shortly after the onset of illumination, ATP accumulates in the chloroplast stroma and, with some lag, there is also a transient increase in the cytosolic ATP level. In contrast, mitochondria respond to light with an initial depletion of ATP and afterwards tend to come back to the original dark level. ADP largely mirrors the changes in ATP, but in the opposite direction. After reaching steady-state rates of photosynthesis (7.5 min of illumination), mitochondria exhibit a slightly higher ATP/ADP ratio compared to dark-treated cells. When comparing dark and light levels (incubation time 9 min each), with their fractionation method, Stitt et al. (1982) reported a considerably higher ATP/ADP ratio in mitochondria from illuminated *Hordeum* mesophyll protoplasts (dark: 0.6; light: 2.6). This difference is not necessarily due to the fractionation method, but more probably to the physiological condition of the plant material usef for protoplast preparation. In our fractionation studies we found considerable differences in the response to illumination of the compartmented adenylate pools during different times of the year, and sometimes obtained considerably higher changes in mitochondrial, but also in cytosolic adenylate ratios between dark and light.

Our data on pyridine nucleotides (Table 1), like that of Takahama et al. (1981), indicates no significant light-mediated conversion of diphospho into triphospho pyridine nucleotides as Muto et al. (1981) reported. Therefore, in each compartment,

Table 1. Levels of metabolites associated with chloroplasts, mitochondria and a cytosolic/vacuolar fraction of *Avena* mesophyll protoplasts during dark/light transition. The values are the mean of 3–6 independent experiments with 3 parallels each (S.E.: 10 to 20% of mean). The sum of oxidized and reduced pyridine nucleotides was constant and averaged at (nmol/10^6 protoplasts) 0.7 [NAD(H)] and 0.35 [NADP(H), chloroplasts]; 0.4 [NAD(H)] and 0.05 [NADP(H), mitochondria]; 1.0 [NAD(H) and 0.5 (NADPH(H), cytosol/vacuole*]. n.d.: value was below our limit of detection (0.05 nmol/10^6 protoplasts). For subcellular volumes see Hampp et al. (1982)

Metabolite	Period of illumination (min)	Chloroplasts (nmol/10^6 protoplasts)	Mitochondria	Cytosol/ vacuole
ATP	0	1.1	0.8	2.7
	0.5	2.4	0.65	3.1
	1.0	2.7	0.85	3.0
	2.5	3.0	0.95	2.7
	7.0	2.4	0.85	2.5
ADP	0	2.1	0.55	0.45
	0.5	1.1	0.70	0.38
	1.0	0.8	0.65	0.40
	2.5	0.7	0.50	0.42
	7.0	1.1	0.50	0.40
Inorganic phosphate	0	50	58	475
	0.5	46	79	476
	1.0	14.8	37	444
	2.5	11	33	277
	7.0	7.9	53	246
NAD*)	0	0.46	0.35	0.90
	0.5	0.54	0.20	0.50
	1.0	0.43	0.32	0.62
	2.5	0.43	0.30	0.70
	7.0	0.40	0.32	0.79
NADP*)	0	0.24	n.d.	0.24
	0.5	0.15	n.d.	0.20
	1.0	0.16	n.d.	0.19
	2.5	0.22	n.d.	0.20
	7.0	0.27	n.d.	0.20
Triose phosphates	0	1.1	2.2	7.0
	0.5	7.5	11.8	50
	1.0	7.9	7.7	41
	2.5	4.5	7.5	15
	7.0	4.2	2.8	13
PGA	0	11.5	3.6	0.8
	0.5	2.6	3.9	1.4
	1.0	3.5	5.5	0.6
	2.5	6.5	5.0	1.6
	7.0	4.3	6.1	1.1

Table 1 (cont.)

Malate	0	1.0	0.8	16.5
	0.5	0.6	2.6	24.5
	1.0	1.9	0.7	26
	2.5	2.1	0.9	38
	7.0	1.9	1.0	35
Aspartate	0	2.0	n.d.	3.9
	0.5	11.5	4.7	2.0
	1.0	12.5	n.d.	11.6
	2.5	2.5	n.d.	11.4
	7.0	4.1	n.d.	12.5
Glutamate	0	11.5	2.8	8.0
	0.5	4.4	2.2	7.0
	1.0	3.2	4.1	8.5
	2.5	4.0	1.9	5.4
	7.0	3.1	2.0	1.7
Citrate	0	n.d.	n.d.	5.7
	0.5	n.d.	n.d.	6.4
	1.0	n.d.	n.d.	4.5
	2.5	n.d.	n.d.	3.7
	7.0	n.d.	n.d.	5.7

changes in, e.g., the oxidized form mirror opposite levels of the reduced molecule. In the resolution of incubation time we were able to handle, the stroma NAD system did not show larger variations upon illumination, although, sometimes we found a decrease in NAD during the initial 30 s. In contrast, there was a pronounced drop in the mitochondrial NAD level, indicative for an immediate increase in the redox state of this system in response to light. Similarly the cytosolic redox charge was higher during the first 30 s of light treatment.

The NADP system, not present in oat mesophyll mitochondria, becomes more reduced in chloroplasts only within the first seconds of illumination, whereas towards a steady state of photosynthesis the NADPH/NADP ratio is lower than in the dark, obviously due to a limitation of CO_2 fixation by NADPH. The redox state of the cytosol (NADPH/NADP + NADPH) assumes a value of about 0.5 in the dark and exhibits some increase upon illumination (about 0.6). As far as a comparison is possible, these results agree to some extent with levels of pyridine nucleotides determined via nonaqueous fractionation of lyophilized leaves. Heber and Santarius (1965) reported less than 80% of the total NADPH and more than 50% of the NADP to be chloroplast associated. In addition, 25% of the NADP(H) system of chloroplasts from dark tissue, and 70 to 100% of the extrachloroplast space were in the reduced state. For the NAD system, however, Heber and Santarius (1965) reported still lower NADH/NAD ratios for chloroplasts and an extrachloroplast fraction. Comparable experiments concerning mitochondria only exist for animal tissue. The low level of the NADP system in oat mesophyll mitochondria is in contrast to results on mitochondria from rat liver cells where this system is predominantly located in these organelles (Sies 1982).

Both, chloroplast-located TP and PGA (3-phosphoglycerate) show a fast response to light treatment. While PGA levels decrease to a lowest value within 30 s of illumination, TP accumulate. In both cases, intermediate steady state levels are reached between 2.5 and 7 min of illumination. Much more interesting, however, are the levels of these compounds found to be associated with oat mitochondria. Similar to chloroplasts, the level of TP transiently increases upon illumination. The actual levels (about 12 nmol TP/10^6 protoplasts; 30 s light) are higher than in chloroplasts, but still considerably below the amount of TP recovered in the cytosol during the onset of illumination. In parallel to the increase in cytosolic TP, there is also some increase in PGA in comparison to the dark level. Again, under prolonged illumination (>60 s), TP largely disappear (50 to 13 nmol/10^6 protoplasts), while PGA levels increase, possibly reflecting steady state levels of photosynthesis. From labeling experiments and pool size determinations Heber et al. (1967) and Urbach et al. (1965) calculated about 75% of the total PGA and about 20–25% of total dihydroxy acetone phosphate to be chloroplast-associated. This compares well with about 70% and 21% (TP) respectively in our fractionation study. To our knowledge there is only one report on the location of PGA in rat liver mitochondria (Heldt 1966) and no translocation of PGA or triose phosphates has been investigated (see Wiskich 1977, 1980).

Values for P_i (inorganic phosphate) show that darkened chloroplasts contain about 50 nmol P_i/10^6 protoplasts (about 20 mM). Upon illumination there is a significant decrease, which is most pronounced between 30 and 60 s, leading to a concentration of about 3.5 mM after 2.5 min. Mitochondria start from a P_i level in the dark of about 50 mM and, after considerable fluctuations (up to 75, down to 30 mM), tend to return to the dark level after prolonged illumination. The remaining cellullar P_i (cytosol + vacuole) shows a response to light which is similar to that of chloroplasts, but somewhat retarded. Starting from an overall concentration of about 25 mM in the dark, a larger decrease in P_i can be observed at a stage, where chloroplasts are already depleted of most of their P_i, levelling off at about 12 mM.

The time course of changes in compartmented levels of malic acid indicates a light-dependent increase in stromal and cytosolic/vacuolar pool sizes. In contrast, mitochondria show a very pronounced but transient increase in malate, which seems to be restricted to a period of about 30 s of illumination, leading back to the original dark level.

The aspartate pool of chloroplasts exhibits a fivefold transient increase within the first 60 s of illumination, whereas that of glutamate starts at a high dark level and decreases immediately upon illumination to about a third (Table 1). Mitochondrial aspartate was above the limit for detection only up to the 30-s stage, whereas glutamate was rather stable at a concentration of 2–3 mM. While extraorganellar levels of glutamate decreased upon prolonged illumination, those of aspartate behaved different in that the amount of this metabolite increased gradually (obviously due to chloroplast export) from 4 to about 12 nmol/10^6 protoplasts (0.2 to 0.6 mM).

Citrate was below our detection limit (0.05 nmol/10^6 protoplasts) as far as mitochondria and chloroplasts were concerned, but showed reasonable levels in the rest of the protoplast (presumably the vacuole). No larger changes due to illumination could be measured. These data are also in contrast to results on animal mitochondria, where citrate and aspartate are mainly matrix located (Sies 1982), possible due to a lack of vacuoles in animal cells. Oxoglutarate and oxaloacetate, finally, were below the limit of detection in all subcellular fractions except between 1 and 3 nmol/10^6 protoplasts of 2-oxoglutarate, which, in some experiments, could be recovered in the chloroplast fraction after 30 s of illumination. The concentration of oxaloacetate reported by others (110–20 μM in spinach leaves; Krause and Heber 1976), equivalent to about 0.2 nmol/10^6 oat mesophyll protoplasts, is near the detection limit of our method and thus could explain the lack of its recovery.

From results given by Lehner and Heldt (1978) on aqueously isolated spinach chloroplasts, glutamate, malate and aspartate concentrations can be calculated that are higher than ours. However, osmotic volume changes in conventionally isolated chloroplasts, in addition to a different plant species, could be the reason for this difference.

Perhaps the most interesting result of the work presented in this paper is the association of triose phosphates and of phosphoglyceric acid with oat mitochondria, although the exact location (matrix or intermembrane space, or both) cannot be given. The fluctuations of these metabolites upon illumination have to be seen in context with the other metabolite pools given in Table 1, and the mitochondrial TCA cycle, which is controlled by a number of regulatory sites (Wiskich 1980). One of the controls is thought to exist on the level of the P_i acceptor ADP which, when limiting, should restrict the rate of electron flow and hence the oxidation of mitochondrial NADH. If this would be the case, then we should find a decrease in mitochondrial ADP as a response to light due to a direct or indirect (TP/PGA shuttle) ATP export from the chloroplast stroma. The results given in Table 1, however, show the opposite: an increase of ADP between 0 and 60 s of illumination (see also Hampp et al. 1982). Only under prolonged illumination the mitochondrial ATP/ADP ratio appears to be higher than in the dark (Fig. 1; 2.5–7 min), a result also reported by Stitt et al. (1982), who, due to the limitations of their method (see above) only compared steady state situations in dark and light (9 min of incubation) and thus were not able to deal with events taking place immediately during dark/light transition.

The alteration in mitochondrial adenylate levels runs parallel to a fast increase in both cytosolic and mitochondrial NADH (decrease of NAD in Table 1). It is known that NADH inhibits the oxidation of malate by malate dehydrogenase (Asahi and Nishimura 1973) as well as its decarboxylation by the NAD-linked malic enzyme (Wedding et al. 1976; Macrae 1971). As a consequence, the mitochondrial malate pool exhibits a steep increase. This is parallel by a change in aspartate. Both events indicate an interruption of the tricarboxylic acid cycle for a limited period of time, as was concluded from experiments with excised mung bean leaves (Chapman and Graham 1974a,b). In addition, the mitochondrial P_i pool shows larger fluctuations, which range from an increase during the first 30 s of illumination to an expressed reduction in level between 30 and 60 s and result in considerable alterations in the phosphorylation potential ($[ATP] \cdot 10^3/[ADP] \cdot [P_i]$); D, 30 s, 2.5 and 7 min of light: 26, 12, 54, 33. Such changes could be due to altered transport processes across the inner mitochondrial membrane, where at least three different systems for P_i transport exist (Wiskich 1980).

Another way to affect mitochondrial phosphorylation and thus NADH oxidation could result from an inhibition of the coupling factor or the adenylate translocator in response to illumination. As a direct effect of light obviously can be excluded (Goller et al. 1982), such a postulated inhibition should be due to one or several compounds increasing their pool sizes immediately upon illumination in all compartments.

The cytosolic ATP/ADP ratio, regarded as a regulator of mitochondrial activites during dark/light transition, obviously is not that important. Experiments with isolated pea and cauliflower mitochondria (Dry and Wiskich 1982) showed that some reduction in respiration only occurred at ratios exceeding 20:1. Normally, such ratios do not exist in the cytosol (Hampp et al. 1982; Goller et al. 1982, Stitt et al. 1982), although the levels of free and bound adenylates are not known. Thus additional mechanisms to control mitochondrial respiration should exist.

As TP (triose phosphates) respond to illumination with an immediate increase in their cytosolic concentration, with high levels bound to mitochondria, we looked for the effects of glyceraldehyde and dihydroxy acetone phosphates on isolated oat mesophyll mitochondria. The values given in Table 2 indicate a significant reduction in the rate of ADP phosphorylation (state 3 respiration) with increasing concentrations of GAP. The effect is most expressed at molarities of GAP which were calculated to exist in the cytosol shortly after starting the illumination and a pH around 7.6.

Table 2. Effect of different concentrations of glyceraldehyde-3 phosphate (GAP) on mitochondrial respiration after addition of ADP (100 nmol, state 3) and its consumption (state 4). Assay conditions: 0.35 M mannitol, 25 mM TES (pH 7.6), 2 mM $MgSO_4$, 5 mM KH_2PO_4, 0.1% (w/v) BSA, 10 mM glutamate, 10 mM malate. The cytosolic concentration of GAP after 60 s of illumination was about 25 to 30 mM (assumption: no GAP in the vacuole; cf. Kaiser et al. 1983)

	GAP (mM)						
	0	2.5	7.5	9.0	13.0	18.5	28.0
State 3/state 4 (respiratory control)	4.1	3.1	2.7	2.5	2.2	2.0	1.8
Time needed for the consumption of ADP in state 3 (% of control)	100	163	188	200	225	394	570

Work under progress is designed to show whether this reduction in the rate of ADP phosphorylation is due to an interaction of triose phosphates with the transport or the phosphorylation of ADP.

Conclusions

Methods are now available for the rapid fractionation of mesophyll protoplasts. These, together with rapid quenching of metabolic reactions and sensitive techniques for metabolite assay, allow for the determination of subcellular pool sizes of metabolites under, e.g., changing environmental conditions, and thus enable the investigation of processes concerning intracellular compartmentation and regulation. Under the aspect of plant mitochondrial respiration in the light, the results on the kinetics of changes in chloroplast, cytosolic and mitochondrial pool sizes of metabolites during dark/light transition indicate that respiration is regulated not only by the cytosolic energy state, but possibly also by photosynthetic intermediates released from chloroplasts immediately upon starting CO_2 fixation.

Acknowledgement. The author is indebted to Dr. M. Stitt, for showing his method of membrane filtration, and to Dr. M. Goller, for critical reading of this paper. This work was supported by a grant of the Deutsche Forschungsgemeinschaft.

References

Arnon DI (1949) Copper enzymes in isolated chloroplasts. Polyphenoloxidase in *Beta vulgaris.* Plant Physiol (Bethesda) 24:1–15

Asahi T, Nishimura M (1973) Regulatory function of malate dehydrogenase isoenzymes in the cotyledon of mung bean. J Biochem (Tokyo) 73:217–225

Chapman EA, Graham D (1974a) The effect of light on the tricarboxylic acid cycle in green leaves I. Relative rates of the cycle in the dark and the light. Plant Physiol (Bethesda) 53: 879–885

Chapman EA, Graham D (1974b) The effect of light on the tricarboxylic acid cycle in green leaves II. Intermediary metabolism and the location of control points. Plant Physiol (Bethesda) 53:886–892

Dry IB, Wiskich JT (1982) Role of the external adenosine triphosphate/adenosine diphosphate ratio in the control of plant mitochondrial respiration. Arch Biochem Biophys 217:72–79

Goller M, Hampp R, Ziegler H (1982) Regulation of the cytosolic adenylate ratio as determined by rapid fractionation of mesophyll protoplasts of oat. Effect of electron transfer inhibitors and uncouplers. Planta (Berl) 156:255–263

Hampp R (1980) Rapid separation of the plastid, mitochondrial and cytoplasmic fractions from intact leaf protoplasts of *Avena.* Determination of in vivo ATP pool sizes during greening. Planta (Berl) 150:291–298

Hampp R (to be published 1984) ADP and AMP, luminometric method. In: Bergmeyer J, Graßl M (eds) Methods in enzymatic analysis, vol 7. Chemie, Weinheim

Hampp R, Ziegler H (1980) On the use of *Avena* protoplasts to study chloroplast development. Planta (Berl) 147:485–494

Hampp R, Goller M, Ziegler H (1982) Adenylate levels, energy charge and phosphorylation potential during dark/light and light/dark transition in chloroplasts, mitochondria and cytosol of mesophyll protoplasts from *Avena sativa* L. Planta (Berl) 62:448–455

Hampp R, Goller M, Füllgraf H (1984) Determination of compartmented metabolite pools by a combination of rapid fractionation of oat mesophyll protoplasts and enzymatic cycling. Plant Physiol (Bethesda) 75:1017–1021

Heber UW, Santarius KA (1965) Compartmentation of pyridine nucleotides in relation to photosynthesis. Biochim Biophys Acta 109:390–408

Heber U, Santarius KA, Hudson MA, Hallier UW (1967) Untersuchungen zur intrazellulären Verteilung von Enzymen und Substraten in der Blattzelle I. Intrazellulärer Transport von Zwischenprodukten der Photosynthese im Photosynthese-Gleichgewicht und im Dunkel-Licht-Wechsel. Z Naturforsch Teil B Anorg Chem Org Chem 22:1189–1199

Heldt HW (1966) The participation of endogenous nucleotides in mitochondrial phosphate-transfer reactions. In: Tager JM, Papa S, Quagliariellö E, Slater EC (eds) Regulation of metabolic processes in mitochondria. Biochim Biophys Acta Library, vol 7, pp 51–63

Kaiser G, Martinoia E, Wiemken A (1983) Rapid appearance of photosynthetic products in the vacuoles isolated from barley mesophyll protoplasts by a new fast method. Z Pflanzenphysiol 107:103–113

Kato T, Berger SJ, Carter JA, Lowry OH (1973) An enzymatic cycling method for nicotine-adenine dinucleotide with malic and alcohol dehydrogenase. Anal Biochem 53:86–97

Krause GH, Heber U (1976) Energetics of intact chloroplasts. In: Barber J (ed) The intact chloroplast. Elsevier North Holland, New York, pp 171–214

Lehner K. Heldt HW (1978) Dicarboxylate transport across the inner membrane of the chloroplast envelope. Biochim Biophys Acta 501:531–544

Lowry OH, Passonneau JV (1972) A flexible system of enzymatic analysis. Academic, New York

Macrae AR (1971) Isolation and properties of a 'malic' enzyme from cauliflower bud mitochondria. Biochem J 122:495–501

McLilley CR, Stitt M, Mader G, Heldt HW (1982) Rapid fractionation of wheat leaf protoplasts using membrane filtration. The determination of metabolite levels in chloroplasts, cytosol and mitochondria. Plant Physiol (Bethesda) 70:965–970

Muto S, Miyachi S, Usuda H, Edwards GE, Bassham JA (1981) Light-induced conversion of nicotinamide adenine dinucleotide to nicotinamide adenine dinucleotide phosphate in higher plant leaves. Plant Physiol (Bethesda) 68:324–328

Neuburger M, Journet E-P, Bligny R, Carde J-P, Douce R (1982) Purification of plant mitochondria by isopycnic centrifugation on density gradients of percoll. Arch Biochem Biophys 217: 312–323

Robinson SP, Walker DA (1979) Rapid separation of the chloroplast and cytoplasmic fractions from intact leaf protoplasts. Arch Biochem Biophys 196:319–323

Sies H (1982) Nicotinamide nucleotide compartmentation. In: Sies H (ed) Metabolic compartmentation. Academic, New York, pp 205–231

Stitt M, McLilley CR, Heldt HW (1982) Adenine nucleotide levels in the cytosol, chloroplasts and mitochondria of wheat leaf protoplasts. Plant Physiol (Bethesda) 70:971–977

Takahama U, Shimizu-Takahama M, Heber U (1981) The redox state of the NADP system in illuminated chloroplasts. Biochim Biophys Acta 637:530–539

Urbach W, Hudson MA, Ullrich W, Santarius KA, Heber U (1965) Verteilung und Wanderung von Phosphoglycerat zwischen den Chloroplasten und dem Cytoplasma während der Photosynthese. Z Naturforsch Teil B Anorg Chem Org Chem 20:890–898

Wedding RT, Black MK, Pap D (1976) Malate dehydrogenase and NAD malic enzyme in the oxidation of malate by sweet potato mitochondria. Plant Physiol (Bethesda) 58:740–743

Wirtz W, Stitt M, Heldt HW (1980) Enzymatic determination of metabolites in the subcellular compartments of spinach protoplasts. Plant Physiol (Bethesda) 66:187–193

Wiskich JT (1977) Mitochondrial metabolite transport. Annu Rev Plant Physiol 28:45–69

Wiskich JT (1980) Control of the Krebs Cycle. In: Davies DD (ed) The biochemistry of plants, vol 2. Metabolism and respiration. Academic, New York, pp 243–278

Protoplast Evacuolation

R.J. GRIESBACH and R.H. LAWSON

Introduction

Plant cells are characterized by a very prominent vacuole. This organelle carries out many essential cellular functions. For example, vacuoles act as both a temporary and permanent storage site for toxic materials. In addition, the solutes found within this organelle are responsible for osmotic pressure. The vacuole is also a rich reservoir for many enzymes, inorganic molecules, metabolic intermediates and secondary metabolites (Matile 1978).

Vacuoles have several properties, however, which can be quite troublesome in protoplast studies. Their autophagic activity can, at times, make it difficult to incorporate foreign material into a cell. The cell usually recognizes this material as foreign and sequesters it within the vacuole. Another bothersome characteristic is the vacuole's involvement in senescence. The onset of senescence triggers changes in the vacuolar membrane which leads to the release of hydrolytic enzymes and autolysis. Many studies involving protoplasts are carried out under stress conditions which could easily trigger senescence (Thomas and Stoddart 1980).

There are numerous procedures for isolating intact vacuoles but none for isolating intact cells minus their vacuoles. Such evacuolated protoplasts should be of great use in studying the effects the vacole has on protoplast regeneration, genetic transformation, cell division and other studies involving protoplasts.

Materials and Methods

Evacuolated protoplasts were prepared as previously described (Griesbach and Sink 1983). Mesophyll protoplasts were isolated from cut leaf tissue (1 x 10 mm sections) which had been exposed overnight to 0.8% cellulysin, 0.4% macerase, 10% mannitol, 5 mM MES (2N-morpholino-ethanesulfonic acid), and 10 mM $CaCl_2$ at pH 5.7. The protoplast suspension was cleaned of debris by layering on top of 30% sucrose and centrifuging at 200 *g* for 15 min. The band of "cleaned" protoplasts at the interface was then washed twice with 5 mM MES, 10 mM $CaCl_2$ and 10% mannitol at pH 6.0 via centrifugation at 100 *g* for 5 min. One half of a milliliter of 2×10^6 protoplasts ml^{-1} were mixed with 4.5 ml of Percoll containing 100 mM $CaCl_2$, 5 mM N-2-hydroxyethylpiperazine-N-2-ethanesulfonic acid (HEPES) (pH 7.0) and 8% mannitol. The Percoll solution was filter sterilized (45-μm millipore filter). The Percoll/proto-

The Physiological Properties of Plant Protoplasts
Edited by P. E. Pilet

plast suspension was pipetted into a 5-ml polycarbonate centrifuge tube and centrifuged for 60 min at 23°C at 40,000 r.p.m. (SW 56 rotor-Beckman). In order to do an ultracentrifuge run under sterile conditions, the following steps were taken. The centrifuge tube, centrifuge adapter and adapter top were sterilized by submersion in 70% ethanol for 10 min and air dried in a sterile laminar air flow hood. The sterilized tube was then loaded and balanced aseptically in a flow hood. After the centrifuge run, the centrifuge adaptor was opened in a flow hood and 70% ethanol pipetted on top of the Percoll gradient (the pressure changes during centrifugation can force bacteria and fungi into the adapter and tube). The band of evacuolated protoplasts was aseptically removed with a pasteur pipette, and washed three times with regeneration medium which consisted of Murashige and Skoogs salts and vitamins supplemented with 2 gm l^{-1} bacto-peptone, 30 gm l^{-1} sucrose, 100 gm l^{-1} mannitol, 1 mg l^{-1} NAA (napthaleneacetic acid) and 2 mg l^{-1} BA (benzylaminopurine) at pH 6.0 (Griesbach and Sink 1983).

For electron microscopy studies the protoplasts were concentrated in a microfuge tube and the supernatant medium was replaced with a mixture of 2% glutaraldehyde and 1.5% acrolein in 0.05 M Na_2HPO_4/KH_2PO_4 buffer, pH 7.0 containing 5% mannitol. After fixing for 1 h the pellets were washed twice with 0.05 M PO_4 containing 5% mannitol. Post-fixation was for 1 h in 1% OsO_4 in the same buffer. The samples were dehydrated in a graded series of ethanol and embedded in Spurr's medium. The tip of the pellet was usually soft and tissue was trimmed close to the interface between the clear plastic and the sample where the plastic was hard. Sections were cut with a diamond knife and stained for 30 min with uranyl acetate and 8 min with lead citrate.

Protoplasts were fixed and embedded immediately after separation from the Percoll gradient, 8 h and 17 h after the initial evacuolation. Samples were processed for embedding and sectioning as described.

Results and Discussion

This procedure resulted in 100% evaculation of mesophyll protoplasts of *Petunia parodii* and *P. hybrida* cv. Pink Magic. Approximately 40% of these protoplasts were viable and underwent further development. When evacuolated protoplasts were placed on regeneration medium, a new vacuole reformed within 17 h and the chloroplasts began to accumulate starch. Evacuolated protoplasts showed some fragmentation when observed in the light microscope immediately after separation from the Percoll gradient. Most of the protoplasts were reduced in size compared to vacuolated controls. In ultrathin section the evacuolated protoplasts showed no evidence of a tonoplast or membrane-bound vacuoles in the 0 time embedding (Fig. 1a). Control protoplasts contained a large central vacuole and normal-appearing chloroplasts. The grana of the chloroplasts in the evacuolated protoplasts appeared normal but the bounding membrane of chloroplasts was evaginated (Fig. 1b). After 8 h incubation some of the evacuolated protoplasts showed developing vacuoles and after 17 h most of the protoplasts formed new vacuoles in the regeneration medium (Fig. 2). In some

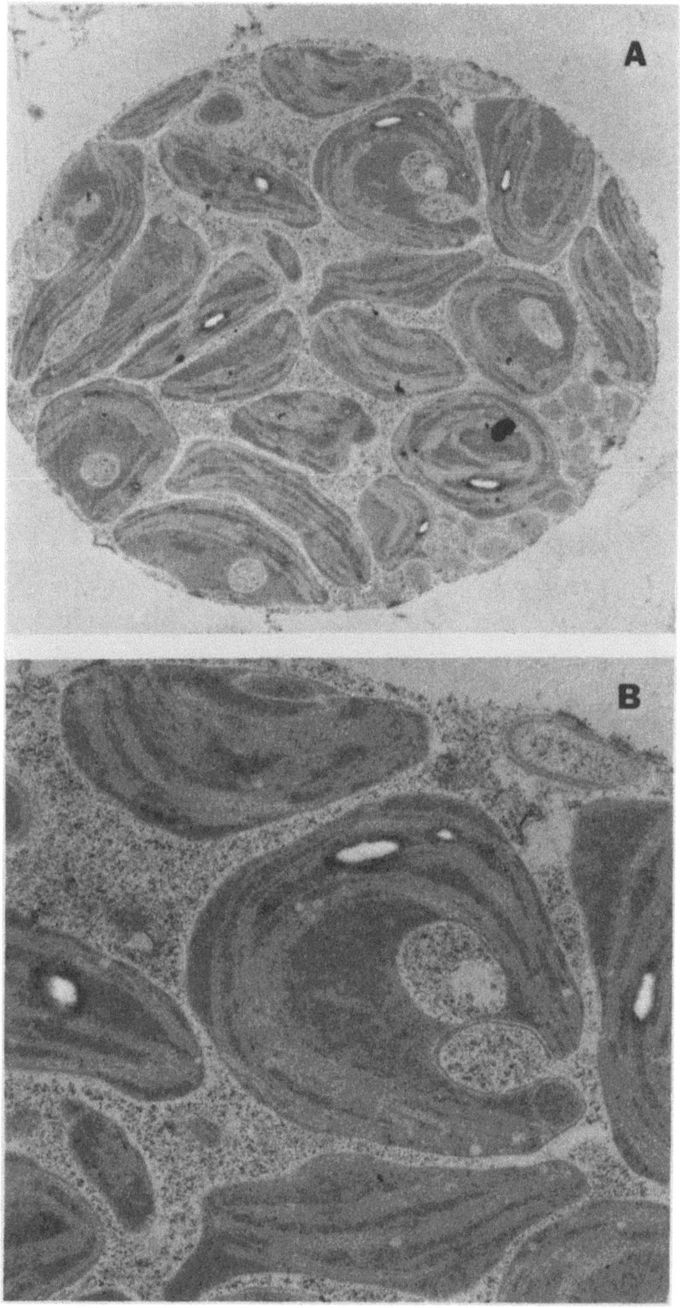

Fig. 1A,B. Evacuolated mesophyll protoplasts of *Petunia hybrida* cv. Pink Magic: **A** x 12,600 **B** x 27,000

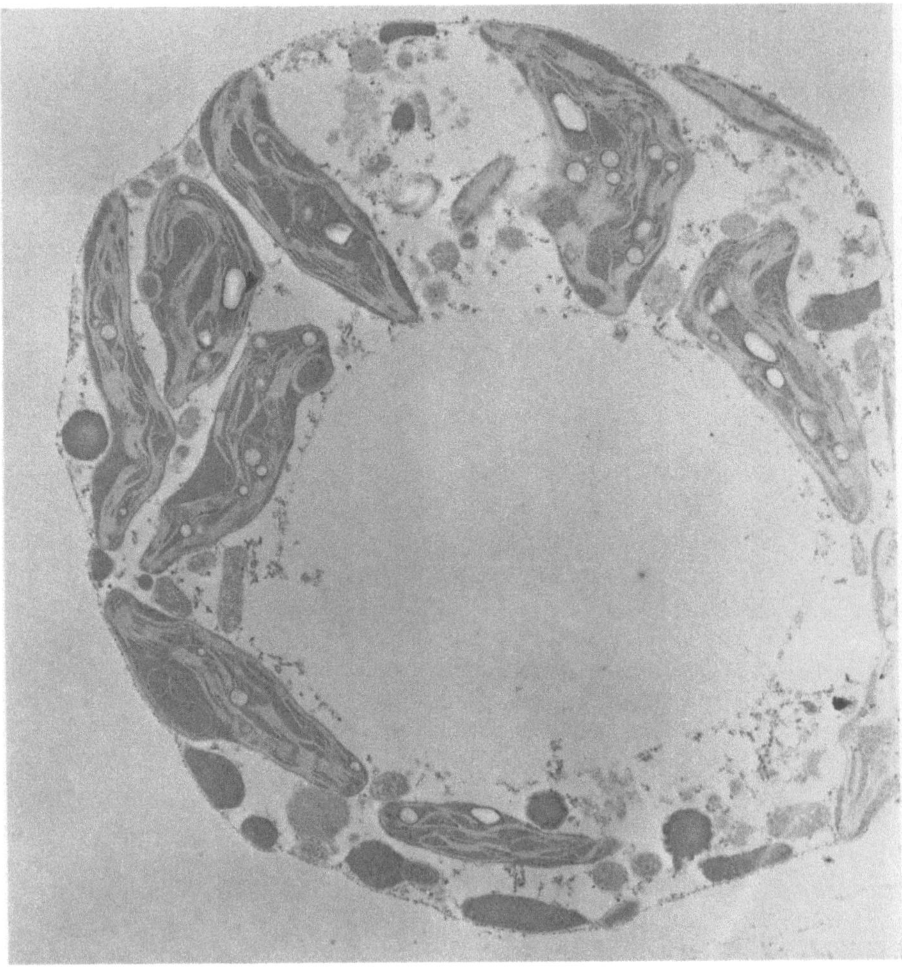

Fig. 2. Mesophyll protoplast of *Petunia hybrida* cf. Pink Magic 17 h after evacuolation (x 8500)

of the treated protoplasts there were only a few smaller vacuoles after 17 h, but in many others the tonoplast appeared to have completely reformed and the large central vacuole appeared the same as vacuoles in the normal untreated control. After 1 to 2 weeks the protoplasts started to divide and grow. Whole plants which have been regenerated from evacuolated protoplasts were morphologically identical to control plants.

This procedure can also evacuolate suspension cultured cells of *Petunia hybrida* cv. Pink Magic and *P. alpicola*. Approximate yields of evacuolated protoplasts from these cells were 35%. The viability of evacuolated suspension cultured protoplasts varied considerably. Because of the large size of most suspension cultured cells, the evacuolation procedure resulted in three types of protoplasts. The most numerous evacuolated protoplasts were quite small and were also enucleate. These protoplasts were the result of continued centrifugal force after the vacuole had been removed.

Small subcellular cytoplasmic regions were pinched off. One can control these "cytoplasts" by changing the total centrifugal force. By either decreasing the time of centrifugation or the centrifugal force, one could decrease the number of cytoplasts. In general, these cytoplasts usually died within 48 h.

Another type of evacuolated protoplast was the second most numerous after centrifugation. This type was quite large and retained the nucleus. These "evacuolated protoplasts" did, however, vary in size. Their size depended upon the original size of the suspension-cultured cells and the number of cytoplasts which were pinched off during the evacuolation procedure.

The last type of protoplast which was produced during centrifugation still retained some form of a vacuole. The vacuole was not as large as in the original protoplast. Large suspension-cultured cells could contain more than one vacuole. Usually there was a very large central vacuole and several much smaller peripheral vacuoles. One or more of these peripheral vacuoles could be retained after centrifugation. The number of these "mini-protoplasts" depended upon the total centrifugal force. A longer centrifugation time or higher centrifugal force resulted in a decrease in mini-protoplasts and an increase in evacuolated protoplasts.

Both the mini-protoplasts and evacuolated protoplasts were viable, unlike the cytoplasts which died within 48 h. Their viability depended greatly upon the centrifugation conditions and could vary from 20 to 50%. The milder the conditions the greater the viability. In general, the mini-protoplasts had a higher viability than the evacuolated protoplasts.

Evacuolated mesophyll protoplasts have been used in microinjection experiments (Griesbach 1983). After injection of 0.1 μl of chromosomes in a $CaCl_2$ buffer, 42% of the intact vacuolated protoplasts and 58% of the evacuolated protoplasts were viable and capable of division. The viability of control, noninjected protoplasts was set at 100%. The reason for the decrease in viability when using vacuolated protoplasts is that during injection the tonoplast membrane is damaged resulting in the release of the vacuole's toxic contents (e.g., waste products, hydrolytic enzymes, etc.) into the cytoplasm. Some investigators have circumvented this problem by using very large vacuolated protoplasts from suspension-cultured cells and carefully placing the microneedle in between the central vacuole and peripheral cytoplasm. Other investigators use protoplasts from meristematic cells which usually contain considerable cytoplasm and a very small vacuole. Neither of these alternatives is as simple or easy to perform as using evacuolated protoplasts.

Evacuolated protoplasts were more sensitive to mutagens and other toxic substances. Evacuolated mesophyll protoplasts of *P. hybrida* cv. Pink Magic were twice as sensitive to mutagenic agents like UV light. To date, mutagens have not significantly increased the nuclear mutation rate of cultured cells. Many investigators suggest that the vacuole's ability to sequester toxic substances precludes the exposure of the genetic material to the mutagen.

There are many enzymes associated with the vacuole and its membrane. Even though evidence suggests that there are no enzymes involved in anthocyanin biosynthesis located on the vacuolar membrane (Jonsson et al. 1983), the vacuolar membrane is very important in the accumulation and storage of these pigments. The vacuoles which reform from evacuolated protoplasts of pigmented mung bean cells do not contain the anthocyanin. This suggests that the resynthesized vacuoles are not identical to the original vacuoles. This is not unexpected, since the vacuoles reform within 17 h. It would be difficult for the cell in such a short period of time to "re-load" a vacuole and its membrane with all the necessary enzymes and metabolites. This is probably the main reason why evacuolated protoplasts do not regenerate into whole plants at the same frequency as control vacuolated protoplasts.

References

Griesbach R (1983) Protoplast microinjection. Plant Mol Biol Rep 1:32–37

Griesbach R, Sink K (1983) Evacuolation of mesophyll protoplasts. Plant Sci Lett 30:297–301

Jonsson L, Donker-Koopmann W, Uitslager P, Schram A (1983) Subcellular localization of anthocyanin methyltransferase in flowers of *Petunia hybrida*. Plant Physiol (Bethesda) 72: 287–290

Matile P (1978) Biochemistry and function of vacuoles. Annu Rev Plant Physiol 29:193–213

Thomas H, Stoddart J (1980) Leaf senescence. Annu Rev Plant Physiol 31:83–111

Protoplasts in Studies of Vacuolar Storage Compounds

G. ALIBERT, A.M. BOUDET, H. CANUT, and P. RATABOUL

Introduction

Plant cells have the ability to store carbon and (or) nitrogen compounds in a reversible or irreversible way. The accumulation of these individual substances, sometimes at high concentrations (up to 15% of the dry weight), often occurs in the large central vacuole of mature cells. This was first observed by microscopic examinations and more recently confirmed by biochemical analysis of isolated vacuoles (Matile 1978; Marty et al. 1980; Boudet and Alibert 1982). These last studies have extended the range of known vacuole-stored compounds and have provided quantitative data on their cytoplasm–vacuole distribution.

As shown in Table 1, most of the organic acids, carbohydrates with storage functions, secondary metabolites and ions are essentially stored in vacuoles. Proteins are also located inside the vacuoles and high concentrations have been reported for specific plant materials: castor bean endosperm, ananas leaves, cell suspension cultures.

Beyond these analytical determinations, mechanistic problems arise in relation to the translocation of these compounds into the vacuoles, their retention in the vacuole or the processes of their remobilization at a given stage in cell life.

Plant protoplasts can be a very useful tool for elucidating such problems:

- firstly they can be used a simplified systems in various experimental approaches,
- secondly they frequently constitute an intermediary step to obtain isolated vacuoles.

Even though protoplasts can exhibit, in comparison with the starting material, considerable changes in their biosynthetic activities (Premecz et al. 1978; Fleck et al. 1982), they also keep most of their basic metabolic abilities and can be used in short-term experiments on mechanisms of synthesis and intracellular transport.

In this way protoplasts offer many advantages: they are relatively homogeneous, easy to handle and can be fed by exogeneous compounds without the obstacle of the cell wall. In addition, they represent a particularly suitable material for the isolation of intact and pure specific organelles. Accordingly protoplasts can be fruitfully used to give a dynamic view of the intracellular transport of metabolites, between compartments. Kinetic studies including different physical and chemical treatments of protoplasts have been particularly developed in order to understand the exchanges occurring between vacuoles and the extravacuolar compartments of the cell (Kaiser et al. 1982; Boller and Alibert 1982).

The Physiological Properties of Plant Protoplasts
Edited by P. E. Pilet

Table 1. Occurrence of organic compounds and ions in plant vacuoles obtained from protoplasts

Compounds	Plant	Amount in 10^6 vacuoles	Vacuolar content in %	References
Organic Compounds				
Malic acid	*Bryophyllum*	22 μmol	100	Buser and Matile 1977
Malic acid	*Sedum*	0.75 μmol	100	Kenyon et al. 1978
Malic acid	*Melilotus*	0.17 μmol	92	Canut 1980
Isocitric acid	*Sedum*	0.12 μmol	100	Kenyon et al. 1978
Ascorbic acid	*Armoracia*	0.16 μmol	100	Grob and Matile 1980
Sucrose	*Saccharum* (cells)	2 μmol	100	Thom et al. 1982
Gentianose	*Gentiana*	2.7 μmol	100	Keller and Wiemken 1982
Fructanes	*Helianthus*	n.d.	100	Frehner et al. 1983
Fructanes	*Hordeum*	130 μg	100	Wagner et al. 1983
Glucose	*Hordeum*	0.62 μmol	100	Wagner et al. 1983
Dhurrin	*Sorghum*	1.1 μmol	100	Saunders and Conn 1978
Nicotine	*Nicotiana*	n.d.	100	Saunders 1979
Glucosinolates	*Armoracia*	n.d.	100	Grob and Matile 1979
Anthocyanins	*Tulipa*	n.d.	100	Wagner 1979
o-Coumaric acid glucoside	*Melilotus*	0.24 μmol	100	Oba et al. 1981
Proteins	*Ricinus*	1 290 μg	25	Nishimura and Beevers 1978
Proteins	*Ananas*	820 μg	36	Boller and Kende 1979
Proteins	*Nicotiana* (cells)	67 μg	12	Boller and Kende 1979
Proteins	*Acer* (cells)	58 μg	24	Alibert et al. 1982
Proteins	*Lycopersicum* (cells)	81 μg	29	Glund et al. 1984

Table 1 (cont.)

Ions					
K^+	*Bryophyllum*	n.d.		60-80	Buser and Matile 1977
	Hippeastrum	n.d.		100	Lin et al. 1977
	Acer (cells)	n.d.		76	Rona et al. 1982
Na^+	*Hippeastrum*	n.d.		100	Lin et al. 1977
	Acer (cells)	n.d.		89	Rona et al. 1982
Ca^{2+}	*Bryophyllum*	n.d.		86-90	Buser and Matile 1977
	Acer (cells)	0.009	μmol	85	Teulieres and Alibert 1984[a]
PO_4^{3-}	*Sedum*	0.008	μmol	60	Black et al. 1981
	Acer (cells)	n.d.		90	Rebeille et al. 1984
NO_3^-	*Hordeum*	1.4	μmol	99	Martinoia et al. 1981

[a]Unpublished results; n.d.: not determined

In this paper we present two examples of the use of protoplasts for these purposes: the transport and the accumulation of o-coumaric acid glucoside (o-CG) in *Melilotus alba* and the degradation of vacuolar proteins in *Acer pseudoplatanus* cell suspension cultures.

Technical Background

Protoplast were isolated from mesophyll cells of *Melilotus* or from sycamore cell suspension cultures by rapid degradation (2–3 h) of the cell wall using pectolytic and cellulolytic enzymes (cellulase: Onozuka R 10; pectinase: Onozuka macerozyme R 10).

Vacuoles were obtained from these protoplasts following the general procedure described by Boudet et al. (1981) which is summarized in Fig. 1. This method, which can be easily adapted, with slight modifications, to different plant materials has the following advantages:

- rapid separation of the vacuoles from the other compartments of the protoplast (less than 1 min)
- no changes in the osmotic pressure of the medium during the isolation procedure
- preparation and purification of the vacuoles in one and the same step.

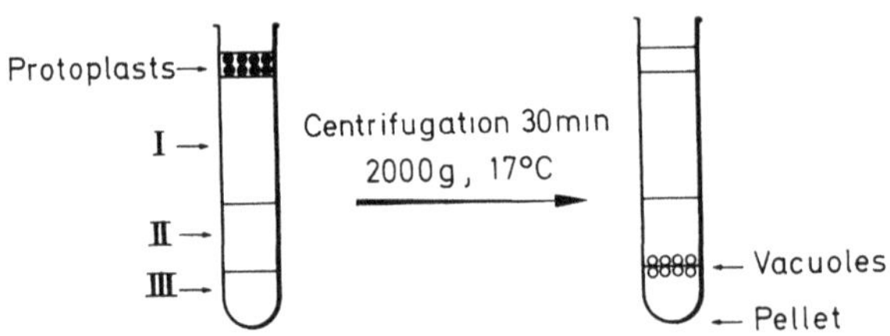

Fig. 1. Procedure for obtaining and purifying vacuoles. *I* 5 ml of 2% Ficoll + DEAE Dextran; *II* 2 ml 5% Ficoll + Dextran sulphate; *III* 2 ml 20% Ficoll. The concentration of DEAE Dextran and Dextran sulphate are adjusted for each experimental plant material

Consequently this method is particularly suited for studying on a kinetic basis, the exchanges of solutes through the tonoplast because it allows the rapid and accurate determination of intravacuolar storage pools.

Transfer and Accumulation of o-Coumaric Acid Glucoside in *Melilotus alba* Mesophyll Cell Protoplasts (Rataboul et al. 1985)

In *Melilotus* mesophyll cells an o-coumaric acid glucoside (Fig. 2) accumulates to constitute up to 8% of the dry weight of the plant. Oba et al. (1981) have shown that this compound is entirely located in the vacuolar compartment of the cell. Thus *Melilotus* appears to be a good material for studying the mechanisms involved in the accumulation of phenolic glucosides by plant cells.

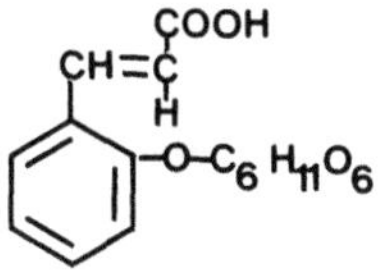

Fig. 2. Trans o-coumaric acid glucoside

o-CG is synthesized from phenylalanine (Kosuge and Conn 1961) and is rapidly labelled when the isolated protoplasts are fed with this radioactive precursor. Figure 3 shows the kinetics of synthesis of o-CG (radioactivity in protoplasts): in our experimental conditions, following a initial rapid rise, the radioactivity in the glucoside levels off after 15 h.

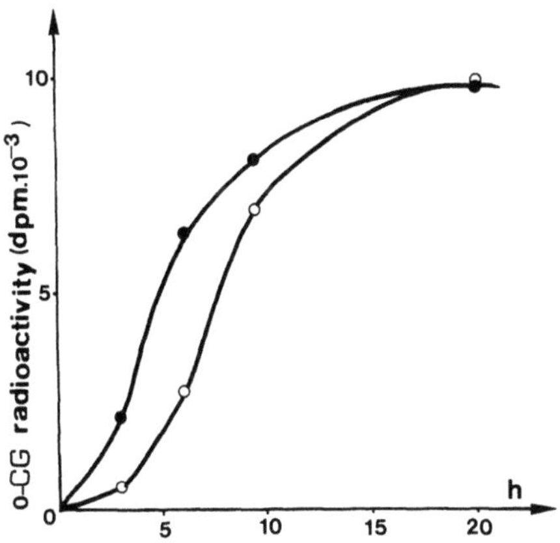

Fig. 3. Time course changes of o-coumaric acid glucoside labelling from [^{14}C] phenylalanine in protoplasts ●——● and in the corresponding vacuoles ○——○. Radioactivity is expressed for 10^6 protoplasts or 10^6 vacuoles. In the material studied one protoplast contains one vacuole

The time course increases of o-CG radioactivity in vacuole samples corresponding to the above protoplast fractions show a delay in comparison with the protoplasts. After short incubation times most of the newly synthesized radioactive glucoside was absent from vacuoles, indicating an initial cytoplasmic location. These results demonstrate that the glucoside molecule is first synthesized in the cytoplasm and then transferred to the vacuoles. These data disagree with previous hypotheses (Fritsch and Grisebach 1975; Marty et al. 1980), suggesting that phenolic compounds could be glucosylated at the tonoplast by a membrane-bound glucosyl-transferase. Moreover in *Melilotus* the glucosyl-transferase active on o-coumaric acid, which is present in protoplasts, has not been detected in vacuoles.

After the cessation of o-CG synthesis, its radioactivity in the vacuoles still increases until all the compound is recovered in the vacuolar compartment. Thus a slow but continuous transfer of o-CG into the vacuole occurs, explaining the nearly complete absence of o-CG in the cytoplasmic fraction.

As other phenolics, this compound does not exhibit a marked turnover (results not shown). The continuous transfer of o-CG in the vacuole, together with the absence of degradation of vacuolar o-CG, explain the accumulation of the compound in the organelle.

As yet nothing is known about the mechanisms (energization, biochemical and physiological controls) involved in the transfer of secondary metabolites into the vacuole. We have studied the effect of light on the transport of o-CG using *Melilotus* protoplasts.

The protoplasts were fed with [^{14}C] phenylalanine for 4 h in the light, then a fraction of the sample was transferred to the dark while the remaining protoplasts were kept in light conditions with or without 5 μM DCMU (an inhibitor of photosynthesis) (Fig. 4).

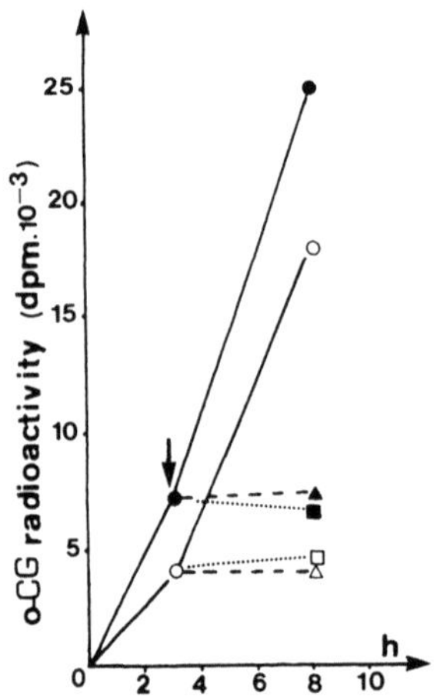

Fig. 4. Effect of light on synthesis and transport into vacuoles of o-coumaric acid glucoside. Protoplasts were fed with [^{14}C] phenylalanine at time 0 in light conditions. *Arrow* indicates time of transfer to dark conditions or addition of DCMU to the incubation medium. Radioactivity is expressed for 10^6 protoplasts or 10^6 vacuoles.

Protoplasts	●——●	light	○——○	Vacuoles
	▲- - -▲	dark	△- - -△	
	■ .... ■	light + DCMU	□ .... □	

In the light, as previously shown, o-CG is synthesized and continuously transported to the vacuole. In contrast, dark prevents both synthesis and vacuolar transfer of o-CG. Similar results are obtained in the presence of DCMU. Thus we can conclude that photosynthesis controls both the synthesis and the transport of o-CG into the vacuole. Light induces a significant acidification of the vacuolar sap which is prevented by addition of DCMU (Fig. 5). In addition, an artificial increase of the vacuole sap pH, obtained by feeding high levels of benzylamine, a lipophilic base which accumulates in the vacuole (results not shown), inhibits the vacuolar accumulation of o-CG without any effect on its synthesis.

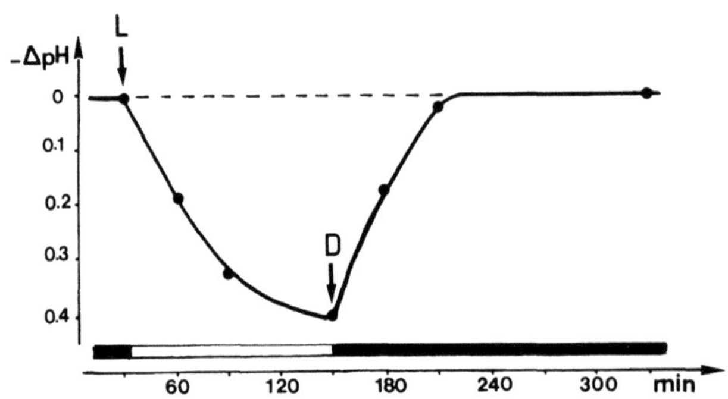

Fig. 5. Reversible changes in vacuolar pH induced by dark/light transitions in *Melilotus* protoplasts. Vacuolar pH was estimated using the distribution of [^{14}C] benzylamine (Kurkdjian 1982). Dark/light transitions ●——● ; light + DCMU ●- - -●

These results suggest that the vacuolar pH is under the control of photosynthesis and that the △pH existing between the cytoplasm and the vacuole is an important factor in the transfer of the glucoside into the vacuole (the lower the pH of the vacuolar sap the easier o-CG is transported).

Protein Degradation in Vacuoles (Canut et al. 1985)

In the early 1960's Matile identified for the first time (Matile and Wiemken 1967) hydrolases in plant cell vacuoles and reported that vacuoles can be compared to animal lysosomes. Further works by Pujarniscle (1968), Boller and Kende (1979) or Canut et al. (1985) confirmed these first results with special emphasis on the presence of proteases in this organelle.

However, the identification of proteolytic activities in vacuoles was essentially performed using artificial substrates (e.g. p-nitroanilide derivatives) or molecules of animal origin (e.g. albumin, globin etc.) and, apart from the work of Nishimura and Beevers (1979) on castor bean endosperm, no demonstration of the in vivo effective proteolytic function of vacuoles has been given.

Using protoplasts from sycamore cell suspension cultures we attempted to show that proteolysis actually does occur inside the vacuoles.

When transferred to a new medium, *Acer* cells accumulate proteins until day 8 of culture, in part in the vacuolar compartment (Fig. 6); subsequently the protein content decreases rapidly between days 8 and 10 (Alibert et al. 1982). We took advantage of this situation to study protein degradation in vacuoles.

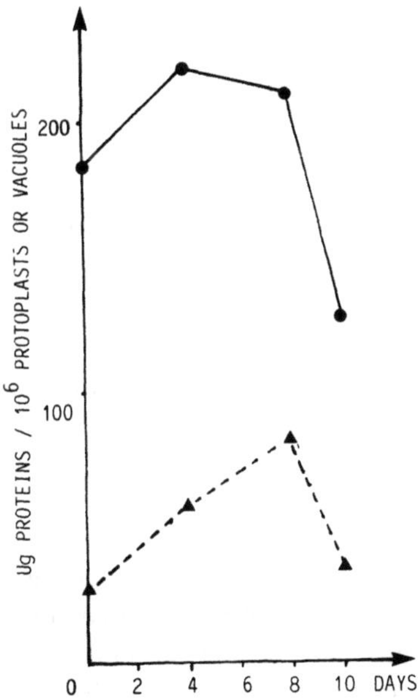

Fig. 6. Protein content of protoplasts •——• and vacuoles ▲- - - -▲ obtained from *Acer* cells, at different times after transfer on a new medium

Seven-day-old cells were supplied with tritiated leucine to label the proteins. After 18 h, the cells were harvested and protoplasts prepared. Radioactivity in the TCA-precipitable material from protoplasts and the corresponding vacuoles maintained in various conditions was followed with time.

Incorporation of [^{3}H] leucine into the TCA-precipitable material occurs when protoplasts are maintained at 20°C showing that isolated protoplasts are able to sustain protein synthesis from endogeneous aminoacids. In contrast, no protein synthesis occurs at 4°C (Fig. 7).

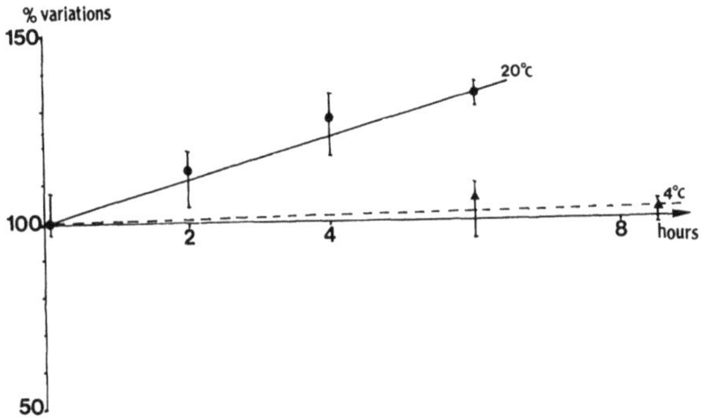

Fig. 7. Time course changes of TCA-insoluble radioactivity in protoplasts obtained from *Acer* cells fed with [^{3}H] leucine. Protoplasts were incubated either at 20°C or at 4°C. At time 0, 100 % corresponds to 19740 dpm for 10^6 protoplasts

Vacuoles contain about 30% of the total labelled proteins of the protoplasts at the time of their isolation. Following incubation of isolated vacuoles, the TCA-insoluble radioactivity decreases with time (Fig. 8). This decrease is temperature-(Fig. 8) and pH-dependent (Fig. 9) like enzymatic proteolysis (Canut et al. 1984a).

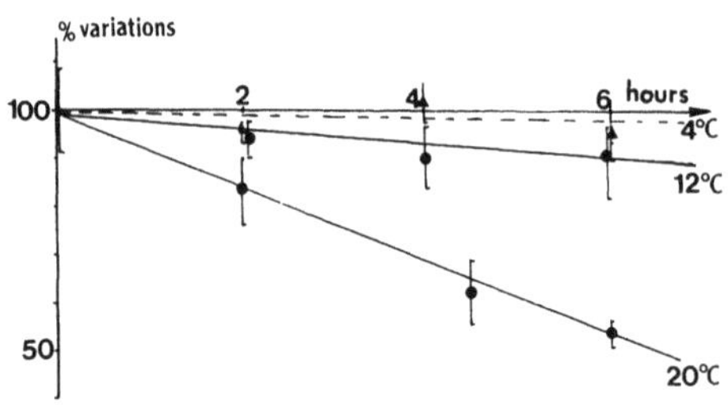

Fig. 8. Degradation of proteins by intact isolated vacuoles – Temperature dependency . At time 0, 100 % corresponds to 1 179 dpm. Incubations were performed at pH 6.5

These results demonstrate that the isolated vacuoles contain endogeneous labelled proteins (i.e. that proteins synthesized in the cytoplasm are transferred to the vacuoles) and that protein breakdown, probably due to proteases, occurs in isolated vacuoles.

Taken together, the results of Figs. 6 and 8 point out the involvement of the vacuole as a transitory storage site for cellular proteins in *Acer pseudoplatanus* cells. Proteins accumulated in vacuoles can be further degraded and the released amino acids are likely to be remobilized for supplying cellular nitrogen metabolism. Similar remobilization of proteins, accumulated in the vacuoles of paraveinal mesophyll cells of soybean, has been recently suggested by Franceschi et al. (1983).

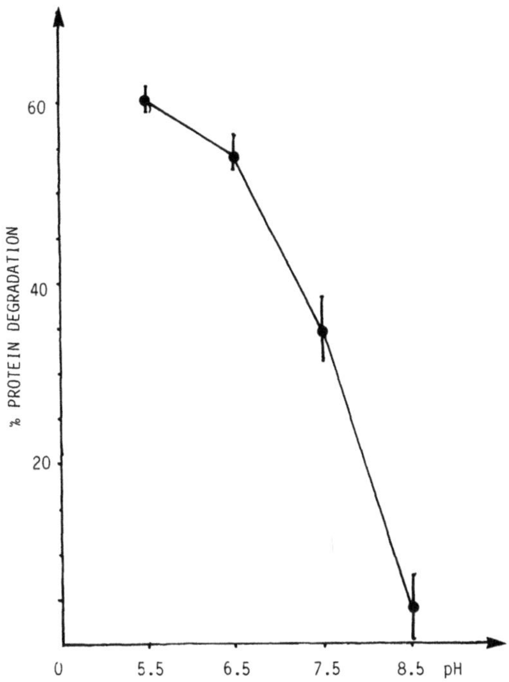

Fig. 9. Degradation of proteins by intact isolated vacuoles – pH dependency. Vacuoles were isolated in medium buffered at different pH values and incubated at 20°C for 6 h

Conclusion

The previous examples concern two types of vacuole-stored compounds:

1. a secondary metabolite apparently irreversibly stored as a glucoside – the most frequent form in which natural substances are accumulated in plants.

2. storage proteins which can be remobilized and participate in the nitrogen supply of the cytoplasm.

The results obtained in both cases show that isolated protoplasts associated with organelle isolation (vacuoles) and labelling studies are unique tools for investigating, at the subcellular level, the dynamics of the exchanges between the different cell compartments. No doubt that improvements in the above individual techniques will increase our knowledge of plant cellular compartmentation and particularly the physiological role of vacuoles.

Acknowledgements. This research was supported by CNRS U.A. No. 241. We express our most sincere thanks to our colleagues for helpful technical assistance and discussions.

References

Alibert G, Carrasco A, Boudet AM (1982) Changes in biochemical composition of vacuoles isolated from *Acer pseudoplatanus* L. during cell culture. Biochim Biophys Acta 721:22–29

Black CC, Carnal NW, Kenyon WH (1982) Compartmentation and the regulation of CAM. In: Ting IP, Gibbs M (eds) Crassulacean acid metabolism. Proc 5th annual symp in botany Riverside, California, pp 51–68

Boller T, Kende H (1979) Hydrolytic enzymes in the central vacuole of plant cells. Plant Physiol (Bethesda) 63:1123–1132

Boudet AM, Alibert G (1982) Vacuoles isolées de plantes supérieures. Techniques d'obtention et principales caractéristiques biochimiques. Bull Soc Bot Fr Actual Bot 128:81–93

Boudet AM, Canut H, Alibert G (1981) Isolation and characterization of vacuoles from *Melilotus alba* mesophyll. Plant Physiol (Bethesda) 68:1354–1358

Buser C, Matile P (1977) Malic acid in vacuoles isolated from *Bryophyllum* leaf cells. Z Pflanzenphysiol 82:462–466

Canut H (1980) Isolement des vacuoles de *Melilotus alba* Desr. Etude préliminaire de leurs caractéristiques biochimiques. Thesis Toulouse

Canut H, Alibert G, Boudet AM (1985a). Intracellular localization of proteases in *Melilotus alba* mesophyll protoplasts (submitted)

Canut H, Alibert G, Boudet AM (1985b) Hydrolysis of intracellular proteins in vacuoles isolated from *Acer pseudoplatanus* cells (submitted)

Fleck J, Dürr A, Fritsch C, Vernet T, Hirth L (1982) Osmotic-shock stress proteins in protoplasts of *Nicotiana sylvestris.* Plant Sci Lett 26:159–165

Franceschi VR, Wittenbach VA, Giaquinta RT (1983) Paraveinal mesophyll of soybean leaves in relation to assimilate transfer and compartmentation III. Immunohistochemical localization of specific glycopeptides in the vacuole after depodding. Plant Physiol (Bethesda) 72: 586–589

Frehner M, Wiemken A, Keller F, Matile P (1983) Subcellular compartmentation of fructan metabolism in tubers of Jerusalem artichoke (*Helianthus tuberosus*)

Fritsch H, Grisebach H (1975) Biosynthesis of cyanidin in cell cultures of *Happlopappus gracilis.* Phytochemistry (Oxf) 14:2437–2441

Glund K, Tewes A, Abel S, Leinhos W, Walther R, Reinbothe H (1984) Vacuoles from cell suspension cultures of tomato (*Lycopersicum esculentum*) – Isolation and characterization. Z Pflanzenphysiol 113:151–161

Grob K, Matile P (1979) Vacuolar location of glucosinolate in horseradish root cells. Plant Sci Lett 14:327–335

Grob K, Matile P (1980) Compartmentation of ascorbic acid in vacuoles of horseradish root cells. Note on vacuolar peroxydase. Z Pflanzenphysiol 98:235–243

Kaiser G, Martinoia E, Wiemken A (1982) Rapid appearance of photosynthetic products in the vacuoles isolated from barley mesophyll protoplasts by a new fast method. Z Pflanzenphysiol 107:103–113

Keller F, Wiemken A (1982) Differential compartmentation of sucrose and gentianose in the cytosol and vacuoles of storage root protoplasts from *Gentiana lutea* L. Plant Cell Rep 1: 274–277

Kenyon WH, Kringstadt R, Black C (1978) Diurnal changes in the malic content of vacuoles isolated from leaves of the crassulacean acid metabolism plant *Sedum telephium.* FEBS Lett 94:281–283

Kosuge T, Conn EE (1961) The metabolism of aromatic compounds in higher plants III. The β-glucosides of o-coumaric, coumarinic and melilotic acids. J Biol Chem 236:1617–1621

Kurkdjian A (1982) Absorption and accumulation of nicotine by *Acer pseudoplatanus* and *Nicotiana tabacum* cells. Physiol Veg 20:73–83

Lin W, Wagner GJ, Siegelman HW, Hind G (1977) Membrane-bound ATPase of intact vacuoles and tonoplasts isolated from mature plant tissue. Biochim Biophys Acta 465:110–117

Martinoia E, Heck U, Wiemken A (1981) Vacuoles as storage compartments for nitrate in barley leaves. Nature (Lond) 289:292–294

Marty F, Branton D, Leight RA (1980) Plant vacuoles. In: Stumpf PK, Conn EE (eds) The biochemistry of plants, vol I. Academic, New York, pp 625–658

Matile P (1978) Biochemistry and function of vacuoles. Annu Rev Plant Physiol 29:193–213

Matile P, Wiemken A (1967) The vacuole as the lysosome of yeast cell. Arch Microbiol 56:148–155

Nishimura N, Beevers H (1978) Hydrolases in vacuoles from castor bean endosperm. Plant Physiol (Bethesda) 62:44–48

Nishimura N, Beevers H (1979) Hydrolysis of proteins in vacuoles isolated from higher plant tissue. Nature (Lond) 277:412–413

Oba K, Canut H, Boudet AM, Conn EE (1981) Subcellular localization of 2-(β-D-glucosyloxy)-cinnamic acids and the related β-glucosidase in leaves of *Melilotus alba* desr. Plant Physiol (Bethesda) 68:1359–1363

Premecz G, Ruzicska P, Olah T, Farkas GL (1978) Effect of "osmotic stress" on protein and nucleic acid synthesis in isolated tobacco protoplasts. Planta (Berl) 141:33–36

Pujarniscle S (1968) Caractère lysosomal des lutoides du latex *d'Hevea brasiliensis* Mull Physiol Veg 6:27–46

Rataboul P, Alibert G, Boller T, Boudet AM (1985) o-coumaric acid glucoside in vacuoles isolated from *Melilotus alba* protoplast mesophyll cells I. Intracellular transport and vacuolar accumulation (submitted to BBA)

Rebeille P, Bligny R, Guern J, Martin JB, Douce R (to be published 1984) Phosphate metabolism in intact plant cells. In: Alibert G, Boudet AM, Marigo G, Lea P (eds) Annual proc of the Phytochemical Society of Europe, vol 24. Oxford University Press, Oxford

Rona JP, Cornel D, Grignon C, Heller R (1982) The electrical potential difference across the tonoplast of *Acer pseudoplatanus* cells. Physiol Veg 20:459–463

Saunders JA (1979) Investigation of vacuoles isolated from tobacco. Plant Physiol (Bethesda) 64:74–78

Saunders JA, Conn EE (1978) Presence of the cyanogenic glucoside dhurrin in isolated vacuoles from sorghum. Plant Physiol (Bethesda) 61:154–157

Thom M, Maretzki A, Komor E (1982) Vacuoles from sugarcane suspension culture I. Isolation and characterization. Plant Physiol (Bethesda) 69:1315–1319

Wagner GJ (1979) Content and vacuole/extravacuole distribution of neutral sugars, free amino acids and anthocyanin in protoplasts. Plant Physiol (Bethesda) 64:88–93

Wagner W, Keller F, Wiemken A (1983) Fructan metabolism in cereals: induction in leaves and compartmentation in protoplasts and vacuoles. Z Pflanzenphysiol 112:359–372

Distribution of Saccharides Between Cytoplasm and Vacuole in Protoplasts

F. KELLER, P. MATILE, and A. WIEMKEN

Introduction

Glucose and fructose have a key function in metabolism. They can be utilized as a source of carbon in most synthetic pathways and can also satisfy the demand for metabolic energy. It is not surprising, therefore, that plant cells always contain a considerable pool of free hexoses and sucrose. It confers to the cell a certain independence on the supply of carbon and energy. Depending on species, type of organ, developmental state, and environmental conditions, pools of hexoses, sucrose, and other saccharides such as oligosaccharides and water soluble polysaccharides (fructans) may be exceedingly high, particularly in storage tissues. When, e.g., barley leaves are excised and illuminated strongly they are not able to export the enormous amounts of fixed carbon. They then store it in the form of fructan which may eventually account for as much as 70% of the dry weight (Wagner et al. 1983). It is obvious that such a large mass of solutes cannot be accommodated in the cytoplasm. Solutes imported or produced in excess are stored in the large internal extraplasmatic space of the plant cell, the central vacuole. This compartment offers to the cells an almost unlimited capacity for the accommodation of saccharides.

Protoplasts are most useful tools for the study of functions of vacuoles such as the storage of metabolites. Not only can vacuoles be isolated from them and distributions of compounds and enzyme activities between cell sap and cytoplasm be determined by direct analysis, but also interactions between cytoplasm and vacuoles can be investigated by following the intracellular partitioning of metabolites in tracer experiments or else, by studying transport across the tonoplast in isolated vacuoles (reviews by Wagner 1982; Leigh 1983).

Subcellular Distribution of Saccharides

In some instances, sugars have been localized exclusively in vacuoles. Yet the distribution of various saccharides appears to depend on species, type of tissue and perhaps also on their developmental state. Wagner (1979) has shown that sucrose is located almost completely in vacuoles isolated from *Tulipa* petals. He recovered most of the monosaccharides in the vacuoles prepared from *Hippeastrum* petals and *Tulipa* leaves but in *Tulipa* petals half of the total fructose was extravacuolar. It is not sur-

The Physiological Properties of Plant Protoplasts
Edited by P. E. Pilet

prising that in an organ like the storage root of beet root (Leigh et al. 1979) or in sugar cane suspension culture (Thom et al. 1982a) most of the sucrose was localized in the vacuoles. But a finding of Nishimura and Beevers (1978) is rather unexpected: in the endosperm cells of germinating castor beans the main product of lipid mobilization, sucrose, is not only secreted into the apoplast (and eventually absorbed by the cotyledons) but also stored in the vacuoles; as much as 62% of the total protoplastic sucrose was recovered in the vacuoles. In immature apple flesh practically all of the hexoses present in the protoplasts were found in the vacuoles, but sucrose appeared to be excluded from the vacuoles; in contrast, the principal compound for long distance transport of carbon in apple trees, sorbitol, was found to be accumulated in the vacuoles (Yamaki 1984). In the grape berry glucose and fructose are the principal sugars accumulated; concentrations as high as ca. 1.5 M have been determined in vacuoles isolated from subepidermal protoplasts (Moskowitz and Hrazdina 1981).

A study on the various saccharides in parenchyma cells of the storage root of *Gentiana lutea* may further illustrate the fact that the distributions between vacuole and cytoplasm may differ markedly from one compound to the other (Fig. 1): the trisaccharide gentianose which is by far the most abundant saccharide in this organ as well as gentiobiose were found exclusively in the vacuoles whereas ca. 50 % of the total sucrose was localized in the extravacuolar space of protoplasts (Keller and Wiemken 1982).

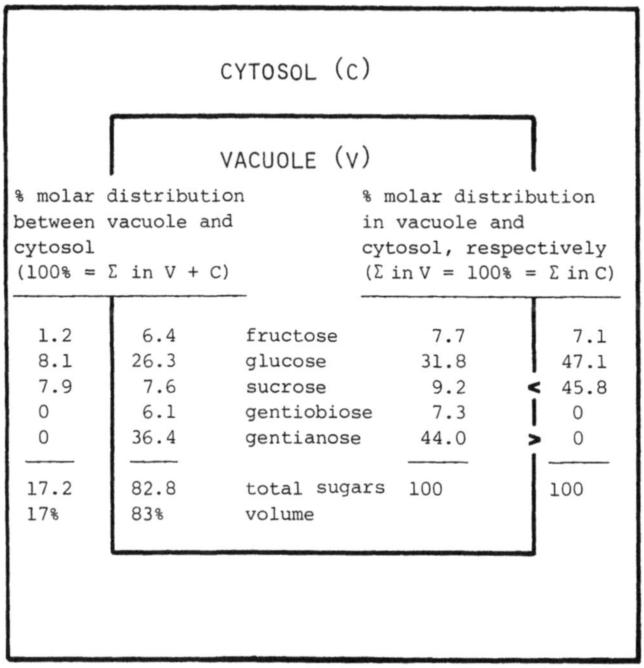

Fig. 1. Distribution of saccharides between cytosol and vacuole in protoplasts of *Gentiana lutea* L. roots. (Keller and Wiemken 1982)

A similar distribution of the tetrasaccharide stachyose and sucrose, respectively, has been determined in protoplasts from the storage tubers of *Stachys Sieboldii* (Keller, unpublished).

Percent distributions between vacuole and cytplasm as calculated by counting numbers of protoplasts and vacuoles in the preparations analysed (e.g. Wagner 1979; Yamaki 1984) or by assessment of vacuolar marker enzymes (e.g. Keller and Wiemken 1982) may cause a wrong impression about the apparent concentrations within the compartments. Since vacuoles may occupy 90 % or even more of the total volume of protoplasts, the presence of a high proportion of a sugar in the isolated vacuoles may merely indicate that concentrations in vacuolar and extravacuolar compartments are about equal. Keller and Wiemken (1982) have calculated the molar distributions of saccharides in quiescent gentian roots under the assumption that the cytosolic and vacuolar fluids are isotonic. It turned out that the 50 % of the total sucrose which are squeezed into the small volume of the cytoplasm produce a concentration five times higher than that of the vacuolar fluid. As far as the molar concentrations in the compartments are concerned, three classes of saccharide can be distinguished in gentian roots: gentianose and gentiobiose being exclusively vacuolar, sucrose having a high concentration in the cytosol, and the monosaccharides which are distributed evenly.

In sugar cane suspension cultured cells an analysis of sugar accumulation and distribution has yielded convincing evidence that the concentrations of hexoses and sucrose in the cell sap may exceed those in the cytoplasm several times (Komor et al. 1982; Thom et al. 1982b). Taken together with the exclusive vacuolar localization of certain saccharides mentioned before, this example clearly demonstrates that accumulation of corresponding solutes is an energy-requiring process. There is little doubt that the tonoplast is a site of active transport of sugars from the cytosol into the cell sap. Moreover, the distinct distributions of sugars as e.g. in the gentian storage parenchyma cells suggest that transport across the tonoplast may not only be active but also selective with regard to the sugar species concerned.

The transport mechanisms at the tonoplast responsible for the pumping of sugars from the cytoplasm into the cell sap are not yet fully understood. Only in the case of fructans has the complete compartmentation in the cell sap been recently explained by the finding that their synthesis takes place within the vacuoles (Wagner et al. 1983; Frehner et al. 1984).

Storage and Metabolism of Fructans

Fructans (polyfructosylsucrose) are water soluble, non structural carbohydrates which are widely distributed in the plant kingdom (for recent reviews see Kandler and Hopf 1982; Meier and Reid 1982). They can occur in all vegetative parts of plants but are often found to accumulate in large amounts (up to 90 % of the dry weight) in storage organs such as roots, bulbs, tubers and rhizomes, in meristematic regions of leaves or in the lower sections of stems of grasses.

Three main classes of fructans have been found which differ in their glycosidic linkages: (a) the inulin type with $(2 \rightarrow 1)$-β-D-fructofuranosidic linkages, (b) the phlein type with $(2 \rightarrow 6)$-β-D-fructofuranosidic linkages and (c) the branched type with a mixture of both linkage types. The fructans found in dicotyledons are all of the inulin type. The fructans of the monocotelydons are of the phlein or the branched type and only rarely of the inulin type.

The degree of polymerisation (DP) varies according to plant species, life cycle and environmental conditions but is always low (between 3 and 250) as compared to for instance starch. Two main physiological roles have been ascribed to fructans: (a) reserve carbohydrate which is obvious considering the high fructan content often found in plants and (b) protecting agent which is less obvious and only poorly studied but is a conclusion of the observation that many fructan-containing plants can survive cold or dry periods during their life cycle quite well.

Despite their importance fructans have received relatively little attention. Most information on their biosynthesis and degradation is due to the pioneering work of Edelman and co-workers (Edelman and Jefford 1968), who studied the metabolism of inulin in tubers of Jerusalem artichoke (*Helianthus tuberosus* L.).

They found four enzymes to be involved in the inulin metabolism (Fig. 2): SST forms the trisaccharide isokestose (1^F-fructosylsucrose) from sucrose, FFT transfers single terminal fructosyl residues from inulins of DP $\geqslant 3$ to sucrose or inulins of DP $\geqslant 3$, and two FEH which are very similar hydrolyse single terminal fructosyl residues off from the chain ends of inulin.

$$GF + GF \xrightarrow{SST} GFF + G \qquad (a)$$

$$GFF_n + GFF_m \xrightarrow{FFT} GFF_{n+1} + GFF_{m-1} \qquad (b)$$

$$GFF_m \xrightarrow[+H_2O]{FEH} GFF_{m-1} + F \qquad (c)$$

$$GF \xrightarrow[+H_2O]{INV} G + F \qquad (d)$$

Fig. 2. Summary of the reactions catalyzed by the main enzymes directly involved in the metabolism if inulin. *F* fructose; *G* glucose; *GF* sucrose; *GFF* isokestose (trisaccharide) GFF_n inulin of DP_{n+2} ($0 \leqslant n < 30$ to 35); GFF_m inulin of DP_{m+2} ($1 \leqslant m \leqslant 30$ to 35); *SST* sucrose-sucrose-1^F-fructosyltransferase; *FFT* fructan-fructan-1-fructosyltransferase; *FEH* fructan exohydrolase; INV invertase. (Frehner et al. 1984)

In a logical continuation of Edelman's studies Frehner et al. (1984) have recently been able to answer some of the open questions regarding the subcellular organization of the inulin metabolism. They prepared protoplasts and vacuoles from tubers of Jerusalem artichoke and compared their saccharide contents and activities of the enzymes involved in the inulin metabolism. Using α-mannosidase and β-N-acetylglucosaminidase as marker enzymes for the vacuoles they concluded that all the inulin saccharides with a DP $\geqslant 3$ and all the inulin enzymes (SST, FFT, FEH) were entirely vacuolar. A new working model was proposed (Fig. 3), in which it is suggested that the key steps of the metabolism of inulin all take place in the large central vacuole which also serves as its storage site. It is obvious that a careful regulation of the actions of anabolic and catabolic enzymes must be postulated if they are present in the same compartment. This is partly fulfilled in that SST and FEH are never active simultaneously in the tubers during their life cycle, and FEH is strongly inhibited by sucrose (Edelman and Jefford 1968). However, more work will be needed in order to further clarify the regulatory aspects of the proposed model as well as to demonstrate transport systems for fructose, glucose, and sucrose across the tonoplast, and the soluble nature of the inulin enzymes.

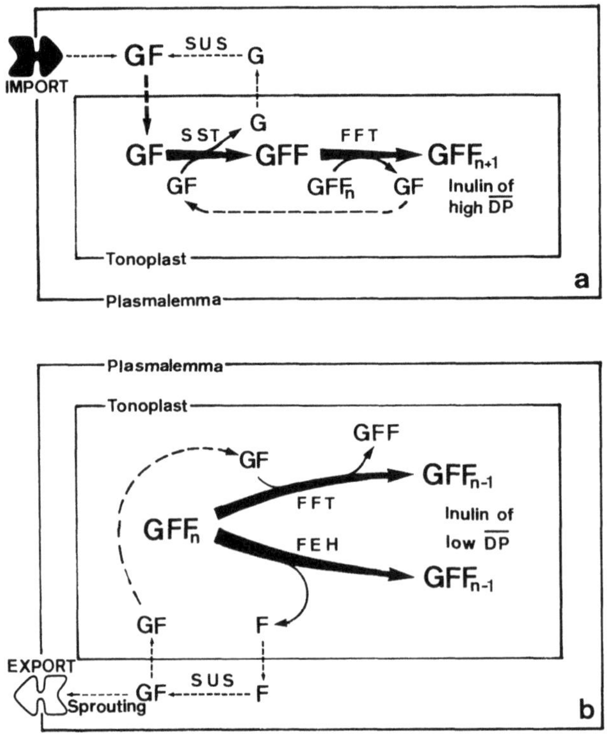

Fig. 3a,b. Model proposed for the compartmentation of the metabolism of inulin in tubers of *Helianthus tuberosus* L.. Developing stage (**a**), resting and sprouting stage (**b**). SUS: sucrose synthesis. Other abbreviations see Fig. 2. (Frehner et al. 1984)

As compared to the dicotelydons the fructans of the monocotelydons do not only differ in their glycosidic linkages and are more complex, but their enzymology is also not understood very well at the present time. However, numerous studies exist which demonstrate the accumulation of fructans in leaves of temperate grasses whenever the supply of assimilates by photosynthesis exceeds the demand. Such a reaction may be induced experimentally by exposing plants to high light intensity and low temperatures (cf. reviews cited). Wagner et al. (1983) made use of this when they set out to study the metabolism of fructans in leaves of barley. They excised primary leaf blades and illuminated them overnight with continuous light of high intensity. This treatment induced a marked accumulation fructans. The subcellular arrangement of some of the enzymes, substrates and products involved in that fructan metabolism was then studied using a similar approach to that of Frehner et al. (1983) by isolating and comparing leaf protoplasts and their vacuoles. In fructan-enriched barley leaves, all the fructans (DP $\geqslant$ 3) as well as all the SST activity were found to be completely associated with the vacuoles. Studies on the subcellular distribution of FEH in the same system also clearly demonstrated a vacuolar localization (Wagner, unpublished). With regards to FFT, the enzyme(s) promoting chain elongation of fructans, only very limited information is available at the moment. Preliminary studies have demonstrated the ability of purified vacuoles to form higher fructans (DP > 3) (Wagner, unpublished). Detailed analyses of the enzymology and the subcellular localization of FFT has not yet been undertaken. However, the data which are presently available for barley leaves allow the presumption that the situation might well be analogous to that found in tubers of Jerusalem artichoke, so that the model proposed by Frehner et al. (1984) (Fig. 3) could be a fairly general one. Much more work will of course be needed to demonstrate its full validity. With the recent development of techniques for the isolation of protoplasts and vacuoles and the inherent experimental possibilities rapid progress should be made towards a better understanding of the metabolism of fructans in higher plants.

References

Edelman J, Jefford TG (1968) The mechanism of fructosan metabolism in higher plants as exemplified in *Helianthus tuberosus*. New Phytol 67:517–531

Frehner M, Keller F, Wiemken A (1984) Localization of fructan metabolism in the vacuoles isolated from protoplasts of Jerusalem artichoke tubers (*Helianthus tuberosus* L.) J Plant Physiol 116:197–208

Kandler O, Hopf H (1982) Oligosaccharides based on sucrose (sucrosyl oligosaccharides). In: Loewus FA, Tanner W (eds) Encyclopedia of plant physiology, new series, vol 13A. Springer, Berlin Heidelberg New York, p 348

Keller F, Wiemken A (1982) Differential compartmentation of sucrose and gentianose in the cytosol and vacuoles of storage root protoplasts from *Gentiana lutea* L. Plant Cell Rep 1:274–277

Komor E, Thom M, Maretzki A (1982) Sugar transport by sugar cane vacuoles. Physiol Veg 20: 277–287

Leigh RA (1983) Methods, progress and potential for the use of isolated vacuoles in studies of solute transport in higher plant cells. Physiol Plant 57:390

Leigh RA, Rees T, Fuller WA, Banfield J (1979) The location of acid invertase activity and sucrose in the vacuoles of storage roots of beet root (*Beta vulgaris*). Biochem J 178:539–547

Meier H, Reid JSG (1982) Reserve polysaccharides other than starch in higher plants. In: Loewus FA, Tanner W (eds) Encyclopedia of plant physiology, new series, vol 13A. Springer, Berlin Heidelberg New York, p 418

Moskowitz AH, Hrazdina G (1981) Vacuolar contents of fruit subepidermal cells from *Vitis* species. Plant Physiol (Bethesda) 68:686–692

Nishimura M, Beevers H (1978) Hydrolases in vacuoles from castor bean endosperm. Plant Physiol (Bethesda) 62:44–48

Thom M, Maretzki A, Komor E (1982a) Vacuoles from sugar cane suspension cultures I. Isolation and partial characterization. Plant Physiol (Bethesda) 69:1315–1319

Thom M, Komor E, Maretzki A (1982b) Vacuoles from sugar cane suspension cultures II. Characterization of sugar uptake. Plant Physiol (Bethesda) 69:1320–1325

Wagner GJ (1979) Content and vacuole/extravacuole distribution of neutral sugars, free amino acids and anthocyanin in protoplasts. Plant Physiol (Bethesda) 64:88–93

Wagner GJ (1982) Compartmentation in plant cells: the role of the vacuole. Recent Adv Phytochem 14:1–45

Wagner W, Keller F, Wiemken A (1983) Fructan metabolism in cereals: induction in leaves and compartmentation in protoplasts and vacuoles. Z Pflanzenphysiol 112:359–372

Yamaki S (1984) Isolation of vacuoles from immature apple fruit flesh and compartmentation of sugars, organic acids, phenolic compounds and amino acids. Plant Cell Physiol 25:151–166

Anthocyanin-Containing Vacuoles Isolated from Protoplasts of Daucus carota Cell Cultures

W. HOPP, W. HINDERER, M. PETERSEN, and H.U. SEITZ

Introduction

The central vacuole is the largest organelle of mature plant cells. The vacuole is a storage compartment for ions, metabolites, secondary products and hydrolytic enzymes involved in various degradative processes.

Among other pigments, flavonoid glycosides are deposited in the vacuolar sap. Techniques recently developed to isolate vacuoles are utilized by several investigators (Hrazdina et al. 1978; Sasse et al. 1979) to study the compartmentation of biosynthesis of these secondary products.

It has been suggested that some of the enzymes involved in anthocyanin biosynthesis are localized at or in the tonoplast (Fritsch and Grisebach 1975). One approach to solve this question is to study the uptake of labelled precursors of anthocyanins by isolated vacuoles. Another possibility is to investigate the localization of enzymes necessary for anthocyanin biosynthesis. The occurrence of three important enzymes from this metabolic pathway of anthocyanins was investigated by Hrazdina et al. (1978) in vacuoles from petals of *Hippeastrum* and *Tulipa*. In these systems chalcone synthase, chalcone isomerase and UDP-glucose: anthocyanidin-3-0-glucosyltransferase were neither present in isolated vacuoles nor in the pellet representing tonoplast membranes. The authors discuss the possibility that biosynthesis of anthocyanins takes place in a multienzyme complex located in or at the endoplasmatic reticulum. In *Petunia hybrida* Jonnsson et al. (1983) studied an anthocyanin methyltransferase which was found to be located in the cytosol of protoplasts.

Pecket and Small (1980) and Small and Pecket (1982) described intensively coloured bodies within the vacuoles of more than 70 species. They suppose that these anthocyanoplasts are membrane-bound and that they are the site of anthocyanin biosynthesis. The phenomenon of anthocyanin-carrying structures within the vacuoles

Abbreviations:
CAH: Trans-cinnamic acid 4-hydroxylase
CCR: NADH cytochrome C reductase
CGT: UDP-glucose: cyanidin-3-0-monoglucosyltransferase
CHI- Chalcone isomerase
CHS: Chalcone synthase
DMSO: Dimethylsulfoxide
IDPase: Inosine diphosphatase
PAL: Phenylalanine ammonia-lyase
PNP: Para-nitrophenyl

The Physiological Properties of Plant Protoplasts
Edited by P. E. Pilet

was also described by other authors (Bünning 1949; Hemleben 1981, Neumann 1983). In order to study transport of anthocyanin into the central vacuole and the localization of enzymes involved in anthocyanin biosynthesis, vacuoles from protoplasts derived from a cell culture of *Daucus carota* (Schmitz and Seitz 1972; Noé et al. 1980) were isolated and characterized with respect to anthocyanin content and several marker enzymes. These vacuoles contain cyanidin as the only aglycone (Schmitz and Seitz 1972); the glycosidic part of the anthocyanins of these carrot cells was determined to be glucose and xylose. More recently, Harborne (1983) reexamined the anthocyanin of these carrot cells; it is described as cyanidin-3-sinapoyl-xylosyl-glucosylgalactoside.

The cell suspension shows a certain heterogeneity with respect to size and anthocyanin content of vacuoles. This might be due to the asynchronous character of the cell culture. The maximum in anthocyanin accumulation is measured at the end of the growth period (Noé et al. 1980). Protoplasts and vacuoles respectively were isolated 7 days after onset of cultivation. At this stage anthocyanin biosynthesis begins.

The vacuolar suspension was characterized with respect to viability, stability and purity. Several putative marker enzymes were tested. Using purified vacuoles it should be possible to study transport of anthocyanin precursors and to localize enzyme systems involved in anthocyanin biosynthesis.

Materials and Methods

Cell Suspension Cultures

An anthocyanin-containing cell culture derived from callus of edible, Afghan carrot roots (*Daucus carota* L.) was cultivated as previously described (Noé et al. 1980).

Preparation of Protoplasts

Cells were harvested 7 days after onset of cultivation. The cells (fresh weight 5 g) were incubated in 50 ml isolation medium containing 1.6 % cellulase (Onozuka R-10, Yakult Biochem. Co., Shigikaucho, Japan) and 0.1 % Rhozyme HP-150 (Pollock and Pool, Reading, United Kingdom) for 18 h at 25°C (Langebartels 1981). Aggregates of protoplasts and debris were removed by filtration through nylon nets (140, 100 and 60 μm). The protoplasts were washed three times with isolation medium by centrifugation (100 g). The protoplasts were resuspended in isolation medium and counted for cell number in a haematocytometer (Fuchs-Rosenthal).

Aliquots from this suspension were used for anthocyanin determinations. For enzymatic studies the volumes were increased by 8–10 fold.

Isolation and Purification of Vacuoles

Lysis of protoplasts in order to release intact vacuoles was performed by a two step procedure. The protoplast suspension (8×10^5 ml^{-1}) was incubated with DEAE-dextran (dissolved in 50 mM Hepes, 0.7 M sorbitol, pH 7.6) for 5 min at room temperature; 1 ng DEAE-dextran per protoplast was applied. After the incubation an equal amount of dextran sulphate was added in order to protect isolated vacuoles. In a final lytic step with 0.3 M NaCl (dissolved in buffer A) vacuoles were released. The purification was carried out according to Grob and Matile (1979). The lysate (15–20 ml) was layered onto a step gradient of Urografin (Schering AG, Berlin, West Germany): The lower layer was composed of a mixture of 25 % Urografin (w/v) with a solution of 0.7 M NaCl (dissolved in buffer A) in a ratio of 1:1 (v/v); the upper layer contained these solutions in a ratio of 1:7 (v/v). The gradients were centrifuged at 1500 *g* (Minifuge, Haereus Christ, West Germany). The vacuoles banded between these two layers and were harvested for further purification with a 5-ml pipette. Therefore the vacuoles were mixed with Urografin/0.7 M NaCl (7:3, v/v) and overlayed with 0.7 M NaCl (dissolved in buffer A). This second gradient was centrifuged at 1000 *g* for 30 min. The purified vacuoles were concentrated at the interphase and then collected with a pipette.

Preparation of Particulate Fractions from Protoplasts and Vacuoles

Protoplasts were centrifuged (100 *g*, 5 min) and the resulting pellet was suspended in buffer A and homogenized in a Potter-Elvehjem homogenizer (5 ml, glass/glass). In order to determine the activities of chalcone synthase and UDP-glucose: cyanidin-3-O-glucosyltransferase we took buffer B respectively buffer C. After counting the vacuoles were homogenized in a Potter-Elvehjem homogenizer. From these resulting homogenates aliquots were taken in order to determine enzyme activities. To get membrane pellets the protoplast/vacuole suspensions were diluted in buffer A (1:2), homogenized and centrifuged at 100,000 *g* for 60 min (+4°C). The pellets containing membranes were resuspended in 1 ml of buffer A (PP = protoplast pellet, VP = vacuolar pellet). Supernatants of both kinds of preparations were concentrated by ultrafiltration (Amicon, PM 10) to a volume of 0.5–1 ml (PS = supernatant of protoplasts, VS = supernatant of vacuolar membranes).

Enzyme and Chemical Assays

Anthocyanin content was determined after extraction of protoplasts or vacuoles respectively (aliquots of 200 μl) with MeOH-HCl (99:1) for 1 min on a whirl-mix. After removal of insoluble material by a 16,000 *g* centrifugation (10 min) absorbance of anthocyanin was measured at 537 nm. The values were multiplied by the total volume of the samples. These units were taken as a measure for anthocyanin content of a fraction (Table 1).

Glucose-6-phosphate-dehydrogenase was assayed according to Bergmeyer (1970). NADH cytochrome C reductase (antimycin A-insensitive) was measured according to Hodges and Leonard (1974). IDPase was assayed according to Ray et al. (1969). Hydrolases were determined using various nitrophenyl derivates: α-Mannosidase (PNP-α-mannopyranoside), β-mannosidase (PNP-β-mannopyranoside), β-N-acetylglucosaminidase (PNP-β-N-acetylglucosaminide), β-glucosidase (PNP-β-glucopyranoside),

Table 1. Comparison of anthocyanin content and α-mannosidase activity during purification of vacuoles from *Daucus carota.* The anthocyanin content and the enzyme activity was determined in the first urografin gradient and in an aliquot of the vacuolar band. In the second purification step analysis of fractions was carried out in the same way. The yield of vacuoles was 3.9 % determined by counting based on the protoplast number before lysis

	Anthocyanin content		α-Mannosidase activity	
	E_{537} ml	%	ΔE_{405} min^{-1} ml	%
Protoplasts	47	100	40.4	100
Lysate	46.9	99.7	42.0	104
First gradient:				
Vacuolar fraction	12.8	28	10.0	27
Rest of the first gradient	33.8	71.2	27.2	65
Total in the first gradient	46.6	99.2	37.2	92
Second gradient:				
Purified vacuoles	2.1	4.6	1.5	4.0
Rest of the second gradient	10.8	23.5	8.5	22.0

acid phosphatase (PNP-phosphate), phosphodiesterase (bis-PNP-phosphate). All substrates were from Sigma (München, West Germany). The assay conditions were as follows: 0.4 ml of 0.1 M sodium citrate buffer (pH 4.5) containing 5–10 mM PNP derivatives as substrates were preincubated for 5 min at 30°C. The reaction was started by adding 20–100 μl of crude extracts. After 30 min the reaction was stopped by adding 0.8 ml 0.2 M borate buffer (pH 9.8). The absorbance of para-nitrophenol was measured at 405 nm. Samples treated with borate buffer before adding the crude extracts were used as blanks. Cinnamic acid 4-hydroxylase and phenylalanine ammonia-lyase activities were determined according to Diesperger et al. (1974) and Noe et al. (1980). Chalcone synthase and chalcone isomerase were measured as previously described (Hinderer et al. 1984). UDP-glucose: cyanidin-3-0-glucosyltransferase activity was measured as follows: 25 μl crude extract, 10 μl cyanidine chloride (Roth, Karlsruhe, West Germany) 5.5 mM in DMSO, 10 μl 7 mM UDP-glucose (Uridine Diphospho-D-[U-^{14}C] glucose 9 x 10^{-2} μCi per sample and 10 μl of buffer C. UDPG was purchased from Amersham Buchler (Braunschweig, West Germany). After incubation for 15 min at 30°C the reaction was stopped by adding 5 μl 6 % (v/v) HCl. The test assay was chromatographed on paper (solvent system: acetic acid – HCl – water 3:1:8). The spot of cyanidin-3-0-monoglucoside was cut out and counted in Unisolve 1 (Zinsser, Frankfurt, West Germany).

Labelling of Protoplasts with Concanavalin A

N-acetyl-[^{3}H]-Con A was purchased from Amersham Buchler (Braunschweig, West Germany). Purified protoplasts were incubated for 5 min with [^{3}H]-Con A at 25°C according to Boss and Ruesink (1979). After washing three times with isolation medium vacuoles were isolated. The radioactivity was determined by scintillation counting in Unisolve 1.

Microspectrophotometric Measurements

Vacuoles were immobilized in alginate. The absorbance at 550 nm was measured in individual vacuoles with an ultramicro-spectrophotometer (UMSP I, Zeiss, West Germany) for 2 h.

Protein Determination

Protein concentrations were determined according to Bradford (1976).

Buffers

The following buffers were used: Buffer(A) 50 mM Hepes, 5 mM EDTA, pH 7.6; Buffer(B) 100 mM Potassium phosphate, 1.4 mM 2-Mercaptoethanol, pH 8.0; Buffer (C) 150 mM Potassium phosphate, pH 8.0.

Isolation medium: the isolation medium contained Murashige and Skoog salts, 0.6 mg l^{-1} NAA, 0.1 mg l^{-1} kinetin, 0.25 M mannitol, 0.25 M sorbitol, the calcium chloride concentration was increased to 660 mg l^{-1}, the sucrose content was reduced to 1.5 % (w/v), pH 5.75.

Results and Discussion

Lysis of Protoplasts

In order to get pure vacuolar suspensions it was necessary to modify procedures for lysis of protoplasts described by Wagner and Siegelman (1975), Sasse et al. (1979) and Dürr et al. (1975). Since protoplasts have similar physical properties with respect to sedimentation, it was necessary to optimize lysis of protoplasts. At lysis efficiencies near 100 %, one obtains high yields of pure vacuoles. Different methods were tested. The best results were obtained using a polybase-induced lysis followed by osmotic shock with sodium chloride (Deus 1980). In order to preserve the polybase character of DEAE-dextran we dissolved the polycation in a buffer with a non-ionic osmoticum (Kombrink et al. 1979). Under these conditions 80 % of protoplasts released vacuoles. These vacuoles were less contaminated by cytoplasmic material, contrary to what we found after lysis with phosphate buffer or sorbitol. The crude vacuolar suspension contained only 3–10 % of unruptured protoplasts. High ionic strength in the lytic medium facilitates separation of vacuoles from adherent cytoplasmic material. Similar results were described by Schmidt and Poole (1980) for red beet storage tissue. Grob and Matile (1979) observed similar effects at high ionic strength. Furthermore EDTA is useful for the release of vacuoles from protoplasts. Mettler and Leonard (1979) described an inhibition of aggregation of cytoplasmic material with the tonoplast by EDTA. Yamaki (1982) observed in protoplasts from cotyledons of apple that vacuoles were only released after addition of EDTA. More recently Knuth et al. (1983) described increased lysis and decreased contamination by use of EDTA in the isolation medium. Our observations in carrot cells demonstrate the same effect.

Purification Procedures

Purification of crude vacuolar suspensions was carried out according to Grob and Matile (1979) by using step gradients of urografin. With this procedure 3–15 % pure vacuoles, calculated on the basis of protoplasts at the beginning, were obtained. The yield of vacuoles was in the same range as described by Boller and Kende (1979) and Knuth et al. (1983) using similar procedures.

As shown by light microscopy the vacuolar suspension contained only small amounts of vacuoles with cytoplasmic contamination (2–10 % so-called vacuoplasts). The diameter of these isolated vacuoles varied from 15 to 50 μm. The diameter of vacuoles of intact protoplasts was in the same range.

Viability and Stability

The retention of anthocyanin within isolated vacuoles is a good indication for the integrity of vacuolar membrane. In isolated vacuoles from carrot cells, anthocyanin content remained unaltered during long periods. We have very good hints that during purification procedures only few secondary versicles were formed. After storage at +4°C for 6 h vacuole number decreased only by 20 %. Registration of anthocyanin content in individual vacuoles, immobilized by alginate, in an ultramicro-spectrophotometer at 550 nm showed a constant value for at least 120 min, thus indicating that no efflux of anthocyanin occurred. It is not yet clear whether a pH-gradient over the tonoplast is maintained. As shown by uptake studies with glucose these vacuoles possess an intact transport system (Hopp et al. 1983).

Marker Enzymes

There is a good evidence from several investigations that α-mannosidase is a marker enzyme for vacuoles (Boller and Kende 1979; Boudet et al. 1981; Glund et al. 1984). In order to characterize the anthocyanin-containing vacuoles from *Daucus carota* we looked for a correlation of anthocyanin content and the activity of α-mannosidase. These results were compared to the counting efficiency (Table 1).

The yield of vacuoles determined by counting was 3.9 %. Anthocyanin content and α-mannosidase activity were in good agreement. In the carrot system α-mannosidase and anthocyanin seemed to be good markers for the vacuole. It could also be shown by this experiment that most of the vacuoles released from protoplasts were lost during these two purification steps.

The protein content of purified vacuoles was 19 % (± 5 %) calculated on the basis of lysed protoplasts. A typical distribution of marker enzymes is shown in Table 2.

Glucose-6-phosphate-dehydrogenase is a generally accepted marker enzyme for the cytosol of the cell. As shown by enzyme activity determinations, glucose-6-phosphate-dehydrogenase is present to 6.3 % in purified vacuoles. NADH Cytochrome C

Table 2. Distribution of marker enzymes in protoplasts and purified vacuoles from *Daucus carota*

Marker enzymes	Vacuoles		Protoplasts		Ratios	
	A $\Delta E\ h^{-1}\ mg^{-1}$	B $\Delta E\ h^{-1}/10^6$	C $\Delta E\ h^{-1}\ mg^{-1}$	D $\Delta E\ h^{-1}/10^6$	A/C	B/D
[a]Glucose-6-phosphate-dehydrogenase	5.52	0.38	19.3	5.6	0.29	0.068
[b]NADH Cytochrome C reductase	112	10.3	195	68.5	0.57	0.15
[c]Acid phosphatase	180	10.4	102	18.6	1.8	0.56
[c]β-Mannosidase	23	1.4	11.3	2.4	2.02	0.58
[c]β-Glucosidase	137	8.2	36	12	3.8	0.69
[c]α-Mannosidase	83	5	16	5.1	5.2	0.98
[c]Phosphodiesterase	19	2.6	7.2	2.3	6.1	1.13
[c]β-N-acetylglucose-aminidase	7.8	0.5	1.5	0.5	5.2	1.0

[a]E = 340 nm

[b]E = 550 nm

[c]E = 405 nm

reductase is present in rather high amounts in vacuolar preparations (15 %). This result may indicate that ER membranes stick to the tonoplast by unspecific reactions or points to a biogenetic connection of ER and tonoplasts.

α-Mannosidase, phosphodiesterase and N-acetylglucosaminidase were present up to 100 % in isolated vacuoles. The specific activity of these enzymes is increased 5-6-fold by purification of vacuoles. Very similar results were obtained by Boller and Kende (1979) and Nishimura and Beevers (1978). The portion of the other putative vacuolar enzymes is much lower: acid phosphatase (56 %) β-mannosidase (58 %) and β-glucosidase (69 %). These results indicate that acid phosphatase is an insufficient marker enzyme for vacuoles in our system.

Another kind of experiment was carried out to determine contamination of isolated vacuoles by plasma membrane material. The plasma membrane of isolated protoplasts was labelled with tritiated Con A. The portion of Con A was determined in vacuoles isolated from these protoplasts (Table 3).

Table 3. ^{3}H-Labelled Con A in protoplasts and vacuoles from *Daucus carota.* The protoplasts were incubated with Con A and the corresponding vacuoles were isolated as described

10^6 Protoplasts	30,460 cpm	100 %
10^6 Vacuoles	775 cpm	2,5 %

In isolated vacuoles only 2.5 % of labelled Con A was present. The contamination observed in the light microscope seemed to be higher. The value determined by Con A-labelling is somewhat too low. It might be possible that during lysis of protoplasts Con A was removed from the plasma membrane to a certain extent.

Particulate fractions prepared from protoplasts or vacuoles by ultracentrifugation were analyzed for marker enzymes (Table 4).

The fractions were compared on the basis of specific activities of the enzymes. Phosphodiesterase, α-mannosidase, β-glucosidase, N-acetylglucosaminidase and acid phosphatase were not sedimented in the vacuolar pellet as shown by the comparison of specific activities in the vacuolar pellet and in the supernatant.

These enzymes were also found in the supernatant of protoplasts. CCR and IDPase, a putative marker for Golgi vesicles (Ray et al. 1969) were mainly present in the protoplast pellet and not in the corresponding supernatant. In vacuolar pellets there was only a slight contamination of these enzymes because there was no increase in specific activities as it is shown for the vacuolar marker enzymes. The activity of glucose-6-phosphatedehydrogenase was found in the supernatant of protoplasts and not in the pellet. The specific activity of this enzyme in the corresponding supernatant of vacuoles was low. β-Mannosidase is thought to be a potent marker for the tonoplast membrane (Gross J personal communication). The enzyme activity is present in the pellet of homogenized protoplasts, but did not sediment with the tonoplast membranes (VP). Further investigations are underway to determine whether this enzyme is really tonoplast-bound or is removed by high ionic strength during lysis and purification of vacuoles.

Table 4. Particulate fractions from protoplasts and vacuoles. Comparison of pellets and supernatants. All values are given in specific activities ($\Delta E\ h^{-1}\ mg^{-1}$)

	Enzyme activities $\Delta E\ h^{-1}\ mg^{-1}$			
Marker enzymes	Vacuoles VP	VS	Protoplasts PP	PS
[a]Glucose-6-phosphate-dehydrogenase	0.6	9.0	12	30
[b]NADH Cytochrome C reductase	36	12.6	102	54
[c]Inosine diphosphatase	6	0.6	12.6	1.2
[d]β-Mannosidase	4.2	25	72	9.6
[d]Acid phosphatase	96	480	75	107
[d]β-Glucosidase	84	313	15.6	106
[d]α-Mannosidase	28	198	16.8	45
[d]Phosphodiesterase	48	114	3	16
[d]N-acetylglucosaminidase	0.06	54	0.78	11

[a]E = 340 nm
[b]E = 550 nm
[c]E = 366 nm
[d]E = 405 nm

VP = Vacuolar pellet, VS = Supernatant of vacuolar membranes, PP = Protoplast pellet, PS = Supernatant of protoplasts

Enzymes of Flavonoid Biosynthetic Pathway

According to the model of Fritsch and Grisebach (1975) CAH and other monooxygenases which catalyze the hydroxylation of naringenin to eriodictyol and of dihydrokaempferol to dihydroquercetin were found to be microsomal. Enzyme activity for 3'-5'-hydroxylation of the B-ring of flavanones and dihydroflavonols was found to be localized in the microsomal fraction from *Verbena hybrida* (Stotz and Forkmann 1981). In order to get information of the compartmentation of anthocyanin biosynthesis, isolated vacuoles were tested for several enzyme activities (Fig. 1).

PAL and CHI were only found in protoplasts; in the vacuolar fraction these enzymes were not detectable. CAH is present in minor activity in the vacuoles (6 %). This activity could be due to the contamination of vacuoles by ER membranes as shown by the activity of CCR (15 %). Fritsch and Grisebach (1975) suggested that the glucosylation of cyanidin might occur at the inner side of the tonoplast. In carrot suspension cultures only a small amount (4.6 %) of a CGT was present in the vacuole. Similar results were obtained by Hrazdina et al. (1978).

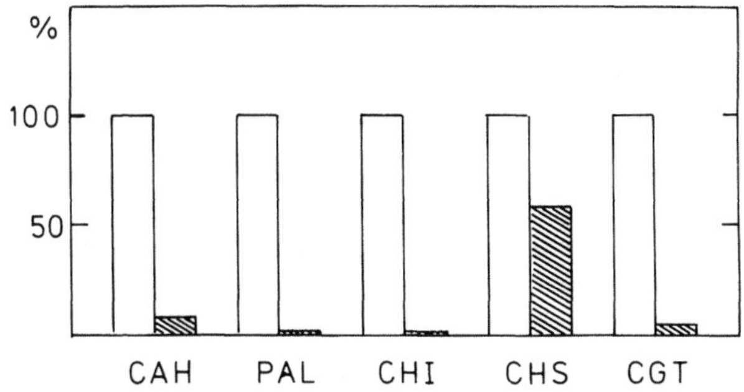

Fig. 1. Enzymes of the flavanoid biosynthesis pathway. CAH (trans-cinnamic-acid 4-hydroxylase, 100%=4.2 x 10^5 cpm min^{-1}), PAL (phenylalanine ammonia-lyase, 100% = 0,85 nkat), CHI (chalcone isomerase, 100 % = 1.5 x 10^{-4} = $E_{385\ nm}\ s^{-1}$), CHS (chalcone synthase, 100 % = 1.33 pkat), CGT (UDP-glucose:cyanidin-3-0-glucosyltransferase, 100 % = 2.5 x 10^6 cpm), □ activity of 10^9 protoplasts, ▨ activity of 10^9 vacuoles

CHS is the first enzyme of the flavanoid pathway. In our system 59 % of the activity of protoplasts could be found in isolated vacuoles. In *Hippeastrum* and *Tulipa* CHS as well as CHI were neither present in isolated vacuoles nor in a pellet representing tonoplast membranes (Hrazdina et al. 1978). The authors discuss the possibility that the conversion of phenylalanine to anthocyanidin may occur at least near the ER by cytoplasmic and ER-bound enzymes acting in close proximity to each other and to the tonoplast. Jonnsson et al. (1983) studied the subcellular location of a methyltransferase involved in the anthocyanin biosynthesis. They suggest that this late step of synthesis takes place in the cytosol of the cell. In the carrots the activity of CHS seems to be associated with the vacuolar membrane. This might be an indication that flavanoid formation and the following steps of anthocyanin synthesis are close to the tonoplast within the cytosol.

Acknowledgements. The help of Professor Dr. D. Ammermann (Tübingen) during ultramicrospectrophotometric measurements is gratefully acknowledged. The work was supported by the Deutsche Forschungsgemeinschaft.

References

Bergmeyer HU (1970) Methoden der enzymatischen Analyse, 2nd edn, vol 1. Chemie, Weinheim

Boudet A, Canut H, Alibert G (1981) Isolation and characterization of vacuoles from *Melilotus alba* mesophyll. Plant Physiol (Bethesda) 68:1354–1358

Boss W, Ruesink A (1979) Isolation and characterization of concanavalin A - labeled plasma membranes of carrot protoplasts. Plant Physiol (Bethesda) 64:1005–1011

Boller T, Kende H (1979) Hydrolytic enzymes in the central vacuole of plant cells. Plant Physiol (Bethesda) 63:1123–1132

Bradford MM (1976) A rapid and sensitive method for the quantitation of microgram quantities of protein utilizing the principle of protein-dye binding. Anal Biochem 72:248–254

Bünning E (1949) Über quellbare Zellsaftkolloide in Irisblättern. Planta (Berl) 37:431–436

Deus B (1980) Thesis, Univ of Bochum

Diesperger H, Müller C, Sandermann H (1974) Rapid isolation of a plant microsomal fraction by Mg^{2+} precipitation. FEBS Lett 43:155–158

Dürr M, Boller T, Wiemken A (1975) Polybase induced lysis of yeast spheroplasts; a new gentle method for preparation of vacuoles. Arch Microbiol 105:319–327

Fritsch H, Grisebach H (1975) Biosynthesis of cyanidin in cell cultures of *Haplopappus gracilis*. Phytochemistry (Oxf) 14:2437–2442

Glund K, Tewes A, Abel S, Leinhos V, Walther R, Reinbothe H (1984) Vacuoles from cell suspension cultures of tomato (*Lycopersicon esculentum*). Isolation and characterization. Z Pflanzenphysiol 113:151–161

Grob K, Matile P (1979) Vacuolar location of glucosinolates in horseradish root cells. Plant Sci Lett 14:327–335

Harborne J (1983) Identification of the major anthocyanin of carrot cells in tissue culture as cyanidin 3-(sinapoylxylosylglucosylgalactoside). Z Naturforsch Sect C Biosci 38:1055–1056

Hemleben V (1981) Anthocyanin carrying structures in specific genotypes of *Matthiola incana*. Z Naturforsch Sect C Biosci 36:925–927

Hinderer W, Petersen M, Seitz HU (1984) Inhibition of flavonoid biosynthesis by gibberellic acid in cell suspension cultures of *Daucus carota* L. Planta (Berl) 160:544–549

Hodges T, Leonard R (1974) Purification of a plasma membrane-bound adenosine triphosphatase from plant roots. In: Colowick CSP, Kaplan NO (eds) Methods of enzymology, vol 32B. Academic, New York

Hopp W, Mock HP, Seitz HU (1983) Isolation of anthocyanin containing vacuoles from protoplasts of a cell suspension culture of *Daucus carota* L. In: Protoplasts 1983, Poster Proc 6th Int Protoplast Symp Basel. Exper Suppl (Basel) 45:188–189

Hrazdina G, Wagner G, Siegelman H (1978) Subcellular localization of enzymes of anthocyanin biosynthesis in protoplasts. Phytochemistry (Oxf) 17:53–56

Jonnsson L, Donker-Koopman W, Uitslager P, Schram A (1983) Subcellular localization of anthocyanin methyltransferase in flowers of *Petunia hybrida*. Plant Physiol (Bethesda) 72:287–290

Knuth M, Keith B, Clark C, Gracia-Martinez J, Rappaport L (1983) Stabilization and transport capacity of cowpea and barley vacuoles. Plant Cell Physiol 24:423–432

Kombrink E, Wöber G, Walker D (1979) Einfluß von DEAE-Dextran auf den Elektronentransport in Chloroplasten von *Dunaliella marina*. Ber Dtsch Bot Ges 92:379–392

Langebartels C (1981) Thesis, Univ of Tübingen

Mettler J, Leonard T (1979) Isolation and partical characterization of vacuoles from tobacco protoplasts. Plant Physiol (Bethesda) 64:1114–1120

Neumann D (1983) Subcellular localization of anthocyanins in red cabbage seedlings. Biochem Physiol Pflanz (BPP) 178:405–407

Nishimura M, Beevers H (1978) Hydrolases in vacuoles from castor bean endosperm. Plant Physiol (Bethesda) 62:44–48

Noé W, Langebartels C, Seitz HU (1980) Anthocyanin accumulation and PAL activity in a suspension culture of *Daucus carota* L. Inhibition by L-α-aminooxy-β-phenylpropionic acid and t-cinnamic acid. Planta (Berl) 149:283–287

Pecket R, Small C (1980) Occurrence, location and development of anthocyanoplasts. Phytochemistry (Oxf) 19:2571–2576

Ray PM, Shininger IL, Ray MM (1969) Isolation of β-glucansynthase particles from plant cells and identification with golgi membranes. Proc Natl Acad Sci USA 64:605–612

Sasse F, Backs-Hüsemann D, Barz W (1979) Isolation and characterization of vacuoles from cell suspension cultures of *Daucus carota*. Z Naturforsch Sect C Biosci 34:848–853

Schmidt R, Poole R (1980) Isolation of protoplasts and vacuoles from storage tissue of beet. Plant Physiol (Bethesda) 66:25–28

Schmitz M, Seitz HU (1972) Hemmung der Anthocyansynthese durch Gibberellinsäure A_3 bei Kalluskulturen von *Daucus carota* L. Z Pflanzenphysiol 68:259–265

Small C, Pecket R (1982) The ultrastructure of anthocyanoplasts in red cabbage. Planta (Berl) 154:97–99

Stotz G, Forkmann G (1982) Hydroxylation of the B-Ring of flavonoids in the 3'- and 5'-position with enzyme extracts from flowers of *Verbena hybrida*. Z Naturforsch Sect C Biosci 37:19–23

Wagner G, Siegelman H (1975) Large-scale isolation of intact vacuoles and isolation of chloroplasts from protoplasts of mature plant tissue. Science (Wash DC) 190:1298–1299

Yamaki S (1982) Distribution of sorbitol, neutral sugars, free amino acids, malic acid and some hydrolytic enzymes in vacuoles of apple cotyledons. Plant Cell Physiol 23:881–889

Vacuolar pH Variability in a Protoplast Population

P. MANIGAULT

Introduction

Different methods to determine the vacuolar pH have been proposed (see Kurkdjian et al. 1984 for a review). Techniques used for plant organs or cells are not always suitable for protoplasts or isolated vacuoles. Furthermore, most of them consider a cell population as a whole (plant organ or suspension of cells in liquid medium) and give a "mean" pH value which in turn depends on the technique used to obtain it (Kurkdjian et al. 1984c).

This report presents two methods of individual pH measurements: one using H^+-ion selective microelectrodes, the other based on 9 aminoacridine (9AA) accumulation measured by microfluorimetry. These measurements enable calculation of the mean vacuolar pH value of a protoplast population but they also reveal the variability of vacuolar pH within that same population. Some characteristics of this variability are described.

Material and Methods

Preparation of Protoplasts

Protoplasts of *Acer pseudoplatanus* were prepared from cultured cells during the exponential phase of growth according to the technique reported before by Kurkdjian and Barbier-Brygoo (1983). Cells of a 7-day-old culture were incubated for 60 min at 26°C in 20 ml of buffer solution Tris-Mes 25 mM pH 5.5 and mannitol 0.7 M containing the following enzymes: cellulase Onozuka RS 2%, macerozyme Onozuka R10 1.5%, (Yakult Honsha Co Ltd, 8–21 Shingikancho, Nishinomyia 662 Japan); and pectolyase Y23 0.25% (Seishin Pharmaceutical Co Ltd 9–500–1.Nagareyama, Nagareyama-Shi, Chiba-keu, Japan). The suspension was succesively filtered through nylon cloth (200, 100 and 30 μm pore diameters) and centrifuged for 30 s at 3900g. The supernatant was discarded and the pelleted protoplasts were resuspended in 10 ml of a buffer solution Tris-Mes 25 mM pH 7.5; mannitol 0.7 M).

The Physiological Properties of Plant Protoplasts
Edited by P. E. Pilet

Measurement of Vacuolar pH with Microelectrodes

Antimony electrodes are sensitive to organic anions such as citrate and consequently they cannot be used to measure vacuolar pH. Glass microelectrodes built according to Thomas (1974) are most difficult to make. We used a new type of microelectrode introduced by Ammann et al. (1981) and adapted to plant material by Kurkdjian and Barbier-Brygoo (1983). These electrodes are easy to make and highly selective to H^+ ions. The H^+ selective liquid is constituted by the ligand tri-n-dodecyl amine with o-nitrophenyl-n-octyl ether and sodium tetraphenylborate (this ready-to-use cocktail is now available from Fluka).

The vacuolar H^+ activities are calculated from the electromotive force recorded for each protoplast and corrected for the transmembrane electrical potential difference.

Measurement of Vacuolar pH with Lipophilic Bases

Lipophilic bases such as benzylamine, nicotine or 9-amino-acridine diffuse freely through membranes in their neutral form. They dissociate inside and outside the cells or protoplasts according to the intra- and extracellular pH. The charged protonated form does not diffuse through the membrane and so does not itself re-equilibrate. The diffusion equilibrium is established for the non-dissociated molecules between the suspension medium and the protoplast. When the equilibrium is reached, the intracellular (Ci) and extracellular (Ce) concentrations of the probe of known pK are measured and the intracellular pH (pHi) is calculated from the equation of Waddell and Bates (1969):

$$Ci/Ce = 1+10^{pK-pHi}/1+10^{pK-pHe} \qquad (1)$$

Equation (1) shows that vacuolar pH can be calculated from the measurement of the accumulation ratio Ci/Ce.

Measurement of the Accumulation Ratio of 9AA

In the case of the fluorescent lipophilic base 9AA the Ci/Ce ratio is calculated from the intra- and extracellular fluorescence according to a procedure already described for cells by Manigault et al. (1983).

The microfluorimeter (Leitz, MPV2) is equipped with high luminance light sources, generally high pressure mercury arcs (Osram HBO 50 W). The most efficient configuration is to use the microscope objective lens both for irradiation of the object and for its observation and collection of emission (epifluorescence). A field diaphragm and a measurement diaphragm reduce the prejudicial effects of fading in the preparation plane and glare at the image level. The protoplasts suspended in the liquid medium are set on a slide such as an hematocytometer whose depth is constant and precisely known. For protoplasts and vacuoles having a diameter between 20 and 80 μm, a 100 μm depth slide may be chosen. For the observation and the measurement, an objective lens 40x NA 0.65 or 25x NA 0.50 may be used. Two

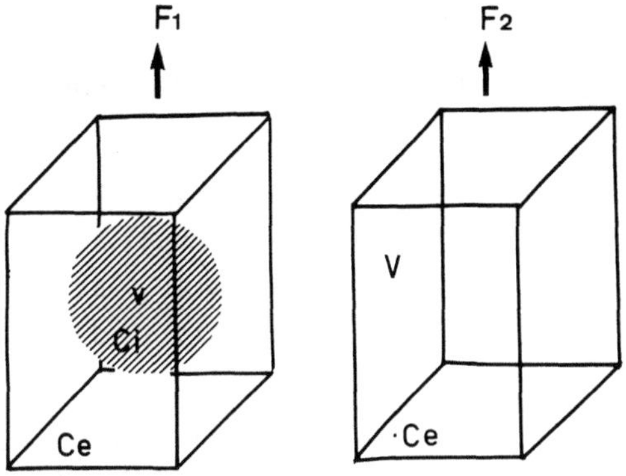

Fig. 1. Schematic representation of F1 and F2 measurements. *F1* is the sum of two signals: one originates from the vacuole compartment and corresponds to the 9AA concentration *Ci* in the volume *v*. The other originates from the surrounding medium whose volume is *(V−v)*, and where the concentration is *Ce*. *F2* is the signal issued from an adjacent parallelepipedic volume of medium alone, *V*, and concentration Ce

fluorescence signals are used: F1 from the object and the surrounding medium; F2 from the medium without object. For a spherical object such as a protoplast or a vacuole the image is framed by a square diaphragm slightly larger than the sphere diameter, and the dimensions of each are measured with a reticular eyepiece. A previous calibration determines the convenient range in which F1 and F2 must be situated so that their fluorescence is proportional to the 9AA concentrations (practically between 0,05 μM and 5 mM) (Fig. 1).

To complete the calculation, it is necessary to measure a third signal, the background F3, coming from the suspension medium with neither protoplast nor fluorochrome.

To summarize:

The fluorescence of the medium around the protoplast is

$$\frac{V-v}{V}(F2-F3)$$

The fluorescence of the vacuole is

$$(F1-F3) - \frac{V-v}{V}(F2-F3)$$

So the accumulation ratio is:

$$Ci/Ce = V/v\,[(F1-F3)/(F2-F3) - 1] + 1 \qquad (2)$$

Results and Discussion

Vacuolar Accumulation of 9AA in Protoplasts and pH Measurement

The vacuoles of protoplasts incubated in a 5 μM solution of 9AA for about 1 h and observed by fluorescence microscopy appear highly fluorescent, showing that the probe is mainly accumulated inside the vacuole (Manigault et al. 1983). Table 1 shows some examples of the fluorescence intensities F1, F2, F3 measured as described

Table 1. Vacuolar accumulation of 9AA in protoplasts and pHv calculations

O (μm)	v (μm)³	D (μm)	V (μm)³	F1	F2	F3	Fp	Fm	Ci/Ce	pHv
30.9	15448	37.1	178933	109.9	21.3	5.2	90.0	1.4	64.3	5.4
20.6	4577	24.8	79955	33.6	10.1	2.4	24.0	0.5	48.0	5.5
26.8	10079	33.0	141570	66.7	19.3	4.1	48.5	1.1	44.1	5.5
24.8	7986	28.9	108577	44.5	15.1	3.2	30.2	0.8	37.7	5.6

Examples of measurements and calculations made on four protoplasts from a protoplast population incubated for 3 h with 5 μM 9AA (pHe: 7.2). O: vacuolar diameter, v: vacuolar volume, D; diaphragm side, V; parallelepiped volume, F1: sum of two fluorescent signals (protoplast, volume v and surrounding medium, volume V–v), F2: medium fluorescence (issued from volume V), F3: background fluorescence (issued from volume V); Fp: protoplast fluorescence (calculated from volume v), Fm: medium fluorescence (calculated for a volume equivalent to v), Ci/Ce: ratio Fp/Fm

above, the corresponding calculated fluorescence signals issued specifically from the vacuole and from the medium, and finally the corresponding accumulation ratio and calculated pH. For a protoplast population in a medium of constant pHe, the mean pHv values does not depend upon the initial 9AA concentration in the range 0.05–7.5 μM. For example, aliquots of the same population of *Acer pseudoplatanus* protoplasts treated with 5, 0.5 or 0.05 μM 9AA, at the same pHe, accumulate the probe with mean Ci/Ce ratios 38.3, 38.9 and 40.2 respectively.

However, calculations made above assume that the fluorescence intensity measured on protoplasts arises entirely from vacuolar 9AA molecules. This assumption appears reasonable as (1) the vacuolar 9AA concentrations is 100 times the cytoplasmic one (vacuolar pH about 5.5 – cytoplasmic pH about 7.5) (2) the volume occupied by the vacuole is 7 to 10 times the cytoplasmic one. Hence the fluorescent signal from the vacuole is several hundred times that from the cytoplasm. Therefore, the pH value calculated can be considered as a good estimate of the vacuolar pH.

However, complexation of 9AA and quenching of its fluorescence render the measurement of vacuolar pH more difficult than it appears from the principle of the technique. The probe 9AA has been frequently used for the determination of chloroplast pH (Schuldiner et al. 1972; Fiolet et al. 1974; De Benedetti and Garlaschi 1977; Haraux and De Kouchkovsky 1980) and also with other organelles (Deamer et al. 1972). These authors measure the decrease of the fluorescence of the organelle suspension induced by the intravesicular accumulation of the probe. They assume that the fluorescence of the 9AA molecules accumulated inside the organelles is totally quenched. According to the same principle and using the same assumption, Weigel and Weis (1984) measured the pH of vacuoles isolated from *Valerianella locusta*. From our results, however, it is clear that the fluorescence of 9AA accumulated inside the vacuoles can be measured, has a rather high intensity compared to that of the incubation medium and consequently cannot be considered as totally quenched. Therefore, the extrapolation to vacuoles (mean diameter around 20 μm) of the assumption made for small organelles has to be considered with caution.

Two observations suggest that the pH difference between the suspension medium of protoplasts and their vacuoles is not the sole force driving 9AA accumulation.

First, the accumulation ratio of 9AA can be calculated either by measuring directly Ci and Ce as described above (method a) or from the decrease of the extracellular concentration (method b). In this second case, the accumulation ratio is given by $Ci/Ce = (Co/Ce - 1) \times V/v$ where Co is the initial external probe concentration. The results show that the accumulation ratios measured by method b (Rb) are always higher than those given by method a (Ra). For example, the accumulation ratios calculated by methods a and b on a *Catharanthus roseus* protoplast population (relative cumulated volume: 0.29%) incubated at pH 7.0 with 9AA (initial concentration Co: 5 μM) were Ra = 41 and Rb = 194. This cannot be explained by a simple self-quenching of the fluorescence of free vacuolar 9AA molecules (the total intraprotoplast 9AA concentration at equilibrium is around 0.6 mM). A model of 9AA accumulation where an important part of the 9AA molecules is trapped as complexes producing a strong fluorescence quenching accounts for these results (Kurkdjian et al. 1984c).

The second indirect evidence for 9AA trapping comes from the comparison of the pHv values (calculated from the accumulation of the probe) with the pH of the cell sap (pHs). The cell sap pH is always higher than pHv by ca. 0.5 pH unit. Though taking pHs as representative of the vacuolar pH raises several objections (see Kurkdjian et al. 1984c), the difference between pHv and pHs is in good quantitative agreement with the difference that can be calculated from the accumulation ratios Ra and Rb given above.

The model of 9AA trapping as quenched complexes inside the vacuole can account for the difference between our results and those of Weigel and Weis (1984) only if there is a linear relation between free vacuolar 9AA (9AA f) and the amount of complexed molecules (9AA b) i.e. 9AA b = k x 9AA f. In such a case, measuring directly the fluorescence signal from the vacuole is measuring 9AA f whereas monitoring the quenching of fluorescence of the total vacuole suspension is measuring k x 9AA f. Both parameters are directly related to the pH difference between the external medium and the vacuole and can be used to study pH variations. The choice between the two methods is more critical when absolute values of pHv are required.

Variability of Vacuolar pH Values Inside a Protoplast Population

The intensity of 9AA accumulation is different from one protoplast to another, as may be simply seen by the observation of a 9AA-treated preparation (Manigault et al. 1983). This reveals a variability of vacuolar pH in the population. The techniques such as 9AA microfluorimetry and H^+-ion selective microelectrodes allowing individual measurements on protoplasts (or cells, or vacuoles) are by definition the only ones suitable to study the vacuolar pH variability within a protoplast population (Kurkdjian et al. 1984b).

The histogram of individual pH values (see for example Fig. 4A) has a wide range of about 1 pH unit, with a bell-shaped distribution, most frequently unimodal. It is characterized by two parameters, the mean pH and the coefficient of variation.

The mean pH value was calculated from the mean of individual accumulation ratios, Ci/Ce. In order to determine the number of measurements necessary to represent the protoplast population the following procedure was used. During a series of measurements, the successive mean values of Ci/Ce are progressively calculated to reach a stable value indicating that (1) protoplasts are equilibrated with their medium (2) a sufficient number of them has measured (Fig. 2).

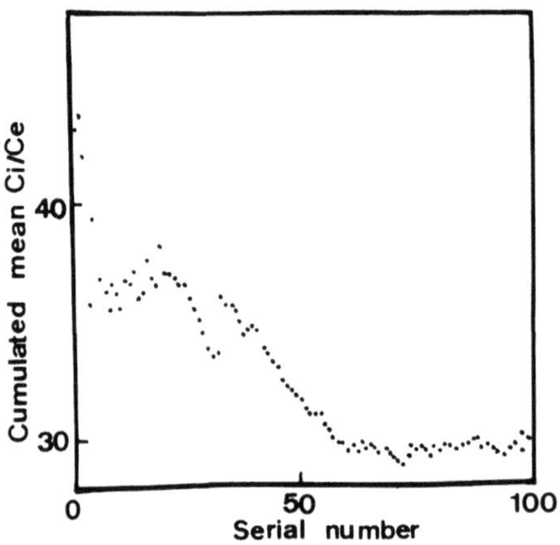

Fig. 2. Cumulated mean Ci/Ce values during 100 successive determinations of pHv. Step by step, the Ci/Ce values were calculated and their means cumulated. In this experiment, after the 50th measurement, the mean was constant and the number of measurements may be considered as sufficient

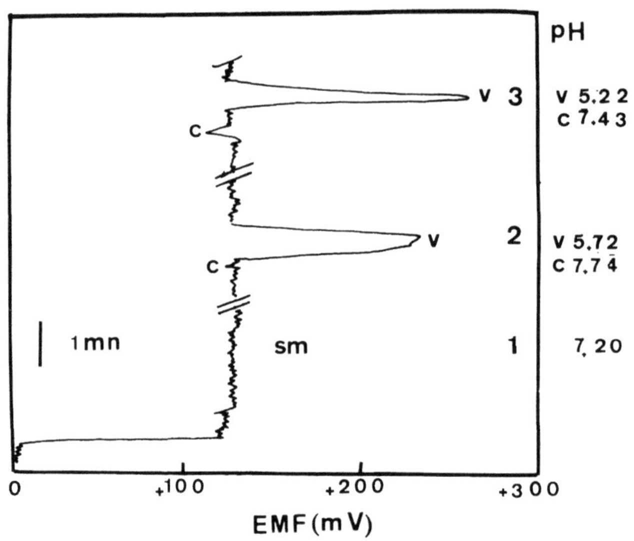

Fig. 3. Measurement of vacuolar pH in protoplasts of *Catharanthus roseus* with a H^+ selective microelectrode.
The H^+ selective microelectrode was calibrated in citrate buffer solution complemented with ions and adjusted to various pH values (citrate 6.7 mM: Na^+ 69 mM; Bo^{3-} 11.4 mM: PO_4^{3-} 10 mM; pH 5.5 to 7.5). The electrode was equilibrated in the suspension medium *(1,sm)*; the protoplasts were impaled *(2,3)* and electron motive force *(EMF)* values were recorded either in the cytoplasm *(c)* or in the vacuole *(v)*. pH values were calculated from these EMF values, the mean values of cytoplasmic and vacuolar electrical potential differences (+ 5.5 mV and + 18.3 mV respectively), and the calibration curve of the H^+-microelectrode

Table 2. Extent of vacuolar pH variability in *Catharanthus roseus* protoplast populations as shown by 9AA microfluorimetry and H^+-selective microelectrodes

	9AA microfluorimetry		H^+-microelectrode
	Ci/Ce	pHv	pHv
Lower	15.31	6.01	6.88
Mean	37.83	5.62	6.23
Higher	93.08	5.23	5.22

The two types of measurement were made on different populations in independent experiments. The pH of the suspension medium was 7.2. The lowest and highest values measured with both techniques are given with the mean value for the populations studied (n = 50 for 9AA measurements and 40 for microelectrode measurements)

The fact that the pH variability revealed by microfluorimetry was not artifactual was established by using a different technique of pH measurement. To illustrate the technique, traces of the electromotive force (EMF) measured on *Catharanthus roseus* protoplasts with H^+-microelectrodes are given in Fig. 3. The derived values of cytoplasmic pH in the two impaled protoplasts were 7.43 and 7.74 and their vacuolar pH 5.22 and 5.72.

A large vacuolar pH variability is also revealed by the microelectrode technique (Table 2). A more extensive study already published (Kurkdjian et al. 1984a) shows that the pH variability exists in cells, protoplasts or isolated vacuoles with about the same characteristics of dispersion and mean values.

Various treatments have been used in order to modify the mean vacuolar pH of protoplast populations and to test their action on its variability. Vacuolar pH can be modified by treating protoplasts with fusicoccin (Kurkdjian et al. 1984b), with FCCP or nigericin or by modifying the pH of the suspension medium. In all cases no significant modification of the pH variability was associated with change in the mean vacuolar pH of the treated populations. This behaviour is exemplified by Table 3 and Fig. 4.

Table 3. Influence of the extracellular pH (pHe) on the mean value and the variability of vacuolar pH in a population of *Acer pseudoplatanus* protoplasts

pHe	Protoplasts					
	Low pHv		Mean pHv		High pHv	
	Ci/Ce	pHv	Ci/Ce	pHv	Ci/Ce	pHv
5.16	5.7	4.41	3.1	4.79	1.5	4.99
7.28	75.6	5.40	39.4	5.68	20.1	5.98
Δ (Ci/Ce)	13		13		13	

Protoplasts were incubated in buffered medium with 5 μM 9AA. The accumulation ratios and pHv values were determined for 100 protoplasts, the mean and extreme values are given. The relative increase of the accumulation ratio (Ci/Ce) induced by the increase of pHe was calculated in each case

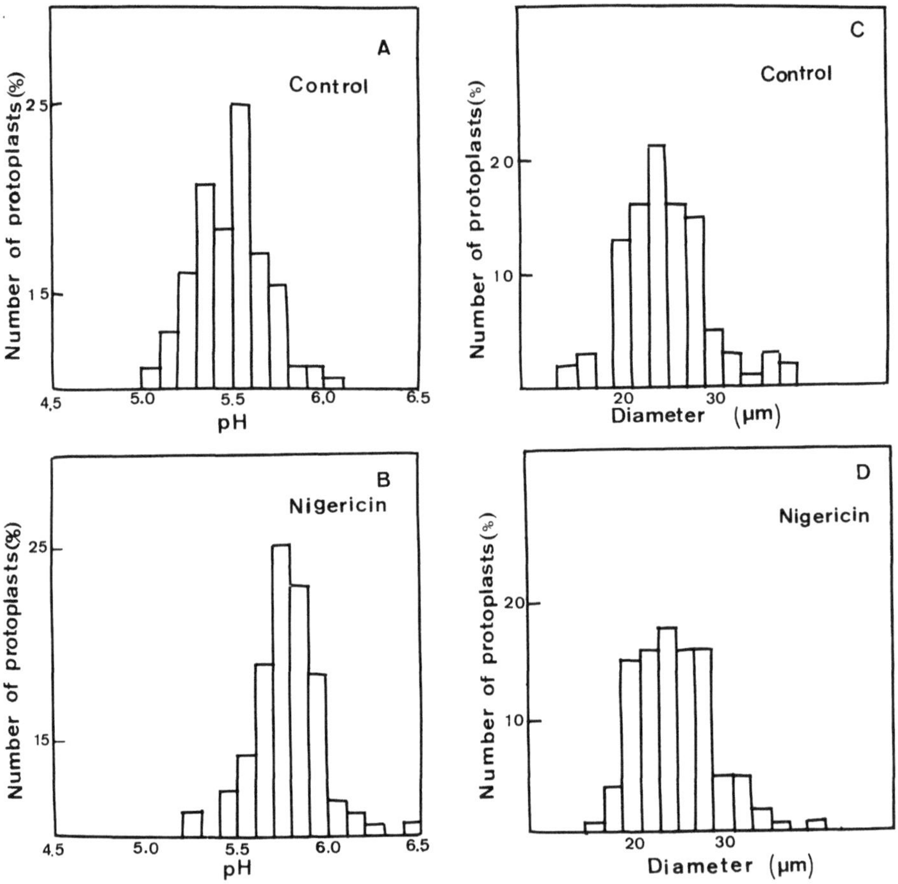

Fig. 4 A–D. Distribution of vacuolar pH values from the accumulation ratio of 9AA in an *Acer pseudoplatanus* protoplast suspension. **A** Control protoplasts incubated for 3 h with 5 μM 9AA (pHe 7.0). Mean pHv 5.53; mean Ci/Ce: 29.5 ± 1.5 (SE). Coefficient of variation 0.49, n=100. *B* Protoplasts treated with nigericin (5 μM, 3 h). Mean pHv = 5.83, mean Ci/Ce = 14.8 ± 0.7 (SE). Coefficient of variation 0.47, n=100. *C* Histogram of diameters for control protoplasts. Each class of the abcissa corresponds to an interval of 4 microns. Mean volume: 9883 $(\mu m)^3$. The mean volume is obtained from the sphere volume for each diameter class, the number of objects of each class, the total volume of the objects, and the total number of objects. **D** Histogram of diameters for nigericin-treated protoplasts. The repartition of the diameters is unchanged. Mean volume: 9577 $(\mu m)^3$

Conclusion

Microfluorimetry with 9AA appears to be a convenient technique to evaluate vacuolar pH. As most other techniques which can be used to measure pHv, the 9AA method is notably suitable to monitor pH changes. One of the adavantages of this technique is that it can be used for vacuolar pH measurement on one protoplast, on a population of protoplasts considered one by one or on a population considered as a whole (as demonstrated by Weigel and Weis). The most original fact revealed by using two very different techniques allowing individual pHv measurements in protoplast populations is the *variability* of vacuolar pH. This observation has been strengthened by recent studies upon populations of *Catharanthus roseus* protoplasts incubated with 9AA and then analyzed and sorted by flow cytometry on their fluorescence intensity, i.e., on their vacuolar pH (Brown et al. 1984).

There is no place here to speculate about the origin of vacuolar pH variability. Of particular interest is the observation reported here that the variability of vacuolar pH in a population is not modified when a shift of the mean value is induced. A model relating this variability to the vacuolar levels of organic acids and strong ions has been presented elsewhere (Kurkdjian et al. 1984c).

Acknowledgements. The author is very greatful to Professor J. Guern for stimulating discussions during the development of this work. He is greatly indebted to Dr. S. Brown for critically reading and to Mrs B. Cervoni for typing the manuscript.

References

Ammann D, Lanter F, Steiner RA, Schulthess P, Shijo P, Simon W (1981) Neutral carrier based hydrogen ion-selective microelectrode for extra and intracellular studies. Anal Chem 53: 2267–2269

Brown S, Renaudin JP, Prevot C, Guern J (1984) Flow cytometry and sorting of plant protoplasts: technical problems and physiological results from a study of pH on alkaloids in *Catharanthus roseus.* Physiol Veg 22:541–554

Deamer DW, Prince RC, Crofts AR (1972) The response of fluorescent amines to pH gradients across liposome membranes. Biochim Biophys Acta 274:323–335

De Benedetti E, Garlaschi FM (1977) On the estimation of proton gradient and osmotic volume in chloroplast membranes. J Bioenerg Biomemb 9:195–200

Fiolet JWT, Bakker EP, Van Dam K (1974) The fluorescent properties of acridines in the presence of chloroplasts or liposomes. On the quantitative relationship between the fluorescence quenching and transmembrane proton gradient. Biochim Biophys Acta 368:432–445

Haraux F, De Kouchkowsky Y (1980) Measurement of chloroplast internal protons with 9 aminocridine. Biochim Biophys Acta 592:153–168

Kurkdjian A, Barbier-Brygoo H (1983) A hydrogen ion selective liquid membrane microelectrode for measurement of the vacuolar pH of plant cells in suspension culture. Anal Biochim 132: 96104

Kurkdjian A, Barbier-Brygoo H, Manigault J, Manigault P (1984a) Distribution of vacuolar pH values within populations of cells, protoplasts and vacuoles isolated from suspension cultures and plant tissues. Physiol Veg 22:193–198

Kurkdjian A, Manigault P, Manigault J, Guern J (1984b) Action of fusicoccin on the vacuolar pH of *Acer pseudoplatanus* protoplasts as evidenced by 9-aminoacridine microfluorimetry. Plant Sci Lett 34:1–5

Kurkdjian A, Quiquampoix H, Barbier-Brygoo H, Pean M, Manigault P, Guern J (to be published 1984c). Critical evaluation of methods for estimating the vacuolar pH of plant cells. In: Marin PM (ed) Biochemistry and function of vacuolar ATPase in fungi and plants. Springer, Berlin Heidelberg New York

Manigault P, Manigault J, Kurkdjian A (1983) A microfluorimetric method for vacuolar pH measurement in plant cells using 9 aminoacridine. Physiol Veg 21:129–136

Schuldiner S, Rottenberg H, Avron M (1972) Determination of pH in chloroplasts II. Fluorescent amines as a probe for the determination of pH in chloroplasts. Eur J Biochim 25:64–70
Thomas RC (1974) Intracellular pH of snail neurones measured with a new pH-sensitive glass micro-electrode. J Physiol (Lond) 238:159–180
Waddell WJ, Bates RG (1969) Intracellular pH. Annu Rev Plant Physiol 30:289–311
Walker NA, Smith FA (1975) Intracellular pH in *Chara corallina* measured by DMO distribution. Plant Sci Lett 4:125–132
Weigel HJ, Weis E (1984) Determination of the proton concentration difference across the tonoplast membrane of isolated vacuoles by mean of 9-aminocridine fluorescence. Plant Sci Lett 33:163–175

Mitotic Cycle of Mesophyll Protoplasts

Y. MEYER

Introduction

Leaf mesophyll cells are differentiated, most of their metabolic activity being concentrated on photosynthesis. In completely developed leaves these cells do not divide. As most differentiated cells, they are arrested in the G_0 phase of the cell cycle, i.e. they contain an amount of DNA corresponding to 2n. After isolation, mesophyll protoplasts rapidly lose their photosynthetic ability during culture and become capable of undergoing mitosis under favourable medium conditions, including the presence of auxin and cytokinin in the culture medium.

This development is not essentially different from the dedifferentiation which mature cells undergo during callus induction in vitro. Nevertheless, the use of mesophyll protoplasts presents certain advantages in comparison with this latter system. The original mesophyll cells represent a homogeneous population, the protoplast suspension is pipettable and their development is synchronous, allowing the study of this transition in cell function at the biochemical level. This article particularly reports modification in gene expression associated with cell dedifferentiation and with the first cell cycle. In addition, the role of auxin and cytokinin in these processes has been studied. Some characteristics of proteins whose synthesis is modified by auxin are described and the level at which auxin regulates the synthesis of these proteins has been determined.

Materials and Methods

Tobacco mesophyll protoplasts (*Nicotiana tabacum* var. Maryland) were prepared and cultivated as described previously (Meyer and Abel 1975a, b), radioactive labelling as in Meyer and Chartier (1981) and Meyer et al. (1984 a, b). Two-dimensional electrophoresis was carried out according to O'Farrell (1975) as modified by Meyer and Chartier (1981, 1983). Cell-free translation was performed in a rabbit reticulocyte lysate (Meyer et al. 1984a).

The Physiological Properties of Plant Protoplasts
Edited by P. E. Pilet

Results

Time Course of the First Cell Cycle and of its Hormonal Control

Mesophyll cells are blocked in the G_0 phase of the cell cycle, as determined by fluorometric measurement of the amount of cellular DNA (Galbraith et al. 1981; Magnien et al. 1982). In culture, protoplasts pass successively through G_0, S and G_2 before undergoing mitosis. Under optimal conditions they enter S phase after 25–30 h in culture, as determined both by thymidine incorporation (Zelcer and Galun 1980, Cooke and Meyer 1981) and by measurement of DNA content (Galbraith et al. 1981). They undergo their first mitosis between 40–48 h. The time of hormone action has been localized by cultivating protoplasts for different times in a medium depleted in one hormone, followed by a transfer into complete medium and measurement of the time that elapses between this transfer and the first mitosis. This approach has shown that auxin is necessary from the beginning of the culture period, while cytokinin intervenes only later, after about 20–25 h i.e. approximately 5–6 h before the entry into S phase (Meyer and Cooke 1979). In addition to their effect on mitotic development, both auxin and cytokinin are necessary for reformation of the cell wall and control processes related to cell permeability and vacuolization (Meyer and Aspart 1983). Due to our present incapacity to measure accurately these additional hormonal effects, the time course of these processes is unknown and we cannot determine whether they depend on a unique primary hormonal impulse or are triggered independently.

Protoplasts in Culture Synthesize Proteins Which are Different from Those Synthesized in the Leaf

Protein synthesis in protoplasts and leaves has been compared by in vivo labelling followed by electrophoretic separation of the proteins (Fleck et al. 1979). This work clearly shows that leaf proteins, for example the two subunits of Rubp carboxylase, are no longer synthesized in protoplasts.

The inhibition of the synthesis of most leaf proteins results primarly from plasmolysis (Fleck et al. 1982). An analysis of mRNA by in vitro protein synthesis followed by electrophoretic separation of the translation products shows the disappearance of the leaf mRNA during protoplast preparation (Fleck et al. 1980, Vernet et al. 1982). In addition, a few protoplast proteins are induced by plasmolysis and have been called "stress proteins" (Fleck et al. 1982). They are different from heat shock proteins, which can be induced in protoplasts by raising the temperature of the culture from 20°–25°C to 40°C (Meyer and Chartier 1983). Nevertheless, most of the new mRNA's characteristic of protoplasts are not induced by plasmolysis alone, but appear only after preparation of the protoplasts. (Fig. 1).

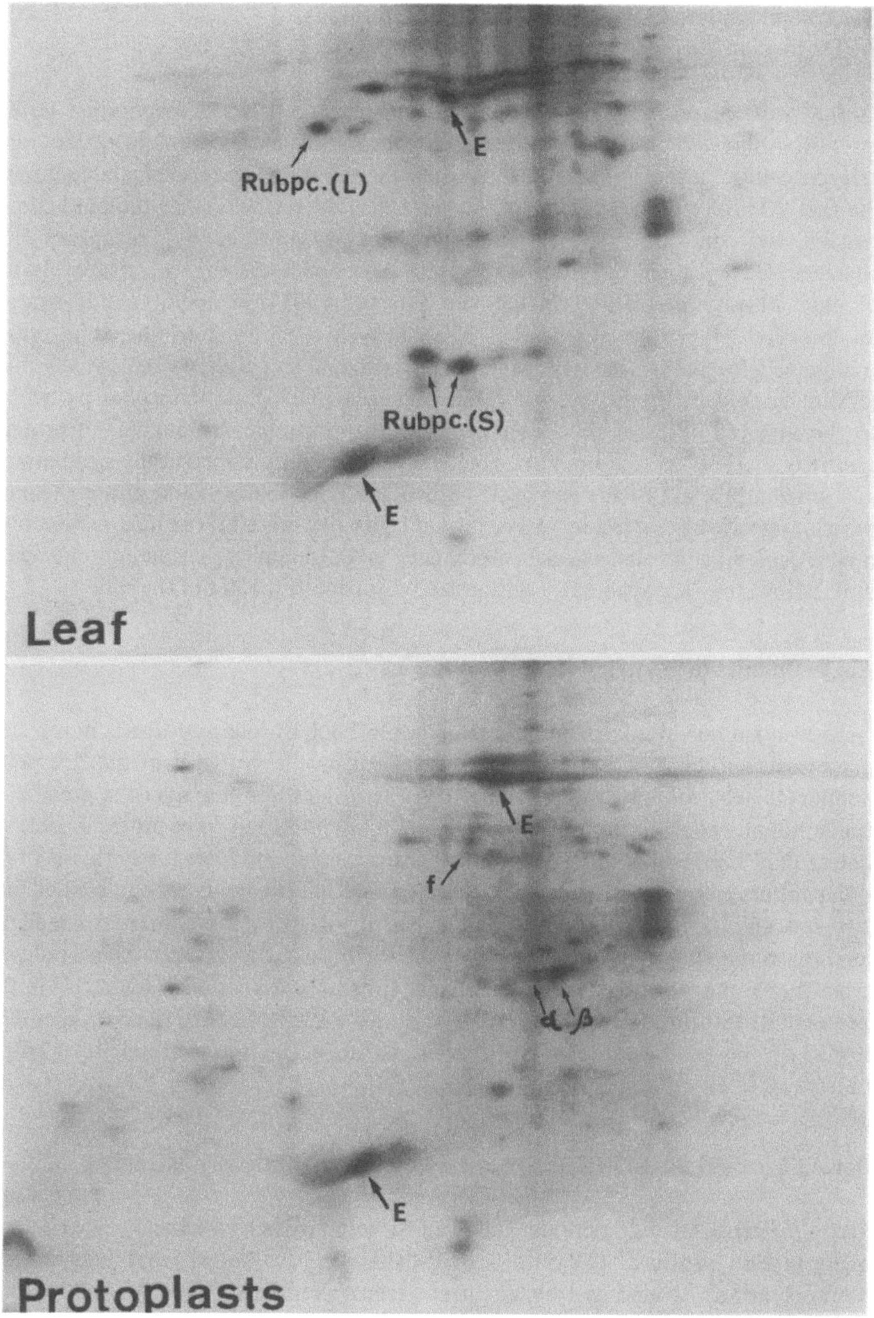

Fig. 1. Comparison of the coding capacities of leaf or protoplast mRNA. In vitro translation products obtained from a rabbit reticulocycte lysate directed by RNA extracted from leaves *(top)* or 6-h-old protoplasts *(bottom)* were separated by two dimensional electrophoresis. The two subunits of Rubp carboxylase, the chloroplast encoded large subunit [*Rubpc. (L)*] and the nucleus encoded small subunit [*Rubpc. (S)*] are indicated on the figure. Spots noted *E* are due to a low endogeneous coding capacity of the rabbit reticulocyte lysate. In addition, three direct translation product *f*, whose level is reduced by auxin and α and β, whose levels are stimulated by auxin, are indicated on the figure. Note the disappearance of leaf translation products in protoplasts and the appearance of new proteins

The Protein Pattern is Very Stable in Culture During the Progression Through the Cell Cycle

Freshly isolated protoplasts are practically devoid of mRNA activity when tested by mRNA-directed in vitro synthesis, but this activity becomes detectable very early in culture, after 30 min to 1 h. Protein synthesis increases considerably during the first 20 h of culture (Cooke and Meyer 1981). Nevertheless, the protein pattern remains very stable. Among about 250 proteins synthesized by protoplasts and characterized by their migration in a two-dimensional system, the relative levels of only 24 vary in culture (Meyer and Chartier 1981). In addition, this limited modification of protein synthesis is not cell-cycle-specific, since the same result is observed for protoplasts cultivated without auxin or cytokinin and which thus remain blocked in the G_1 phase of the cell cycle. This situation is not particular to the first cell cycle of tobacco protoplasts. For example, the analysis of protein synthesis in Hela cells shows that practically no protein is specifically synthesized in a particular phase of the cell cycle (Bravo and Celis 1980). Thus, although some proteins are clearly necessary only during a given phase (for example DNA polymerase and other proteins which participate in DNA replication and chromosome duplication) they are apparently synthesized throughout the life of the cell.

Auxin Modifies the Synthesis of a Few Proteins

Neither auxin nor cytokinin modifies the level of total protein synthesis in tobacco protoplasts. In addition, cytokinin has no effect on the synthesis of the 250 polypeptides detected on two dimensional gels. This does not eliminate the possibility that cytokinin controls mitosis by controlling the synthesis of some proteins, since we detect only the most abundant ones. In contrast, auxin modifies the synthesis of 12 polypeptides among the 250 detected on the gels. These proteins can be classified into two sets (Fig. 2): 10 polypeptides, indicated by Roman letters, present a reduced level of synthesis in the presence of auxin, their level of synthesis being reduced from 2 to 5 times, while two polypeptides (Greek letters) are only detectable in the presence of this hormone. Taking into account the background of the corresponding region of the gel we can estimate that auxin stimulates the synthesis of α and β at least 10 to 50 times.

Some Characteristics of Peptides Whose Synthesis is Modified by Auxin

Using different labeling periods, followed in certain cases by chases, as well as in vitro protein synthesis (Meyer et al. 1984a), we have shown that polypeptides whose synthesis is inhibited by auxin are either short-lived, direct translation products or slowly matured products, suggesting a precursor/product relatinship between these two classes of polypeptides. This slow maturation is reminiscent of the synthesis pathway of proteins which are secreted, for example storage proteins, the maturation process being associated with the secretion through the membrane (Sengupta et al. 1981). Thus, we have looked for a possible cell compartmentali-

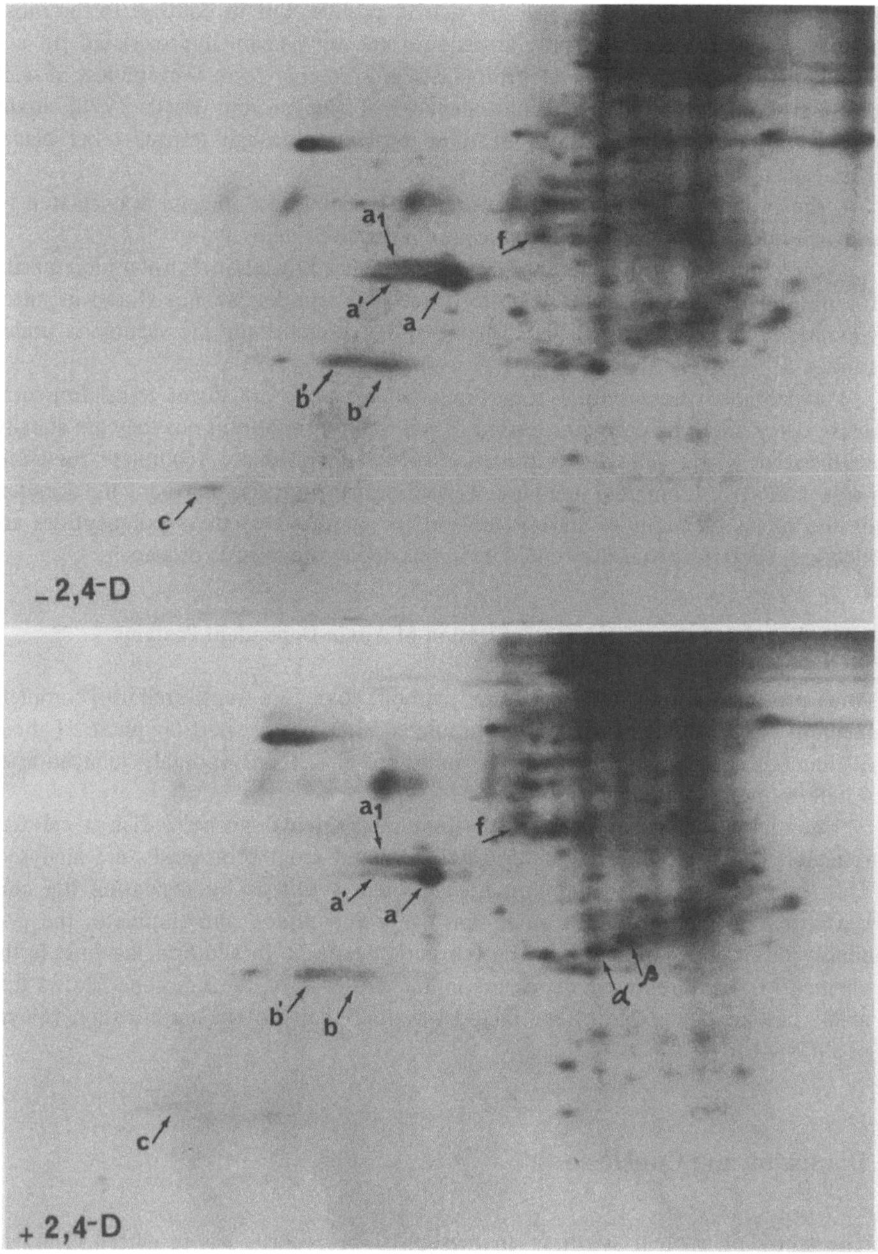

Fig. 2. Auxin effect on protein synthesis in protoplasts. Protoplasts cultivated for two days in the presence of auxin *(+2,4–D)* or in its absence *(–2,4–D)* were labelled for 18 h using ^{35}S methionine. Soluble proteins were extracted and separated by two dimensional electrophoresis. Proteins whose synthesis is reduced by auxin are indicated by *Roman letters.* Due to the long labelling time used, some short-lived proteins whose synthesis is reduced by auxin are not detectable on these gels. Proteins whose synthesis is only detectable in protoplasts cultivated with auxin are marked by *Greek letters.* Note that the synthesis of most proteins is not modified by the hormone

zation of theses polypeptides. All are rich in proline, but in contrast to the most abundant wall proteins the proline residues are not heavily hydroxylated. In addition, these proteins remain in protoplasts in a soluble form. Nevetheless, if wall reformation is inhibited by dichlorobenzonitril (Meyer and Herth 1978) auxin no longer reduces the synthesis of these proteins although mitosis takes place. This leads to two conclusions:

– the proteins whose synthesis is reduced by auxin are in some way related to cell-wall reformation. Nevertheless, they are not wall proteins.

– the reduction in their level of synthesis, induced by auxin, is not a prerequisite for mitotic development. In addition, these polypeptides are not stored in mitochondria, chloroplasts or nuclei. Their possible presence in the vacuole is under study.

Polypeptides whose synthesis is stimulated by auxin are direct translation products. They are apparently not related to cell-wall reformation since they are equally synthesized when cell-wall reformation proceeds normally (complete medium) or is inhibited (complete medium + dichlorobenzonitril). Further, the kinetics of the effect of auxin on mitosis and on the synthesis of these polypeptides are identical, suggesting that they could be implicated in the mitotic process.

Auxin Regulated the mRNA Concentration of Auxin-Dependent Proteins

When protoplasts, previously cultivated without auxin, are transferred into complete medium the synthesis of α and β is stimulated after a very short lag phase of about 30 min. In contrast, synthesis of the basic proteins is only notably reduced after 6 h of hormone treatment.

Essentially the same kinetics are found if proteins synthesized in a cell-free reticulocyte lysate, directed by mRNA extracted from protoplasts, are analysed. This demonstrates that auxin modifies protein synthesis by regulating the concentration of the mRNA of auxin-sensitive polypeptides, and eliminates the possibility of an effect of auxin on the translation process. In addition, labelings in the absence of auxin followed by chases in the presence of the hormone, shows that auxin has no detectable effect on post-translational protein modification (Meyer et al. 1984b).

Discussion and Conclusions

The study of protein synthesis in mesophyll protoplasts has produced clear and original information on two key points in their development:

– dedifferentiation takes place during protoplasts preparation.

– auxin very rapidly regulates the synthesis of a few polypeptides.

Protoplast dedifferentiation, i.e. reprogramming of their gene expression, takes place at two levels. On the one hand, they lose their ability to synthesize proteins present in mesophyll cells, largely as a result of plasmolysis. On the other hand, they synthesize a new set of proteins which remains practically unchanged from the beginning of culture to the first mitosis. This stability allows us to consider that these proteins are characteristic of the meristematic state newly acquired by mesophyll protoplasts. Only a few of these proteins are synthesized by plasmolysed leaves, most of them being detectable only in protoplasts. Their synthesis is practically independent of the culture conditions and culture medium. In particular, hormones do not modify the synthesis of the majority of these proteins. At the molecular level, protoplast preparation, which triggers their synthesis, can be described by two types of event. On the one hand, isolation releases protoplasts from influences coming from the rest of the leaf. On the other hand, enzymaticdigestion of the cell wall places mesophyll cells in contact with substances present in the enzymatic preparations or with products released from their own cell wall by maceration. One may ask if the entry into the cell cycle shown by mesophyll protoplasts results from the disappearance of an inhibition present in the leaf or from a stimulation by a mitogene during maceration. As indicated in the introduction, mesophyll protoplasts are only a particular case of the more general situation of callus induction in a differentiated non-mitotic tissue. It is important to recall that calli develop in vitro only at the wounded face of the explant and it is probably acceptable to consider that the growth of a callus in vitro consists in a hypertrophy of the cicatrization callus which arises spontaneously at the level of the wound. Taking into account our knowledge of the chemical determinant of certain wound reactions (Ryan et al. 1981, Hahn et al. 1981, Albersheim et al. 1983), the hypothesis of mitogenic endogeneous elicitors is very attractive. In addition, this hypothesis can be tested experimentally using the protein pattern as a marker of the differentiated state. The understanding of the mechanism responsible for the dedifferentation of tobacco mesophyll cells during protoplast preparation may lead to new insights into the problem of mesophyll cereal protoplasts: despite many attempts, it has never been possible to induce these cells to divide (Dale 1983). It is interesting to note that cereals do not respond to wounding by callus formation and that all in vitro cultured tissues have been obtained from growing tissues, i.e. without complete dedifferentiation of the cells.

The study of the effect of auxin has entered a new phase since the development of two-dimensional electrophoretic separation of proteins. Up to this point, the effect of auxin on total protein synthesis or on certain enzymatic activities had been described. Nevertheless, these effects were detectable only very late after auxin application, i.e. they were very indirect, and it was not possible to use them to characterize the primary impact of auxin. Only very recently rapid effects of auxin on protein synthesis have been characterized in biological systems responding by cell elongation to the hormone application. (Baulcombe et al. 1980, Zurfluh and Guilfoyle 1980, 1982 a, b, c, Theologis and Rey 1982). As in tobacco protoplasts, auxin very rapidly stimulates the synthesis of a few polypeptides (within 10–30 min) and reduces the synthesis of some others more slowly (after a few hours). In both cases, auxin modifies protein synthesis by regulating the concentration of the corresponding mRNA.

Two research directions are presently open: on the one hand the nature and function of proteins whose synthesis depends on auxin remains to be established, in order to determine to what extent auxin regulates physiological processes such as mitosis, cell wall extension, cell wall reformation, cell permeability etc. by modiying protein synthesis. On the other hand, the biochemical mechanism by which auxin regulates protein synthesis will be analysed. The recent cloning of cDNA corresponding to proteins whose synthesis is stimulated by auxin (Walker and Key 1982) opens the way to the determination of the level of regulation (transcription, mRNA maturation or transport), as well as to the characterization of the structural particularity of the genes coding for auxin-sensitive proteins. This may lead to an understanding of the role of auxin receptors (Van der Linde et al. 1984, Libenga, this book) in the effect of auxin.

Acknowledgments. I should like to thank Y. Chartier for technical assistance and Dr. R. Cooke for translating this paper.

References

Albersheim P, Darvill AG, McNeil M et al. (1983)Oligosaccharins: naturally occuring carbohydrates with biological regulatory functions. In: Ciferri O, Dure L (eds) Structure and function of plant genomes III. Plenum, New York

Baulcombe DC, Giorgini J, Key JL (1980) The effect of auxin on the polyadenylated RNA of soybean hypocotyls. In: Leaver CJ (ed) Genome Organization and expression in plants. Plenum, New York

Bravo R, Celis JE (1980) A search for differential polypeptides synthesis throughout the cell cycle of HeLa cells. J Cell Biol 84 795–802

Cooke R, Meyer Y (1981) Hormonal control of tobacco protoplast nucleic acid metabolism during in vitro culture. Planta (Berl) 52:1–7

Dale PJ (1983) Protoplast culture and plant regeneration of cereals and other recalcitrant Crops. In: Potrykus J et al. (ed) Protoplasts 1983, Poster Proc 6th Int Protoplast Symp Basel. Exper Suppl (Basel) 46:31–41

Fleck J, Durr A, Lett M, Hirth L (1979) Changes in protein synthesis during the initial stage of life of tobacco protoplasts. Planta (Berl) 145:279–285

Fleck J, Durr A, Fritsch C, Lett M, Hirth L (1980) Comparison of proteins synthesized in vivo and in vitro by mRNA from isolated protoplasts. Planta (Berl) 148:453–454

Fleck J, Durr A, Fritsch C, Vernet T, Hirth L (1982) Osmotic-shock "Stress proteins". Plant Sci Lett 26:159–165

Galbraith DW, Mauch TJ, Shields BA (1981) Analysis of the initial stages of plant protoplasts using 33258 Hoechst: reactivation of the cell cycle. Physiol Plant 51:380–386

Hahn MG, Darvill AG, Albersheim P (1981) Host-pathogen interaction XIX. The endogeneous elicitor, a fragment of a plant cell wall polysaccharide that elicits phytoalexin accumulation in soybean. Plant Physiol (Bethesda) 68:1161–1169

Magnien E, Dalschaert X, Faraoni-Sciamanna P (1982) Transmission of a cytological heterogeneity from the leaf to the protoplasts in culture. Plant Sci Lett 25:291–303

Meyer Y, Abel WO (1975a) Importance of the wall for cell division and in the activity of the cytoplasm in cultured tobacco protoplasts. Planta (Berl) 123:33–40

Meyer Y, Abel WO (1975b) Budding and cleavage division of tobacco mesophyll protoplasts in relation to pseudo-wall and wall-formation. Planta (Berl) 125:1–13

Meyer Y, Aspart L (1983) The first mitotic cycle of mesophyll protoplasts. In: Potrykus J et al. (ed) Protoplasts 1983, Poster Proc 6th Int Protoplast Symp Basel. Exper Suppl (Basel) 46: 93–100

Meyer Y, Chartier Y (1981) Hormonal control of mitotic development in tobacco protoplasts. Two-dimensional distribution of newly-synthesized proteins. Plant Physiol (Bethesda) 68: 1273–1278

Meyer Y, Chartier Y (1983) Long-lived and short-lived heat-shock proteins in tobacco mesophyll protoplasts. Plant Physiol (Bethesda) 72:26–32

Meyer Y, Cooke R (1979) Time course of hormonal control of the first mitosis in tobacco mesophyll protoplasts cultivated in vitro. Planta (Berl) 147:181–185

Meyer Y, Herth W (1978) Chemical inhibition of cell wall formation and cytokinesis but not of nuclear division, in protoplasts of Nicotiana tabacum cultivated in vitro. Planta (Berl) 142: 253–262

Meyer Y, Aspart L, Chartier Y (to be published 1984a) Auxin-induced regulation of protein synthesis in tobacco mesophyll protoplasts cultivated in vitro I. Characteristics of auxin-sensitive proteins. Plant Physiol (Bethesda)

Meyer Y, Aspart L, Chartier Y (to be published 1984b) Auxin-induced regulation of protein synthesis in tobacco mesophyll protoplasts cultivated in vitro II. Time course and level of auxin control. Plant Physiol (Bethesda)

O'Farrell PH (1975) High resolution two-dimensional electrophoresis of proteins. J Biol Chem 250:4007–4021

Ryan CA, Bishop P, Pearce CE, Darvill AG, McNeil M, Albersheim P (1981) A sycamore cell wall polysaccharide and a chemically related tomato leaf polysaccharide possess similar proteins inhibitor-inducing activities. Plant Physiol (Bethesda) 68:616–618

Sengupta C, Deluca V, Bailey DS, Verma DPS (1981) Post-translational processing of 7S and 11S components of soybean storage proteins. Plant Mol Biol 1:19–34

Theologis A, Ray PM (1982) Early auxin-regulated polyadenylated mRNA sequences in pea stem tissue. Proc Natl Acad Sci USA 79:418–421

Van der Linde PLG, Borman H, Mennes AM, Libbenga KR (1984) A soluble auxin-binding protein from cultured tobacco tissues stimulates RNA synthesis in vitro. Planta (Berl) 160:102–108

Vernet T, Fleck J, Durr A, Fritsch C, Pinck M, Hirth L (1982) Expression of the gene coding for the small subunit of ribulose biphosphate-carbuxylase during differentiation of tobacco plant protoplasts. Eur J Biochem 126:489–494

Walker JC, Key JL (1982) Isolation of cloned cDNAs to auxin-responsive poly$(A)^+$ RNAs of elongating soybean hypocotyl. Proc Natl Acad Sci USA 79:7185–7189

Zelcer A, Galun E (1976) Culture of newly isolated tobacco protoplasts: precursor incorporation into protein, RNA and DNA. Plant Sci Lett 7:331–336

Zelcer A, Galun E (1980) Culture of newly isolated tobacco protoplasts: cell division and precursor incorporation following a transient exposure to coumarin. Plant Sci Lett 18:185–190

Zurfluh LL, Guilfoyle TJ (1980) Auxin-induced changes in the patterns of protein synthesis in soybean hypocotyl. Proc Natl Acad Sci USA 77:357–361

Zurfluh LL, Guilfoyle TJ (1982a) Auxin-induced changes in the population of translatable messenger RNA in elongating sections of soybean hypocotyl. Plant Physiol (Bethesda) 69: 332–337

Zurfluh LL, Guilfoyle TJ (1982b) Auxin-and ethylene-induced changes in the population of translatable messenger RNA in basal sections and intact soybean hypocotyl. Plant Physiol (Bethesda) 69:338–340

Zurfluh LL, Guilfoyle TJ (1982c) Auxin-induced changes in the population of translatable messenger RNA in elongating maize coleoptile sections. Planta (Berl) 156:525–527

The Use of Guard Cell Protoplasts to Study Stomatal Physiology

J.D.B. WEYERS and P.J. FITZSIMONS

Introduction

Guard cell protoplasts (GCP's) were first studied by Zeiger and Hepler (1976), although Davey et al. (1974) had earlier experimented with isolated epidermal cell protoplasts from tobacco leaves. The microchamber method pioneered by Zeiger and Hepler allowed the observation of small numbers of GCP's adhering to the remnants of strips of leaf epidermis, but did not allow a full range of microbiological techniques to be employed (Zeiger 1981). The full potential of the protoplast technique could only be realised when methods were developed for obtaining high yields of GCP's and for their purification (Schnabl et al. 1978, Outlaw et al. 1981, Fitzsimons and Weyers 1983).

GCP's are now widely studied by stomatal physiologists because their use confers a large number of potential advantages for experiments (Weyers et al. 1983). Among the most important of these are the following:

a) *The possibility of applying novel methods* to investigate stomatal physiology and biochemistry. GCP's represent a refined in vitro system for studying stomata. Dense populations of guard cells can be treated like microbial suspensions, allowing numerous experiments impossible or difficult to carry out with intact tissue.
b) *The ability to isolate guard cells from other cells* and to examine their responses separately. This allows unequivocal assignment of the capability of each cell type to respond to stimuli. The protoplast method can give a very high degree of cell purity compared with other methods. Preparation of epidermal or subsidiary cell protoplasts is the only method currently available to study these cell types in isolation.
c) *The chance to exert greater control over the microenvironment* of the guard cell and to control parameters which cannot otherwise be manipulated accurately.
d) *The possibility of obtaining accurate and precise estimates of volume* due to protoplast sphericity, so that GCP osmotic parameters can be more easily measured.
e) *The removal of the guard cell wall,* which simplifies tracer flux experiments and may aid investigations of the role of the wall in stomatal movements by comparison with more "intact" material, e.g. epidermal strips.
f) *The ability to isolate guard cells from new species.* GCP's can be obtained from digests of whole leaves (Outlaw et al. 1981) and this offers the prospect of removing the bias inherent in experiments which have utilised epidermal strips from plants which "peel" easily (Raschke 1979).

The Physiological Properties of Plant Protoplasts
Edited by P. E. Pilet

Against these important advantages must be set some disadvantages, including the time, expense and expertise required to isolate pure GCP's. As with all protoplast preparations, there is a distinct chance of damage to the guard cells due to both plasmolysis and adverse effects of the enzyme used (see Fitzsimons and Weyers, Chap. 3 this Vol.), and these should be carefully assessed. We must also remain aware that we risk neglecting an important facet of stomatal function by removing the guard cell wall.

In a previous review (Weyers et al. 1983) we considered the methods in use to obtain, purify and manipulate GCP's. We also summarised previous studies of the structure, physiology and metabolism of GCP's. In this chapter, we intend primarily to discuss the potential of the GCP system as a model of stomatal operation. Guard cells appear to be specialised for receiving appropriate stimuli and transducing them via the processes of solute transport and metabolism into stomatal movements (Raschke 1979, Zeiger 1983). At an elementary level, we expect to see this function reflected qualitatively in the osmotic behaviour of GCP's. However, the quantitative aspect of the relationship between GCP solute relations and stomatal aperture necessarily requires a more complex analysis. Despite the specific nature of the comparison between GCP's and stomata, it is hoped that the discussion given here will be of general relevance to all protoplast research in which changes in solute levels occur.

Materials and Methods

We have described the methods used for peeling leaf epidermis (Weyers and Travis 1981), obtaining and purifying GCP's (Fitzsimons and Weyers 1983) and determining the osmotic characteristics of protoplasts (Weyers and Fitzsimons 1982, Fitzsimons and Weyers 1983). Deviations from these procedures and specific information about experiments are given in the figure and table legends.

Results and Discussion

Stomatal movements have been interpreted as part of the general maintenance of homeostasis in plants (Zeiger 1983), an important example of which is the control of water loss while optimising CO_2 uptake (Farquhar and Sharkey 1982). Such theories arose primarily from observations of stomatal responses to various environmental parameters. Even before precise details were known of the mechanism of stomatal movements, much had been discovered about their reactions to light, CO_2 levels, plant water status, humidity and temperature (Meidner and Mansfield 1968). More recently, these responses have been thought of as feedback and feedforward loops of a complex control system (Raschke 1979).

It was thus appropriate that the initial physiological experiments with GCP's should seek to show that they were responsive to the same stimuli as stomata in vivo (e.g. Zeiger and Hepler 1977, Schnabl et al. 1978). The results were not only an important test of GCP viability, but also provided unambiguous information about the site of perception of the stimuli (Weyers et al. 1983). Figures 1–3 show our results with *Commelina* GCP's in experiments where light intensity CO_2 levels and ABA concentration were altered, and are discussed below in the context of other reports from the literature.

GCP Responses to Light

Zeiger and Hepler (1977), working with *Allium*, first demonstrated that GCP's could respond to light by swelling, They found that blue wavelengths were effective at low photon flux densities, whereas red wavelengths resulted in small responses only. With white light, the response could be enhanced substantially by the addition of KCl to the medium. The latter finding has been confirmed by Gotow et al. (1982) (*Vicia* GCP's) and Jewer et al. (1982) (*Pisum* GCP's), and is illustrated clearly in Fig. 1, where white-light-treated *Commelina* GCP's showed a volume increase of over 30% in the presence of 10 mol m^{-3} KCl but an insignificant change in the absence of the salt. After 50 min in the light + KCl treatment, a high proportion of the protoplasts were observed to burst, indicating that the decline in their mean volume thereafter was an artefact. Dark-treated GCP's showed little or no response over the first 50 min, either with or without KCl, although the presence of the salt resulted in a slight volume increase in the long term. It is tempting to speculate from these results that light stimulates a K^+ uptake system (Zeiger 1983), but caution is required because Gotow et al. (1982) showed that only 13% of the increase in *Vicia* GCP volume resulting from a white light + 10 mol m^{-3} KCl treatment was due to uptake of K^+ salts.

GCP Responses to CO_2 Levels

The level of dissolved gases in protoplast incubation media is difficult to control. In our hands, the method normally empolyed with epidermal strips, that of bubbling a gas supply through the medium, led to injury and bursting of GCP's. Others have passed a moist gas supply over incubation media (Gotow et al. 1982, Jewer et al. 1982) and we have developed an experimental apparatus in which two iso-osmotic media are linked by dialysis membrane to minimise disruption of the GCP suspension (Fig. 2a). Using this method, we have observed increases in *Commelina* GCP volumes at low CO_2 tensions (Fig. 2b). The response was only observed in the presence of KCl and, once again, an initial volume increase was accompanied by protoplast lysis which presumably depressed the mean volume of the remaining viable protoplasts. These results extend those of Gotow et al. (1982) who showed volume decreases at levels of CO_2 higher than those of ambient air, but who were unable consistently to find the expected volume increases at lower CO_2 concentrations.

GCP Responses to ABA

Abscisic acid (ABA) causes stomatal closure and is thought to be involved in the responses of plants to water deficits (see Zeiger 1983). Schnabl et al. (1978) investigated the effects of ABA and fusicoccin, a fungal toxin which causes stomatal opening, on GCP's of *Allium* and *Vicia*. The protoplasts showed rapid shrinking and swelling responses analogous to stomatal movements found with epidermal strips. Jewer et al. (1982) found that *Pisum* GCP's treated with ABA did not swell in response to 10 mol m^{-3} KCl, light and CO_2 free air whereas the control GCP's increased in volume by over 50%. Kinetin had no significant effect on this volume increase. Figure 3 illustrates the time course of *Commelina* GCP shrinkage in response to two concentrations of ABA. The effect of the compound appears to continue for at least 2 h at both 10^{-2} and 10^{-3} mol m^{-3} ABA. These results are in contrast to those of Iversen et al. (1983) who found that *Vicia* GCP's decreased significantly in size after 30 min exposure to 10^{-5} mol m^{-3} ABA, but then recovered over the next 90 min to a volume close to their initial size. Iversen et al. (1983) also noted that these responses to ABA, measured under green light, were absent if red light was used.

In contrast to the above results, experiments with epidermal cell protoplasts (ECP's) have not revealed significant responses either to light or ABA (Zeiger and Hepler 1977, Iversen et al. 1983). However, we have pointed out (Fitzsimons and Weyers 1983) that the levels of solutes lost or gained by GCP's during profound changes in their volumes would have little effect on ECP volumes because they are so much larger. The expected effects would probably not be significant with respect to the natural variation in ECP size, and more sensitive methods of analysis of solute exchange are clearly required to investigate the reactions of these cells.

We can conclude from the results so far obtained that GCP's of several species respond to various stimuli by exhibiting volumes changes: guard cells clearly possess the means both to perceive and react to external stimuli and may be the only cells of the stomatal complex which do so. Such GCP reactions are analogous to stomatal movements occuring by *hydroactive* mechanisms (sensu Raschke 1975). That is,

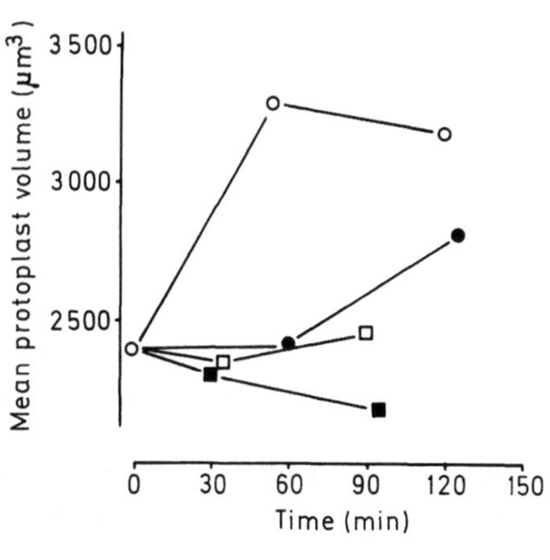

Fig. 1. Effect of light and KCl on GCP volume. *Commelina* GCP's were isolated and purified (Fitzsimons and Weyers 1983) in the evening and maintained in 300 mol m^{-3} mannitol at 23°C for 10 h in the dark, extending the normal dark phase by 2 h. The treatments were carried out at 30°C. Square symbols: 300 mol m^{-3} mannitol only; circles: 10 mol m^{-3} KCl added. *Open symbols* white light (fluorescent tubes) at 25 W m^{-2} PAR; *closed symbols* dark. Volumes are the means of 25 protoplasts. At 60 min, the light + KCl and dark + KCl values were significantly different (Mann-Whitney test, $Z = 4.43$, $p < 0.001$) but not at 120 min. The light – KCl and dark – KCl volumes were never significantly different

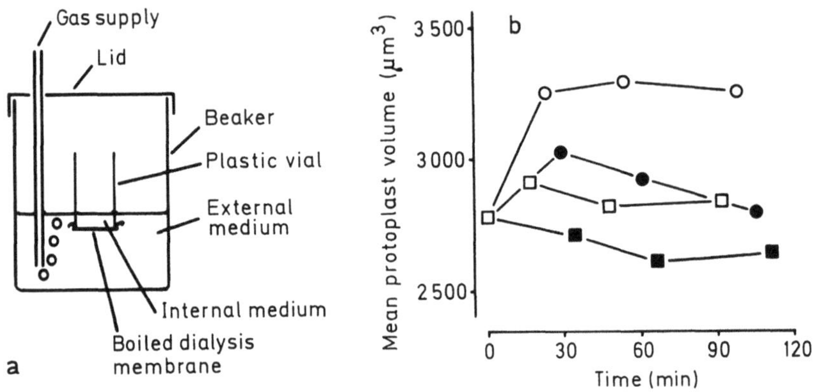

Fig. 2 a Apparatus used for equilibrating GCP's at different CO_2 levels. **b** Effect of CO_2 level on GCP volume. *Commelina* GCP's (Fitzsimons and Weyers 1983) were suspended in 250 mol m^{-3} mannitol, pH 6.5 (NaOH) under white light (as Fig. 1). *Open symbols* equilibrated with CO_2-free air (< 1 vpm); *closed symbols* equilibrated with lab. air (ca. 325 vpm CO_2). *Square symbols* mannitol alone; *circles* 10 mol m^{-3} KCl added. The gas flow rate was 650.10^{-6} m^3 min^{-1} and the system was equilibrated for 1 h before adding protoplasts to the internal medium. Volumes are means from 20–40 protoplasts. The low-CO_2 + KCl and high-CO_2 + KCl values were significantly different at 60 min ($Z = 2.25$, $P < 0.05$) and 95 min ($Z = 2.78$, $P < 0.01$) but not before

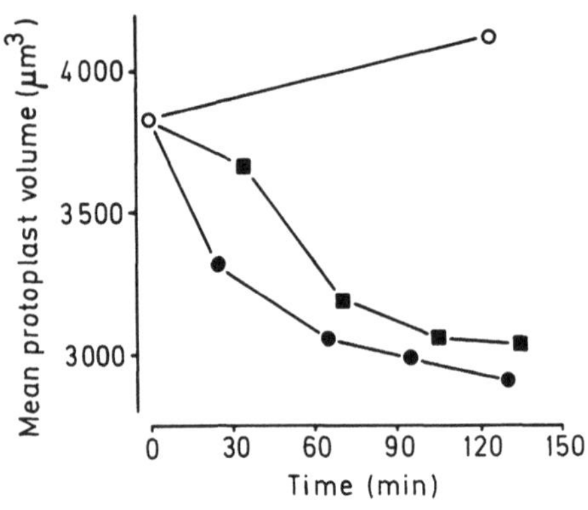

Fig. 3. Effect of ABA on GCP volume. *Commelina* GCP's (Fitzsimons and Weyers 1983) were suspended in 300 mol m^{-3} mannitol plus 10 mol m^{-3} PIPES buffer (pH 6.8) with or without (±) ABA at 30°C in white light (as Fig. 1). *Open circles* mannitol alone; *closed circles* 10^{-2} mol m^{-3} ABA added; *closed squares* 10^{-3} mol m^{-3} ABA added. Volumes are means of 25 protoplasts. At 130 min the ABA-treated protoplasts were significantly smaller than the control ($Z > 3.5$, $P < 0.001$)

they are due to the influx, efflux or metabolism of solutes. Stomata also respond to the water status of the leaf and the humidity of the air by *hydropassive* mechanisms (Raschke 1975). These are due to fluxes of water because of changes in the external water potential. Protoplasts exhibit analogous volume changes when the concentration of osmoticum in the incubation medium is altered. This type of experiment allows the measurement of osmotic parameters of GCP's, knowledge of which permits us to make more quantitative treatments of physiological data and to make quantitative comparisons between GCP's and intact stomata.

If a membrane-bound body like a protoplast acts as a true osmometer, its behaviour in different strengths of impermeant osmotica ought to conform to the

Boyle-van't Hoff equation (Nobel 1983). That is, a plot of protoplast volume (V) against the reciprocal of the solute potential of the medium (ψ_s^{-1}) should be linear with slope nRT (n = number of osmotically active "ideal" solutes in the protoplast, R = the gas constant, T = absolute temperature), and y-intercept b (non-osmotic volume, NOV). This equation, $V = nRT\,(-\psi_s^{-1}) + b$, can be elaborated to take account of permeant osmotic species, protoplast turgor, and matric potential (Nobel 1983), but a simplified approach is probably justified for GCP research where impermeant solutes are generally used as osmotica and turgor and matric potentials seem minimal.

Figure 4 shows plots of this function for GCP's from five different plant species. As shown by Gotow et al. (1982, *Vicia*), Salhani et al. (1982, *Vicia*) and Weyers and Fitzsimons (1982, *Commelina*), GCP's do seem to act as true osmometers within limits. At one extreme, sphericity is lost at high external concentrations of osmoticum (i.e. low ψ_s) as the plasmalemma wraps around the chloroplasts, and at the other, significant lysis occurs at low external concentrations (high ψ_s). Both these conditions lead to errors in estimating the mean volume of the protoplast population. The high degree of linearity obtained within these limits indicates that the opposing pressure and matric potentials within GCP's are either negligible or equal (Nobel 1983).

A corollary of the Boyle-van't Hoff relation worth emphasising is that volume is the appropriate parameter with which to express changes in the solute relations of protoplasts, since at any given external ψ_s, the solute content is directly related to volume. GCP volume are thus to be preferred to diameters, the form in which data are usually collected (in fact, osmotic volume, V minus the NOV, is the best value to use, since this takes account of the "dead space" inside the protoplast) Assuming sphericity, volumes can be calculated simply from diameter measurements using $V = 4/3\ \pi r^3$, but mean volumes must be calculated from the mean of the individual volumes (see Outlaw et al. 1981). In our experience, volume frequency distributions for GCP's are rarely symmetrical so statistical analyses used should avoid requirements for normality.

Figure 4 shows that there is a considerable difference between GCP solute levels and NOVs between species. With *Commelina* we have also noted significant within-species variations, as did Gotow et al. (1982) with *Vicia*. For 22 batches of *Commelina* GCP's given similar post-harvest treatments, we found a negative correlation between the solute content n and the NOV. The least-squares fit of the data was NOV (μm^3) = 850n (pmol) + 1400 μm^3, $r = 0.669$, $P < 0.001$. A possible explanation for this relationship is that GCP solute content is related to the stomatal aperture of the starting material and that a portion of the NOV is guard cell starch involved in starch-solute interconversions associated with stomatal movements (Weyers and Fitzsimons 1982). This hypothesis has been tested by altering the pretreatments given to plant material prior to GCP isolation and osmotic analysis. Figure 5 shows Boyle-van't Hoff plots for GCP's obtained from source plants pretreated with two regimes of light, humidity and CO_2 levels. The closed stomata in one case were reflected in a lower solute content and a higher NOV and vice versa for open stomata. Similar results were obtained with GCP's from *Vicia* and *Pisum*.

The NOV of guard cells appears to be relatively high as a proportion of the normal cell volume encountered in vivo. For *Commelina* mesophyll cells, an estimated NOV of 2545 μm^3 was approximately 10% of cell volume (Fitzsimons and Weyers 1983), and the extremes of values we have found in the course of over 50 determinations on *Commelina* GCP's were 12.5 and 44% of guard cell volume at 0 μm (4000 μm^3, MacRobbie and Lettau 1980). In *Vicia*, the mesophyll cell NOV was found to be 2000 μm^3 (Salhani et al. 1982), and estimates of GCP NOVs have varied from 500 μm^3 (Salhani et al. 1982) to 2230 μm^3 in spring plants (Gotow et al. 1982). The NOV should clearly be taken into account in quantitative studies of guard cell osmotic relations, and the intriguing concept of a *variable* NOV dependent on stomatal aperture must complicate such investigations. At present, protoplasts seem to offer the simplest means of estimating this neglected parameter.

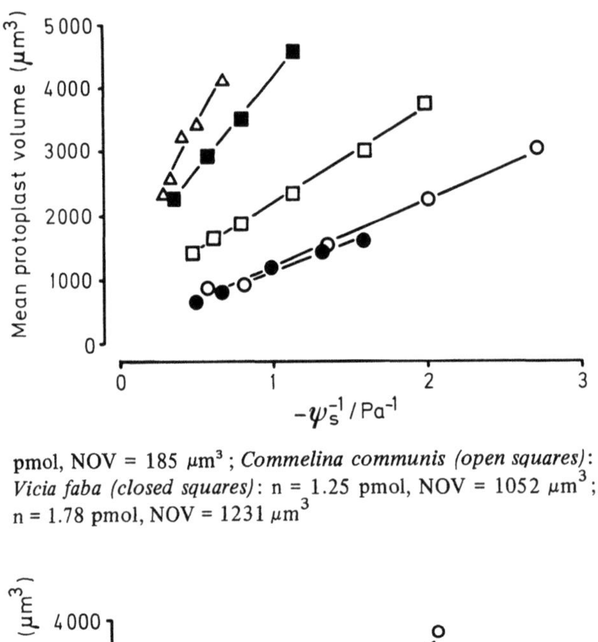

Fig. 4. Boyle van't Hoff analysis of GCP's from five species. GCP's were prepared (Fitzsimons and Weyers 1983) using appropriate mannitol concentrations determined by plasmolysis. Procedures thereafter were according to Weyers and Fitzsimons (1982) except that sorbitol was used for *Allium*. The lines drawn are least-squares linear regressions yielding the following values. *Pisum sativum (closed circles)*: n = 0.36 pmol, NOV = 234 μm^3; *Kalanchöe daigremontiana (open circles)*: n = 0.43 pmol, NOV = 185 μm^3; *Commelina communis (open squares)*: n = 0.59 pmol, NOV = 732 μm^3; *Vicia faba (closed squares)*: n = 1.25 pmol, NOV = 1052 μm^3; *Allium porrum (open triangles)*: n = 1.78 pmol, NOV = 1231 μm^3

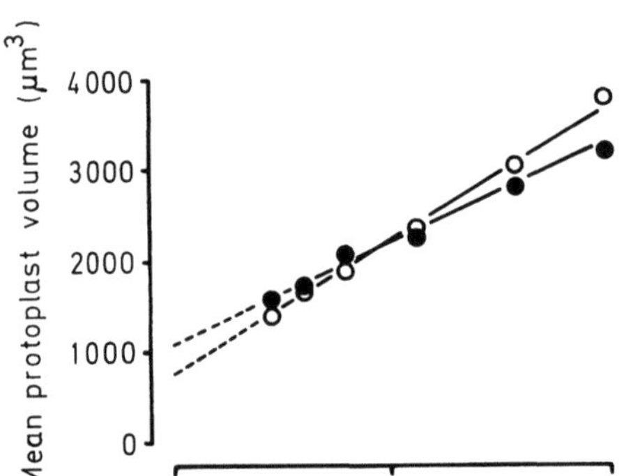

Fig. 5. Effect of pretreatments on GCP osmotic characteristics. *Commelina* GCP's were prepared from pretreated plants as described by Weyers and Fitzsimons (1982). *Open circles* light + high humidity + low CO_2 pretreatment – stomatal aperture at peeling 23.8 μm – n = 0.59 pmol, NOV = 732 μm^3. *Closed circles* dark ambient humidity + ambient CO_2 pretreatments – stomatal aperture at peeling 9.0 μm – n = 0.42 pmol, NOV = 1112 μm^3. Both the n and NOV values obtained were significantly different (covariance analysis, $t > 4.8$, $p < 0.002$)

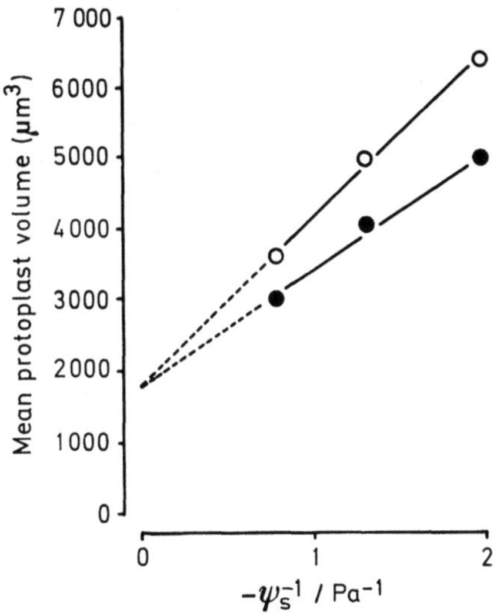

Fig. 6. Effect of ABA on GCP osmotic characteristics. *Commelina* GCP's (Fitzsimons and Weyers 1983) were incubated for 120 min in 350 mol m^{-3} mannitol pH 6.5 (NaOH) with or without 5.10^{-2} mol m^{-3} (±) ABA, then subjected to Boyle van't Hoff analysis (Weyers and Fitzsimons 1982). *Open circles* mannitol alone; n = 0.94 pmol, NOV = 1775 μm^3; *closed circles* plus ABA; n = 0.66 pmol, NOV = 1782 μm^3

Use of the Boyle-van't Hoff relation can thus yield estimates of two parameters, n and the NOV, which are of great relevance to osmotic studies of protoplasts. Net solute uptake or efflux from protoplasts in vitro can be quantified accurately with this approach and the fact that volumes are determined at several osmotic strengths adds to the precision of the estimate. An example of the value of the technique is shown in Fig. 6. ABA-treated GCP's had a lower solute content (as expected from Fig. 3), but the NOV remained unchanged. These results reveal interesting facts about the nature and extent of the ABA response. Further investigations have shown that the decrease in solute content observed is dependent on the availability of metabolic energy (Weyers et al. 1982) and that loss of K salts from the protoplasts may account for some, but not all, of the osmotic volume change (Fitzsimons and Weyers, unpublished).

Armed with the appropriate means for characterising GCP osmotic relations, it should be possible to speculate on the quantitative relevance of any data obtained to stomatal movements in vivo, although at present there have been few comparative studies which could confirm any conclusions made.

We certainly have no guarantee that the values of n obtained bear an absolute correspondence to those existing prior to protoplast isolation, purification, etc. With guard cells, considerable solute leakage can occur during plasmolysis (Raschke 1979). The range of values we have found for *Commelina* GCP solute contents varies between 0.33–0.85 pmol per protoplast after 4–6 h incubation in enzyme medium. We regularly use 300 mol m^{-3} mannitol to "preplasmolyse" guard cells prior to this incuation, and given a guard cell volume of 4000 μm^3 at 0 μm, this sets an upper limit of 1.2 pmol on the solute content at the start of such isolations (this ignores the NOV which would decrease this value). The value of n for our protoplasts is therefore of the same order of magnitude as that in vivo, but some solutes are probably lost during plasmolysis and isolation. It is not possible to say at this stage whether the identities and proportions of the solutes present in GCP's are the same as those in vivo, or whether GCP's can regulate their volume and replace lost solutes by synthesis or uptake.

Even if GCP solute levels were equivalent to those existing in vivo, there would be complications in equating them to stomatal apertures. Guard cell volumes can be related to stomatal aperture by simple equations (see Raschke 1979). For instance, in *Vicia*, Raschke estimated guard cell volume from $V(\mu m^3) = 3150 + 215A$, where A = stomatal aperture in μm, and MacRobbie and Lettau (1980) employed $V = 4000 + 200A$ for *Commelina* guard cells. However, at very high apertures, the relationship between volume and aperture may deviate from lineartiy, and it is clear that volumes less than 3150 and 4000 μm^3 respectively will result in zero aperture. Furthermore, in these equations the effects of subsidiary cell turgor and mechanical advantage are not taken into account.

Such problems probably mean that the direct substitution of GCP volumes for guard cell volumes in these equations would not give meaningful results. Nevertheless, it ought to be possible to predict the stomatal movement expected at a given stomatal aperture due to net or proportional solute changes estimated with GCP's. Also, the rates of volume change in GCP's could be converted into solute fluxes for comparison with data obtained from other systems. We anticipate few problems due to differences in water permeability since the half-times for water exchange for both GCP's and guard cells are rapid (Raschke 1979; Salhani et al. 1982).

The above "physical" analysis does not take into account possible physiological differences between experimental systems. Protoplasts may be damaged due to the effects of plasmolysis and enzymic digestion (see Fitzsimons and Weyers, Chap. 3 this Vol.). Epidermal strips, if used, may suffer artefacts due to cell death and the composition of incubation media (Weyers and Travis 1981). Weyers et al. (1983) pointed out possible errors of this type which may have occurred in the comparison of responses of GCP's and stomata on epidermal strips by Schnabl et al. (1978). Care should be taken to optimise the response to a simulus in all experimental systems used in this type of study.

Conclusions

a) GCP's show responses to light, CO_2 and ABA which qualitatively mimic the reactions found with more intact systems.

b) GCP's act as true osmometers within certain limits of volume and the Boyle-van't Hoff relation can be used to estimate their solute contents and non-osmotic volumes. Both these parameters vary according to the physiological state of the stomata prior to isolation.

c) Protoplast solute content and non-osmotic volumes are required for accurate descriptions of GCP responses in vitro.

d) Guard cell protoplast osmotic parameters can be related to projected stomatal movements but this type of quantitative analysis is subject to many problems and possible artefacts.

Acknowledgement. We thank the U.K. Science and Engineering Research Council for financial support.

References

Davey MR, Frearson EM, Withers LA, Power JB (1974) Observations on the morphology, ultrastructure and regeneration of tobacco leaf epidermal protoplasts. Plant Sci Lett 2:23–27

Farquhar GD, Sharkey TD (1982) Stomatal conductance and photosynthesis. Annu Rev Plant Physiol 33:317–345

Fitzsimons PJ, Weyers JDB (1983) Separation and purification of protoplast types from *Commelina communis* L leaf epidermis. J Exp Bot 34:55–66

Gotow K, Kondo N, Syono K (1982) Effect of CO_2 on volume change of guard cell protoplast from *Vicia faba* L. Plant Cell Physiol 23:1063–1070

Iversen T-H, Johnsson A, Baggerud C (1983) Effect of light and abscisic acid on leaf cell protoplasts. Z Pflanzenphysiol 110:293–300

Jewer PC, Incoll LD, Shaw J (1982) Stomatal responses of *Argenteum* – a mutant of *Pisum sativum* L with readily detachable leaf epidermis. Planta (Berl) 155:146–153

MacRobbie EAC, Lettau J (1980) Potassium contant and aperture in "intact" stomatal and epidermal cells of *Commelina* communis L. J Membr Biol 56:249–256

Meidner H, Mansfield TA (1968) Physiology of stomata. McGraw-Hill, New York

Nobel PS (1983) Biophysical plant physiology and ecology. Freeman, San Francisco

Outlaw WM, Mayne BC, Zenger VE, Manchester J (1981) Presence of both photosystems in guard cells of *Vicia faba*. Plant Physiol (Bethesda) 67:12–16

Raschke K (1975) Stomatal action. Annu Rev Plant Physiol 26:309–310

Raschke K (1979) Movements of stomata. In: Haupt W, Feinleib ME (eds) Encyclopedia of plant physiology, vol 7. Springer, Berlin Heidelberg New York, pp 383–441

Salhani N, Schnabl H, Zimmermann U (1982) The hydraulic conductivity as a criterion for the membrane integrity of protoplasts fused by an electric field pulse. Planta (Berl) 155:140–145

Schnabl H, Bornman CH, Ziegler H (1978) Studies on isolated starch-containing *(Vicia faba)* and starch-deficient *(Allium cepa)* guard cell protoplasts. Planta (Berl) 143:33–40

Weyers JDB , Fitzsimons PJ (1982) The non-osmotic volume of *Commelina* guard cells. Plant Cell Environ 5:417–421

Weyers JDB, Travis AJ (1981) Selection and preparation of leaf epidermis for experiments on stomatal physiology. J Exp Bot 32:837–850

Weyers JDB, Paterson NW, Fitzsimons PJ, Dudley JM (1982) Metabolic inhibitors block ABA-induced stomatal closure. J Exp Bot 33:1270–1278

Weyers JDB, Fitzsimons PJ, Mansey GM, Martin ES (1983) Guard cell protoplasts – aspects of work with an important new research tool. Physiol Plant 58:331–339

Zeiger E (1981) Novel approaches to the biology of stomatal guard cells: protoplast and fluorescence studies. In: Jarvis PG, Mansfield TA (eds) Stomatal physiology. Cambridge Uni Press, Cambridge, pp 103–107

Zeiger E (1983) The biology of stomatal guard cells. Annu Rev Plant Physiol 34:441–475

Zeiger E, Hepler PK (1976) Production of guard cell protoplasts from onion and tobacco. Plant Physiol (Bethesda) 58:492–498

Zeiger E, Hepler PK (1977) Light and stomatal function: blue light stimulates swelling of guard cell protoplasts. Science (Wash DC) 196:887–889

Regulation of Volume Changes in Guard Cell Protoplasts

H. SCHNABL

Introduction

The volume changes of guard cell protoplasts which correlate with stomatal movements (Schnabl et al. 1978) are accompanied by a reversible starch-malate transformation. The process of starch breakdown is paralleled by a concomitant malate synthesis during the K^+-induced swelling, while the disappearance of malate is followed by starch accumulation during shrinkage (Schnabl 1980).

Malate synthesis in stomata has been shown to be catalysed be PEP carboxylase which fixes CO_2 to oxaloacetate (Willmer et al. 1973, Thorpe et al. 1978, Outlaw and Kennedy 1978, Schnabl 1981). The substrate for this reaction, PEP (phosphoenolpyruvate), is derived from the degradation of starch (Outlaw and Manchester 1979). In contrast, malate decarboxylation by malic enzyme has been demonstrated to be coupled to gluconeogenesis resulting in starch accumulation (Schnabl 1980).

Therefore, a further aim was to investigate the key enzymes involved in the synthesis and degradation of metabolites on the basis of the reversible starch-malate transformation in swelling guard cell protoplasts. Determinations of volume kinetics paralleled with enzyme activities of PEP carboxylation and malate decarboxylation, of glycolysis and gluconeogenesis and of oxidative pentose phosphate pathway (OPPP) were used to elucidate the process of volume regulation (Schnabl et al. 1982). The results of enzymic measurements indicate rapid changes of activities during the volume kinetics of GCP (guard cell protoplast) which should be answered by different ways.

On the one hand, the characterization of detailed properties of enzymes, such as PEPC (phosphoenolpyruvate carboxylase) should be used for indicating a possible regulation of enzyme properties in relation to physiological states of protoplasts (Schnabl and Kottmeier 1984). On the other hand, the determination of cellular pool sizes of pyridine and adenine nucleotides as possible regulators of enzymes should offer a valuable answer with respect to the enzyme regulation during the volume changes of GCP (Hampp and Schnabl 1984).

Material and Methods

Seeds of *Vicia faba* L. (Weißkernige Hangdown, purchased from Samenhandlung Pursch, München, FRG) were germinated and grown for 3 weeks in peat moss (TKS I)

The Physiological Properties of Plant Protoplasts
Edited by P. E. Pilet

in a greenhouse with 12 h photoperiod (Osram HQS-lamps, 9 W m^{-2}) at 20°C in light and 17°C in dark. Before starting the experiments, the plants were maintained in a growth chamber in the dark for 24 h at 17°C.

Preparation of Guard Cell Protoplasts

The suspension of guard cell protoplasts was prepared as described by Schnabl et al. (1978) and purified using silicon oil filtration (Schnabl 1981). After storage on ice (in mannitol 0.6 M, lmM $CaCl_2$), the GCP were incubated in KIDA (K^+ iminodiacetate, 10 mM, pH 5.6; mannitol 0.6 M) or in ABA-solutions (10^{-5} M; pH 5.6; mannitol 0.6M). Aliquots were taken at intervals to measure the volume changes microscopically and to determine in parallel samples the enzyme activities after homogenisation for 5 s on ice (Branson sonifier). For determination of PEPC properties, the homogenate was centrifuged at 16,000 g for 15 s in a Beckman microfuge. The supernatant (= crude extract) was either used immediately, frozen at – 80°C or passed through a Sephadex G 25 column (7x0.5 cm, 30μl of the crude extract) and eluted with TRIS/HEPES buffer (50 mM; pH 7.4) under pressure to remove small molecules. The first 600 μml obtained after 60 s were wasted, the subsequent 700 μl taken after 70 s were collected for measuring the enzyme activity (= desalted extract). It required 130 s to produce the desalted extract of enzyme freed to 93–96% from low molecular weight components (Schnabl and Kottmeier 1984). The determination of enzyme activities was calculated on the basis of the number of protoplasts counted by a Fuchs-Rosenthal haemocytometer.

Determination of Enzyme Activities

The PEP carboxylase (EC 4.1.1.31) was carried out spectrophotometrically measuring NADH oxidation by the decrease in extinction of 340 nm (using a Kontron Uvikon model) and radiochemically, measuring the incorporation of $^{14}CO_2$ into oxaloacetate (Schnabl and Kottmeier 1984). Malic enzyme (EC 1.1.1.40) was measured as described by Willmer et al. 1973). PFK (phosphofructokinase) (EC 2.7.1.11) was assayed according to Stitt and ap Rees (1979). Cytoplasmic fructose bisphosphatase (EC 3.1.3.11) was measured as described by Charles and Halliwell (1981). Glucose 6-P-dehydrogenase was assayed according to Ben-Basset and Anderson (1981).

Determination of Adenine and Pyridine Nucleotides

The swelling of GCP was induced by KIDA (10 mM). At different time intervals aliquots were taken, mixed in the ratio 1:1 with 0.2 N HCl, NaOH or 20% $HClO_4$.

The samples were kept on ice for subsequent analysis (NaOH extracts) or frozen in liquid nitrogen and stored at –80°C (HCl and $HClO_4$ extracts) as described by Hampp and Schnabl (1984). Adenine nucleotides were assayed in 10 μml aliquots of neutralized $HClO_4$ extracts directly (ATP) or after enzymatic conversion to ATP by bioluminescence (Hampp et al. 1982). Oxidized and reduced pyridine nucleotides were determined from neutralized HCl or NaOH extracts, brought to pH 6.5 and 7.8 respectively, by enzymatic cycling (Lowry and Passonneau 1972; Hampp et al. 1984).

Results and Discussion

Changes in Enzyme Activities During K^+-Induced Swelling of Guard Cell Protoplasts

The post-harvest purification method employing a mannitol-sucrose-silicon oil phase turnover system (Schnabl 1981) results in the release of about 1.3–2x10^6 highly purified GCP totally free of any contamination by mesophyll or epidermal cell protoplasts using about 40–50 *Vicia faba* plants. In order to quantify the biochemical data, the purity and the viability of GCP have to be checked by determinating the rates of photosynthesis and of K^+-induced swelling. The rate of O_2 evolution normally obtained was between 200–300 μmol mg^{-1} chl h^{-1} using a special O_2 microcuvette (constructed by R. Hedrich and K. Raschke, Göttingen, FRG, unpublished data). Since the GCP are spherical after the isolation process (Fig. 1), the determination of volume increase induced by K^+-treatment is simplified to a great extent by established microscopical methods. The K^+-induced swelling of GCP has been calculated to be between 15 and 20% on the basis of counting about

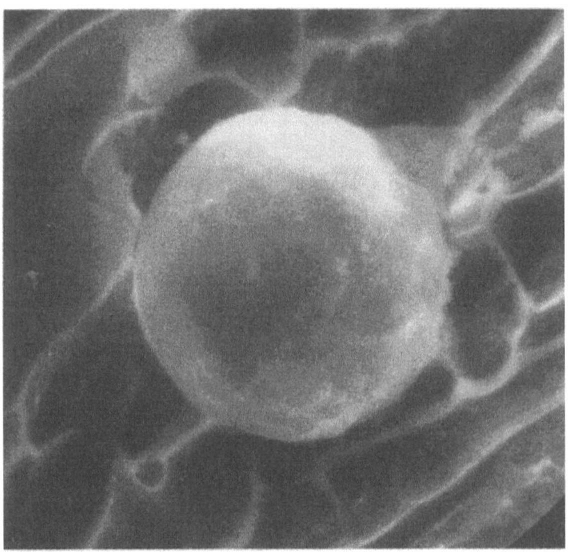

Fig. 1. Scanning electron micrograph of an isolated guard cell protoplast *(Vicia faba)*; 4000:1, done with a Jeol JSM-35 as cryoscan, in an airlock chamber, in which freeze-etching (at -140°C for 5 min) and metal coating (20 s at 20 A) was performed

100 protoplasts over a time range of 10 min. The measurements of enzyme activities were performed only when rates of O_2 evolution and K^+-induced swelling were optimal. In order to investigate the starch malate balance in relation to volume changes of GCP (Schnabl 1980, 1981) in more detail, enzyme activities associated with these processes have been determined. The effects of K^+-ions on the activities of malate-synthesizing and degrading enzymes, PEPC and malic enzyme (both localized in the cytoplasm, Schnabl 1981), have been analyzed in swelling GCP. Moreover, the starch-degrading enzymes – PFK, as the key enzyme of glycolysis, G-6-PDH, that of OPPP – and starch accumulating enzymes – FBPase, as the key enzyme of gluconeogenesis – have been measured during the K^+-incubation of GCP (Schnabl et al. 1982). The rapid increase in PEPC activity during the K^+-treatment of GCP in 30–60 s has been proved to be about 60%, from 650 to 1050 fmol $GCP^{-1} h^{-1}$ (Table 1). Moreover, G-6-PDH has been proved to be another enzyme which is regulated very rapidly within 30–60 s after K^+-addition: in contrast to PEPC, its activity was inhibited from 1050 to 600 fmol $GCP^{-1} h^{-1}$, about 40% (Table 1). Within the following period between 1 and 3 min of K^+-treatment, the activity of G-6-PDH was increased again to an optimum of 850 fmol $GCP^{-1} h^{-1}$ (Fig. 3), whereas PEPC decreased further to 700 fmol $GCP^{-1} h^{-1}$.

From Table 1, it is obvious that other enzymes measured, such as malic enzyme, PFK and FBPase (fructose bisphosphatase) have altered activities after at least 3 min. In contrast to PEPC, the decarboxylating malic enzyme was inhibited from 560 to 350 fmol $GCP^{-1} h^{-1}$ during GCP swelling. A correlation between stimulation and inhibition of carboxylating (PEPC) and decarboxylating steps (malic enzyme) may be involved, i.e. high malate levels and low pH-values are known to inhibit the activity of PEPC, malic enzyme being stimulated. Thus, the cytoplasmic malate pool may be suggested to be controlled by the adversely directed behaviour of PEPC and malic enzyme at different pH-values or at different conditions of feedback inhibition or substrate stimulation.

The pathways of glycolysis and gluconeogenesis are represented by measurements of PFK and FBPase activities. PFK was stimulated by 50% (from 380 to 580 fmol $GCP^{-1} h^{-1}$), whereas the activity of FBPase was reduced by 36% (from 1100 to 700 fmol $GCP^{-1} h^{-1}$) after about 3 min after K^+-addition.

Table 1. Enzyme activities (fmol $GCP^{-1} h^{-1}$) during the K^+-treatment of guard cell protoplasts (0–10 min)

Time of swelling (min)	0	1	3	5	10
GCP (diameter; μm)	15.3 ± 0.8	15.0 ± 1.2	14.5 ± 1.2	16.0 ± 1.0	17.5 ± 1.1
PEP carboxylase	650	1050	700	500	200
Malic enzyme	550	550	560	350	200
PFK	400	390	380	500	580
G-6-PDH	1050	600	850	500	–
FBPase	1000	1000	1100	800	700

The results indicate that changes in enzyme activities appear rapidly during volume increase of GCP and that the enzymic behaviour of PEPC and G-6-PDH differs from that of the other enzymes, measured, with respect to kinetics. On the basis of these data, the question of what triggers the changes in enzyme activities, especially that of PEPC, arises. This is dealt with in the following sections.

Changes in Properties of PEP Carboxylase During K^+-Induced Swelling of Guard Cell Protoplasts

The dramatic and rapid increase in activity (60% within 30 and 60 s after K^+-addition to GCP suspension) enables us to assume that PEPC acts as a potential trigger in volume increase of GCP. PEPC is known to play an important role in providing malic acid yielding protons for H^+/K^+ exchange and anions for balancing positive charges imported during the swelling of GCP. The central role of PEPC at a metabolic branch point (Davies 1979, O'Leary 1982) and its function as a regulatory enzyme in many systems (Smith 1977) suggests a similar role in controlling stomatal movements. Although the overall distribution among different cell types (Outlaw and Kennedy 1978) and the cellular location within the guard cells (Schnabl 1981) is known, the regulation of its activity during volume changes of guard cells is still unresolved.

In order to elucidate the function of PEPC during the volume changes of guard cells, it was necessary to collect kinetic data from the purified enzyme and to understand its properties. Since the activity of PEPC should be determined as a function of PEP- and malate concentrations, the removal of the low molecular weight components from crude extract was a prerequisite. Therefore, a method was worked out in which the endogeneous malate and PEP were removed to 93 to 96%.

Since the stability of this desalted extract decreased in a short time range – 50% of the activity had been lost when stored for 20 min on ice – an accelerated determination was required. After purifying the crude extract within 2.5 min (Schnabl and Kottmeier 1984), PEPC activity was measured in relation to increasing PEP and malate concentrations. The maximum PEPC activity was obtained in the presence of 0.75 -1 mM PEP (1.5 pmol $GCP^{-1} h^{-1}$), and decreased rapidly with higher PEP concentrations to 30%. A K_M (PEP) for PEPC has been calculated to be 0.21 mM (pH 8.3), and a V_{MAX} of 1.54 pmol $GCP^{-1} h^{-1}$. The presence of malate has been shown to increase the K_M from 0.21 to 0.42 (2.5 mM malate) and 0.56 mM (5 mM malate) and to decrease V_{MAX} from 1.54 to 0.71 (2.5 mM malate) and 0.28 pmol $GCP^{-1} h^{-1}$ (5 mM malate) (Schnabl and Kottmeier 1984). In contrast, on the basis of preliminary data, K^+-ions (10 mM) seem to decrease the K_M value from 0.21 to 0.03 mM and to increase V_{MAX} from 1.54 to 9.3 pmol $GCP^{-1} h^{-1}$ when added to the GCP suspension for 10 min (unpublished data). Furthermore, it was demonstrated that K^+-ions were effective in stimulating PEPC activity and in reversing the malate inhibition. Thus, K^+-concentrations of 10 mM increased the V_{MAX} by 40% (without the presence of malate) and relieved the feedback inhibition caused by 1 mM malate (30%) to a value approaching the uninhibited rate (Schnabl and Kottmeier 1984). Hence, K^+ may be regarded as a possible effector of PEPC by inducing a change in K_M, V_{MAX} and malate sensitivity.

A 50% inhibition of PEPC activity was measured with a malate concentration of 0.7 mM. However, the data with respect to malate sensitivity could not be confirmed by repeating the experiments over the range of one year: thus, the results obtained in April gave 50% inhibition by 0.7 mM malate, whereas in July, only 30% and in December 5% were found. The values of malate inhibition seem to vary according to a range of different external and internal parameter of GCP regulating PEPC activity.

The data provide an indication that the properties of PEPC (K_M, V_{MAX} and malate sensitivity) may vary in relation to malate-, PEP and K^+-concentrations which reflect, however, the physiological state of the tissue. Thus, it appears reasonable that PEPC properties are in accordance with the approximate activation state of the enzyme. The data offered support for the hypothesis that the variations in enzyme activities measured during the volume changes of GCP may be related to different enzymic properties dependent on the physiological states of the protoplasts. Further investigations are necessary to confirm this hypothesis.

Changes in the Pool Sizes of Adenine and Pyridine Nucleotides During K^+-Induced Swelling of Guard Cell Protoplasts

Since some of the key enzymes are regulated by either the cellular energy status (PFK) or by the redox state of the pyridine nucleotide system (G-6-PDH), the analysis of cellular pool sizes of pyridine and adenine nucleotides has been used in order to elucidate the enzyme regulation during the K^+-induced swelling of GCP (Hampp and Schnabl 1984). The results show a logical relationship between ATP, NAD^+/NADPH-level and the K^+-induced swelling of GCP. During the first min after K^+-treatment of GCP, the ATP/ADP ratio was decreased from 9 to 6 (Fig. 2), whereas an increase of about 15 was observed during the period between 1 and 3 min. Then, the ratio exhibits again an intensive decrease when GCP begin to swell (after 3 min, Fig 2). These changes may be interpreted on the basis of K^+-import mechanism: When K^+-ions are initially imported, ATP is required, a fact which is reflected by the decrease in the ATP/ADP ratio appearing during the first

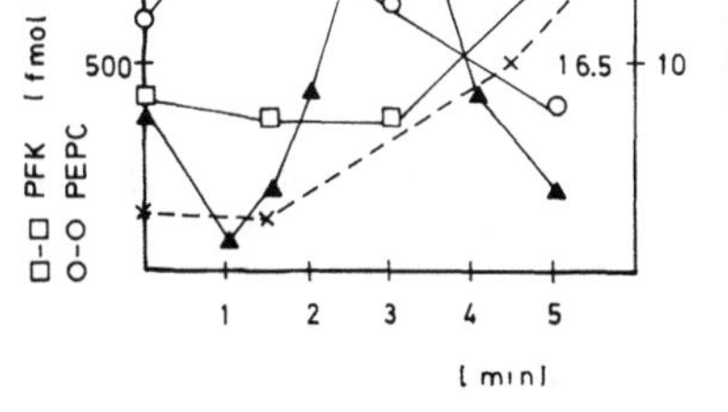

Fig. 2. Levels of ATP/ADP ratio (▲——▲), PEPC (o—— o) and PFK activity (□——□) in swelling guard cell protoplasts (x——x) after different periods of K^+-treatment

minute of K^+-treatment (Fig. 2). The K^+-import is coupled with a stimulation of PEPC activity within 60 s (Fig. 2) (Schnabl et al. 1982) resulting in a subsequent delivery of malate as a possible substrate for respiration. This may be indicated by the higher energy ratio during 1 and 3 min which is followed by an immediate decrease in the ATP/ADP ratio due to an ATPase-dependent K^+-uptake.

The combination of redox pairs and swelling GCP seems to be rather abnormal: The levels of NADH and $NADP^+$ remain nearly constant, whereas those of NAD^+ and NADPH show characteristic oscillations. During the first min of K^+-treatment, the decrease in NAD^+ and the increase in NADPH change into the opposite pattern within the next 1 to 3 min before reversing again after 3 min (Hampp and Schnabl, 1984).

In swelling GCP, some prerequisites have to be maintained with respect to malate synthesis and PEP delivery. Both are known to be provided by starch during swelling. One of the first intermediates of starch breakdown is G-6-P, being metabolized either via glycolysis or via OPPP. The switch of this branch point can be regulated by PFK (via glycolysis) being inhibited by high ATP levels, or by G-6-PDH (via OPPP) being sensitive to high $NADPH/NADP^+$ concentrations.

On the basis of these considerations, K^+-treatment is suggested to activate the conversion of NAD^+ to $NADP^+$ (possibly by an NAD kinase) which is then reduced to NADPH. The increase in NADPH in contrast to $NADP^+$ inhibits G-6-PDH activity after 1 min (from 1050 to 600 fmol $GCP^{-1} h^{-1}$, Fig. 3). During the period between 1 and 2.5 min of K^+-treatment, when ATP is accumulated (Fig. 2), the NADPH/ $NADP^+$ level decreases, G-6-PDH being stimulated (Fig. 3). In the following period, the behaviour of G-6-PDH and $NADPH/NADP^+$ is reversed again.

On the basis of this mechanism, the hexose phosphates are directed from G-6-P, either into the glycolytic pathway at the expense of the function of OPPP or vice versa. When the high ATP level required for K^+-import is reduced after 3 min, PFK is activated (Fig. 2) and the pathway of glycolysis is switched on. In parallel, OPPP is blocked by the increase of $NADPH/NADP^+$ and the reduction in G-6-PDH activity.

The purpose of the alternative switches of glycolysis and OPPP in swelling GCP in relation of adenine and pyridine nucleotides may be highly complex and is difficult to estimate at present. The requirement for PEP and malate during K^+-import may stimulate the glycolytic pathway at the expense of OPPP. If OPPP is activated, it may be an additional source of NADPH and endogeneous CO_2 production. However,

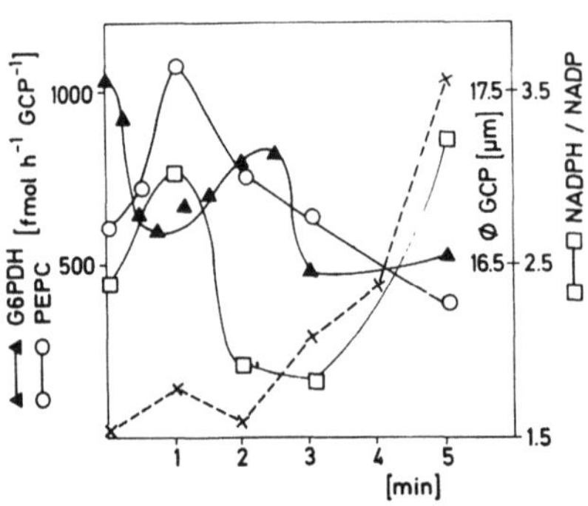

Fig. 3. Levels of $NADPA/NADP^+$ ratio (□——□), PEPC (o——o) and G-6-PDH activity (▲——▲) in swelling guard cell protoplasts (x—— x) after different periods of K^+-treatment

CO_2 has been assumed to lower the membrane permeability for ions (Mansfield et al. 1981), thus inhibiting K^+-import. By this mechanism, it should be possible to combine different ion permeability during stomatal movement with biochemical data with respect to enzyme regulation. More experimental details are required to support this hypothesis.

Concluding Remarks

Both methods examined in order to elucidate the mechanism of enzyme regulation offer interesting aspects in the future. Further kinetic details of purified enzymes involved in starch malate balances are necessary to enable us to describe the enzymic behaviour, to find out possible regulators and effectors and to detect a correlation of enzyme activities with the physiological state of stomatal tissue. The analysis of adenine and pyridine nucleotides should be expanded to investigations of metabolites involved in the glycolytic and oxidative pentose phosphate pathways in order to support the hypothesis with respect to the switch between the two pathways, of glycolysis and OPPP, and to value its consequences.

Acknowledgment. This work was supported by grants from the Deutsche Forschungsgemeinschaft.

References

Ben-Basset D, Anderson LE (1981) Light-induced release of bound glucose-6-phosphate dehydrogenase to the stroma in pea chloroplasts. Plant Physiol (Bethesda) 68:279–283
Charles SA, Halliwell B (1981) Light activation of fructose bisphosphatase in isolated spinach chloroplasts and deactivation by hydrogen peroxide. Planta (Berl) 151:242–246
Davies DD (1979) The central role of phosphoenolpyruvate in plant metabolism. Annu Rev Plant Physiol 30:131–158
Hampp R, Schnabl H (1984) Adenine and pyridine nucleotide status of isolated *Vicia* guard cell protoplasts during K^+-induced swelling. Plant Cell Physiol 25(7):1233–1239
Hampp R, Goller M, Ziegler H (1982) Adenylate levels, energy charge, and phosphorylation potential during dark/light transition in chloroplasts, mitochondria and cytosol of mesophyll protoplasts from *Avena sativa* L. Planta (Berl) 62:448–455
Hampp R, Goller M, Füllgraf H (1984) Determination of compartmented metabolite pools by a combination of rapid fractionation of oat mesophyll protoplasts and enzymatic cycling. Plant Physiol (Bethesda) 75:1017–1021
Lowry OH, Passonneau JV (1972) A flexible system of enzymatic analysis. Academic, New York
Mansfield TA, Travis AJ, Jarvis RG (1981) Responses to light and carbon dioxide. In: Jarvis PG, Mansfield TA (eds) Stomatal physiology. Cambridge Uni Press, Cambridge, pp 119–135
O'Leary MH (1982) Phosphoenolpyruvate carboxylase: an enzymologist's view. Annu Rev Plant Physiol: 298–315
Outlaw WH, Kennedy J (1978) Enzymic and substrate basis for the anaploertic step in guard cells. Plant Physiol (Bethesda) 62:648–652
Outlaw WH, Manchester J (1979) Guard cell starch concentration quantitatively related to stomatal aperture. Plant Physiol (Bethesda) 64:79–82

Schnabl H (1980) CO_2 and malate metabolism in starch-containing and starch-lacking guard cell protoplasts. Planta (Berl) 149:52–58

Schnabl H (1981) The compartmentation of carboxylating and decarboxylating enzymes in guard cell protoplasts. Planta (Berl) 152:307–313

Schnabl H, Kottmeier C (1984) Properties of phosphoenolpyruvate carboxylase in desalted extracts from isolated guard cell protoplasts. Planta (Berl) 162:220–225

Schnabl J, Bornman CH, Ziegler H (1978) Studies in isolated starch-containing and starch-deficient guard cell protoplasts. Planta (Berl) 143:33–39

Schnabl H, Elbert C, Krämer G (1982) The regulation of the starch-malate balances during volume changes of guard cell protoplasts. J Exp Bot 33:996–1003

Stitt M, ap Rees T (1979) Capacities of pea chloroplasts to catalyse the oxidative pentose phosphate pathway and glycolysis. Phytochemistry (oxf) 18:1905–1911

Smith TE (1977) Escherichia coli phosphoenolpyruvate carboxylase: studies on the mechanism of multiple allosteric interactions. Arch Biochem Biophys 183:538–552

Thorpe N, Brady CJ, Milthorpe FL (1978) Stomatal metabolism: primary carboxylation and enzyme activities. Aust J Plant Physiol 5:485–493

Willmer CM, Pallas JE, Black CC (1973) Carbon dioxide metabolism in leaf epidermal tissue. Plant Physiol (Bethesda) 52:448–452

Wall Regeneration in Protoplasts of Higher Plants

G. FRANZ and W. BLASCHEK

Introduction

Isolated higher plant protoplasts will regenerate a new cell wall if cultured in suitable media. After cell-wall formation the cells have the potential for cell division, callus formation and the regeneration of complete plants.

In forming a new cell wall from point zero, protoplasts should be an ideal tool to study the process of biosynthesis and secretion of cell-wall material, where our knowledge is still far from being complete. Furthermore, microscopic and electron microscopic observations of the surface of the naked protoplasts are facilitated. It was quite natural, that plant biochemists, physiologists and cytologists became interested in this very promising system about 15 years ago. However, if one critically overlooks the results obtained in this field, it becomes evident that protoplasts are very useful for some specific questions about cell-wall formation, but they also present some intrinsic problems mainly caused by the artificial process of protoplasts isolation and culture.

The aim of this chapter is to report our present knowledge about wall formation obtained with the protoplast system. It will be discussed especially with respect to the completeness of the regenerated wall, the aspects of wound response and the possible perturbation of wall synthesis caused by protoplast isolation; further, the biosynthesis of cellulose microfibrils during wall regeneration and finally a comparison of the regenerated protoplast wall with the native cell wall. Some of these topics have been discussed in previous reviews as, for example, by Cocking 1970, 1972, Willison 1976, Fowke 1978, Fowke and Gamborg 1980, Galun 1981, Willison and Klein 1982, Meyer and Herth 1982.

Early Events and Time Course of Wall Formation with Protoplasts

Protoplasts have been isolated from a great variety of higher plants, and more recently also cell suspension cultures have been utilized in order to overcome sterility problems (Reinert and Bajaj 1977). The most frequently used sources which are actually taken for studies on cell-wall regeneration, are tobacco mesophyll, sycamore and carrots cells in culture, maize root tissue, soybean and *Vinca rosea* tissue and cell cultures. Based on this variety of objects, the results obtained so far are difficult to compare.

The Physiological Properties of Plant Protoplasts
Edited by P. E. Pilet

Mostly more or less complete cell walls are regenerated within one week of culture (Willison 1976). The failure of some protoplasts (e.g. those of some grasses) to form cell walls in an adequate time, might be the result of inadequate culture conditions rather than any inherent quality of the protoplasts. The rate of wall synthesis depends on the source of protoplasts. Protoplasts isolated from meristematic cells or from rapidly growing tissue cultures start immediately after removal of cell-wall-degrading enzymes with cell-wall regeneration, whereas mesophyll protoplasts exhibit lag periods up to 1 day. Soybean (Williamson et al. 1976) and *Vicia* protoplasts (Willisamson et al. 1977) from suspension cultures generated a new cell wall imediately after isolation. Protoplasts from young leaves of *Allium* initiated microfibrils within 1.5 h compared to 4 h for protoplasts from mature leaves (Prat and Williamson 1976). For tobacco mesophyll protoplasts, lag periods from 3 to 24 h have been reported (Willison and Cocking 1975, Burgess and Linstead 1979, Blaschek et al. 1981). If regenerated mesophyll protoplasts of tobacco were treated with cellulase, the newly formed wall was digested (Burgess et al. 1978). These protoplasts produced new fibres visible by scanning electron microscopy within minutes of their return to culture medium. Most probably the process of protoplast isolation with more or less impure enzyme preparations does not disturb the wall formation, but the situation of differentiation of the starting tissue seems to influence the rapidity of wall regeneration. For this reason also some strange cell walls may relate to the origin of the plant tissue used for protoplast isolation, as it was observed for tomato locule protoplasts forming a multilamellar cell wall before microfibril synthesis (Gregory and Cocking 1965, Willison and Cocking 1972).

One can, of course, imagine that the composition of the culture medium also might influence the time course and composition of the regenerating protoplast wall. This, however, is difficult to examine, because exactly defined media have to be used for a successful culture in the sensitive protoplast system. In investigations concerning the relation of cell wall formation as a prerequisite for cell division, tobacco protoplasts were cultured in saline media (Herth and Meyer 1977). An altered, non-rigid pseudo wall was formed compared to protoplasts cultured in media, where the osmotic value was maintained by the traditionally used sugar alcohols. The events of mitosis, cytokinesis and cell-wall formation, under certain conditions, do not seem to be necessarily connected, although in most cases a rather strict coupling of wall formation and cell division seems to exist (Meyer and Herth 1978, 1982). Cytoplasmic protrusions or budding in protoplasts is another example for local changes in the cell-wall composition of protoplasts, possibly due to the loss of matrix components (Hanke and Northcote 1974). Budding is frequent mainly under suboptimal culture conditions (Willison 1976).

Like suspension cultured cells, protoplasts secrete polysaccharides into the extracellular milieu (Hanke and Northcote 1974, Takeuchi and Komamine 1978, Blaschek et al. 1981a). During the early stages of wall formation, much of the wall polysaccharides disintegrate from the surface of the cells. With prolonged culture the relative amount of these extracellular polysaccharides decreases compared to the amount of regenerated cell-wall material, as was calculated after GLC-analysis of the different polysaccharides formed by tobacco protoplasts (Blaschek et al. 1981) (Fig. 1). This loss of wall material must be taken into account, if comparisons between

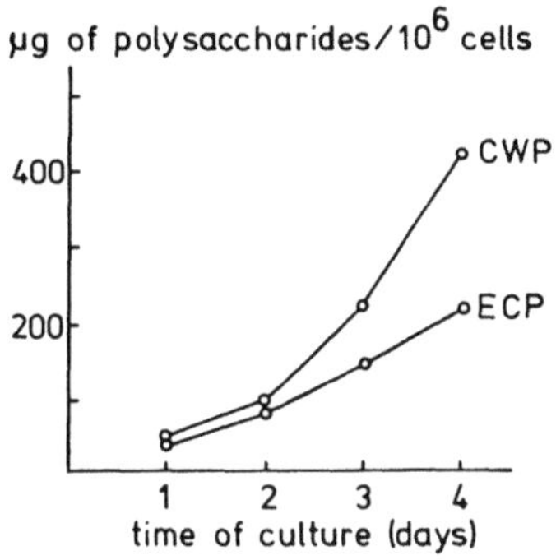

Fig. 1. Time course of cell-wall polysaccharides *(CWP)* and extracellular polysaccharides *(ECP)* synthesis in regenerating tobacco protoplasts. (Blaschek et al. 1981)

protoplast walls and the walls of the native tissues are made. When *Vinca rosea* protoplasts were cultured on agar instead of liquid media, their regenerated walls after some days of culture showed significantly higher amounts of uronic acids, arabinose and galactose, indicating a prevention of the loss of pectic and hemicellulosic material (Takeuchi and Komamine 1982). Together with the accelerated wall regeneration, cell divison was stimulated. This suggests an important role of the physical environment of the protoplasts for the assembly of wall material, which must be considered in investigations about the linkage of newly synthesized wall material to the existing network of polysaccharides surrounding already regenerating protoplasts. Enzymes engaged in this process and secreted by the protoplasts might be lost into the medium at early stages of wall regeneration.

Cytological Aspects

The ultrastructure and the cytological events of cell-wall formation have been studied by light microscopy and different electron microscope techniques; the advantages and disadvantages of these techniques have been summarized by Willison and Klein (1982). Most of the observations were concentrated on the nature and origin of the fibrillar material which appeared at the surface of the protoplasts after the onset of culture. The surfaces of freshly isolated protoplasts usually are devoid of fibrillar material. After various lag periods, first loosely organized microfibrils become visible, and with prolonged culture time, the quantitiy of microfibrils increases rapidly, forming a dense network surrounding the protoplasts (Willison and Klein 1982).

In general it is assumed that these microfibrils are composed of cellulose. But other components besides cellulose may appear as fibrils in plant cell walls (Preston 1974; Herth et al. 1974, Herth and Meyer 1977). In combined ultrastructural and biochemical investigations, the cellulosic nature of the main part of the microfibrils formed by cultured tobacco (Herth and Meyer 1977) and soybean (Klein and Delmer 1979) protoplasts has been investigated. The X-ray diffraction analysis indicated that either non-crystalline chemically different material is associated with part of the native fibrils or the fibrils are not uniformly crystalline. The fibrillar material had dimensions in the range of subelementary fibrils. Whether these results are valid for the fibrillar structures of all higher plant protoplasts, remains uncertain.

The size of an elementary cellulose fibril is believed to be composed of about 32 glucose chains with an average diameter of about 4 nm (Mühlethaler 1967). The microfibrils formed by regenerating protoplasts varied in diameter from 12 to 150 Å (Willison and Cocking 1972, Prat and Williamson 1976, Herth and Meyer 1977, Burgess and Linstead, 1979) and bundles of microfibrils with diameters up to 400 Å have been observed (Willison and Grout 1978, Burgess and Linstead 1979). The chain length of the early formed microfibrils in tobacco protoplasts was about 60 nm (Burgess and Linstead 1979), which would correspond to a degree of polymerization of about 120, provided that unbranched, unfolded glucose chains form the microfibril. These findings will be discussed later in connection with data obtained by chemical analysis. It is assumed that with an increase of the quantity of microfibrils also the length of microfibrils increases, which is, however, difficult to estimate in the dense fibrillar tangle formed with prolonged culture. Most probably branch points are existent in microfibrils connecting one microfibril with another.

Current models for the synthesis and deposition of cellulose microfibrils by plant cells involve the presence of mobile enzyme complexes in the plasma membrane (Staehelin and Giddings 1982). There exists no evidence from studies with protoplasts for an intracellular assembly of microfibrils. Such complexes of particles associated with microfibril ends have been observed in algal cells and secondary plants cell walls (Brown and Montezinos 1976, Mueller et al. 1976, Mueller and Brown 1980, Giddings et al. 1980). In *Spirogyra* the complexes consisted of rosettes with 5–6 particles with a diameter of 25 nm (Herth 1983). The rosettes were supposed to be involved in the formation of elementary cellulose fibrils, whereby one particle is synthesizing about 6 glucan chains. In *Skimmia japonica* protoplasts, regular arrays of particular systems were observed on the plasmalemma (Robenek and Peveling 1977). These findings, however, were also interpreted as artifacts induced by glycerol treatment or simply by plasmolysis (Davey and Mathias 1979). It remains uncertain whether any type of particle array is essential to microfibril formation in protoplasts, since several types of protoplasts are continously forming cellulose in the absence of ordered particular structures (Willison and Grout 1978). If a lot of particles are visible in or on the plasmalemma, which is covered by a network of microfibrils, it is mainly a statistic problem, how much microfibril tips (or imprints of them) are ending at a particle. It seems very difficult to prove, that this very particle is engaged in the synthesis of a special microfibril.

Recent reports suggest that coated vesicles might play a role in the process of endocytosis and exocytosis by plants (Bonnet and Newcomb 1966, Franke and Herth 1974, Ryser 1979, Van der Valk and Fowke 1981, Fowke et al. 1983). It is now well established that matrix components of the wall are secreted via the Golgi apparatus and perhaps also the ER is involved in the synthesis of wall material or precursors of it (Willison 1981). Coated vesicles have been observed to be present on the inner surface of tobacco protoplast membranes (Van der Valk and Fowke 1981, Fowke et al. 1983) and it is suggested that they are produced by the Golgi apparatus, then fuse with the plasma membrane and release cell-wall material and/or enzymes, their coats remaining attached to the inner surface of the plasmalemma. On the other hand, coated vesicles might function by recycling plasma membrane pieces and/or enzymes engaged in cell-wall formation like the cellulose synthase.

The role of microtubules in the orientation of cellulose microfibrils has been studied with protoplasts as well. It has been proposed, that the cellulose-synthesizing complex moves in the plane of the plasma membrane. Either microtubules and/or membrane flow might in some way influence the direction of this movement (Brown and Montezinos 1976, Staehelin and Giddings 1982, Mueller and Brown 1982). Protoplasts provide a useful tool for investigating the inner surface of plant plasmalemmata and the associated organelles which might be involved in the synthesis of wall material (Doohan and Palevitz 1980, Van der Valk et al. 1980, Lloyd et al. 1980, Fowke et al. 1983). In protoplasts, however, cellulose microfibrils are randomly arranged and therefore with this system it will be difficult to prove a direct connection between microfibril and microtubule orientation.

Kinetics of Cellulose Formation

As has already been pointed out, chemical methods have to be employed in order to demonstrate clearly the cellulosic nature of the wall material regenerated by protoplasts. Mainly three methods of hydrolyzing cell-wall polysaccharides withhout degrading cirstalline cellulose have been used in such investigations:

1. boiling the wall material in acetic-nitric reagent (Updegraff 1969)
2. hydrolysis with 2 M TFA (Ring and Selvendran 1978, Mankarios et al. 1979)
3. the nitration procedure for chain length measurements of cellulose (Spencer and MacLachlan 1972).

In most cases the wall polysaccharides of regenerating protoplasts are labelled by feeding ^{14}C-glucose in the medium. Labelled material not degraded by one of the methods mentioned above is taken as an indication for cellulose synthesis.

The amount of cellulose produced in the first days of wall regeneration estimated for soybean, *Vinca rosea,* carrot, and tobacco protoplasts was rather low (about 5% compared to maximal 20% of total cell-wall material after prolonged culture time) and increased in parallel with the amount of non-cellulosic wall material (Klein et al. 1981, Takeuchi and Komamine 1981, Asamizu et al. 1977, Herth and Meyer 1977, Blaschek et al. 1981, 1982). The cellulosic nature of this wall material was confirmed by methylation analysis for soybean and tobacco protoplasts (Klein et al. 1981, Blaschek et al. 1981).

The fluorescent brightener Calcofluor White ST, if added to cell-walls, is connected through H-bonding with β-lined glucan chains (Maeda and Ishida 1967). Calcofluor White was shown to bind with the β-1.4-glucan chains of cellulose, as they are synthesized by the respective plasmalemma bound enzymes, thereby separating the process of polymerization from crystallization (Haigler et al. 1980, Benziman et al. 1980, Quader 1981, Roberts et al. 1982). Calcofluor White was also used for a fluorimetric quantitation of cellulose biosynthesis in tobacco protoplasts

(Galbraith 1981). From these data it was concluded that a protoplast will require 6 to 15 days before the amount of cellulose formed is comparable to that of a mature cell. This implies that the first cell divisions in the protoplast system occur before the cell-wall formation is complete.

First investigations concerning the molecular weight or the degree of polymerization (DP) of cellulose synthesized by regenerating protoplasts were performed with carrot (Asamizu et al. 1977). In the early stage of wall regeneration nearly 70% of the newly formed cellulose had a DP of less than 200. The proportion of higher DP-cellulose (DP> 1000) gradually increased during the cultivation of protoplasts reaching about 50% after 9 days of culture. It remained uncertain whether the short-chain cellulose molecules were still growing or already were assembled in microfibrils. Native primary cell-wall cellulose normally exhibits a broad DP-spectrum in the range of 2000–6000, but also contains glucan chains with low DP (< 500) (Schulz and Marx 1954, Spencer and MacLachlan 1972, Asamizu et al. 1977, Delmer 1977, Blaschek et al. 1982).

At the onset of cell-wall formation in tobacco protoplasts, low molecular weight cellulose (DP< 500) was predominant (Blaschek et al. 1982). With continuous culture time longer cellulose chains up to a DP of 3000 were synthesized. Mainly two classes of chain length were found, either with low (< 500) or high (2000–3000) DP-values (Fig. 2). The relative proportions of these two groups of macromolecules showed only slight variations with culture time, indicating that the short chain cellulose only in part is a precursor for the high DP-cellulose. These findings were confirmed by pulse-chase experiments demonstrating clearly the contribution of low molecular weight glucan chains to the cellulose microfibrils produced by regenerating protoplasts.

The length of the early formed microfibrils of about 60 nm estimated by electron microscopy at the onset of wall regeneration in protoplasts (Burgess and Linstead 1979) is consistent with the occurence of short chain cellulose mentioned above and might also explain the diffuse X-ray diffraction patterns observed with this cellulosic material (Herth and Meyer 1977). The occurrence of more amorphous regions along primary wall cellulose fibrils and a reduced microfibril length might help to expand the primary wall during cell growth.

time of protoplast culture	amount of cellulose/10^6 cells (cpm)	DP-distribution of cellulose		
		< 500	500-2000	2000-3000
1 d	13 539	98 %	2 %	-
2 d	70 638	64 %	-	36 %
3 d	102 595	68 %	1 %	31 %
4 d	210 342	72 %	-	28 %

Fig. 2. Degree of polymerization (DP)-distribution pattern of cellulose in regenerating protoplasts. Tobacco protoplasts were cultured in ^{14}C-glucose containing medium and harvested after 1–4d. The cellulose was nitrated and cellulose nitrate was fractionated according to differences in DP. The DP-distribution is expressed as percent of total synthesized cellulose. (Blaschek et al. 1982)

Chemical Analysis of Non-Cellulosic Wall Material

Comprehensive analysis of cell walls from many sources have been performed in order to get a conception of the three-dimensional organisation of the native primary wall, and a valuable model has been developed by Albersheim and his collaborators (McNeil et al. 1979). Characterisation of the process of wall regeneration by isolated protoplasts may contribute useful insights into the complex assembly process of wall formation as the cell wall is newly built up. One great disadvantage of the protoplast system, however, was already mentioned: the loss of some wall material into the culture medium.

This phenomenon was first described for onion root protoplasts (Prat and Roland 1971), which released callose and pectic substances to the medium. Analysis of soybean protoplasts after feeding labelled glucose showed that up to 40 h of culture the regenerated wall was devoid of pectin, which was found in the medium (Hanke and Northcote 1974). Some pectins apparently were not synthesized by the protoplasts and those which were formed were not assembled into the developing wall.

In carrot protoplasts a 5-h pulse label after varying periods of culture led to a continuous change of the distribution pattern of radioactivity in the pectin-, hemicellulose- and cellulose fractions (Asamizu and Nishi 1980). Most of the radioactivity appeared in extracellular polysaccharides (ECP). The neutral sugar composition of the regenerated wall and of the ECP was distinct from that of the suspension cultured cells (Fig. 3).

In similar investigations on soybean protoplasts, the cell wall and the ECP were analysed after 3 h of culture (Klein et al. 1981). The sugar pattern in these two polysaccharide fractions was quite different, indicating that at least some other polymers were forming the new cell wall than those released into the medium. The composition of the protoplast polysaccharides was different from that of suspension cultured soybean cells. Methylation analysis showed that callose was formed by the protoplasts and partially released into the medium, whereas the native walls were free of 1.3-linked glucans. The synthesis of callose may be induced as a wound response by the act of protoplast isolation.

The regenerated cell wall of *Vinca rosea* protoplasts was mainly composed of non-cellulosic glucans having 1.3- and 1.4-linkages, as was determined by solubility characteristics and methylation analysis (Takeuchi and Komamine 1978). Most of the pectic and hemicellulosic components were released into the medium. The ECP of protoplasts was similar to that of suspension-cultured cells consisting mainly of polyuronide and 3.6-linked arabinogalactan. When the cell wall of *Vinca rosea* protoplasts after 5 days of culture was fractionated according to solubility in EDTA- and KOH-solutions (Takeuchi and Komamine 1981), only about 7% cellulose could be obtained, whereas most glucans were soluble in KOH. They consisted of unbranched β-1.3-glucan chains with a DP of about 290.

Callose, normally occurring in minor amounts in cell walls (Huwyler et al. 1978, Maltby et al. 1979) is rapidly synthesized after wounding of plant cells and is found in a variety of specialized tissues (Currier 1957, Clarke and Stone 1963). Most likely β-1.3-glucans are also synthesized in cell plates formed during cytokinesis

Cell wall material from	*Vinca rosea*[1] (w%)			Carrot[2] (w%)		Tobacco[3] (mol%)		Maize[4] (w%)	
	Susp. cult. cells	Protopl.	Protopl.	Susp. cult. cells	Protopl.	Leaf	Protopl.	Root	Protopl.
Culture conditions		7 days liquid	7 days agar		7 days liquid		4 days liquid		7 days liquid
Sugar composition									
Rhamnose + fucose	2	tr.	1	13	7	4	6	5	1
Arabinose	18	2	15	33	8	9	6	13	14
Xylose	17	0.5	5	11	6	7	1	16	34
Mannose	1	tr.	tr.	2	6	4	5	15	25
Galactose	13	3	15	12	20	11	10	7	13
Glucose	28	91	37	22	44	54	68	38	10
Uronic acids	47	4	28	9	6	11	6	6	4

Fig. 3. Comparative sugar composition of cell walls from regenerating protoplasts and their corresponding parent tissue.

The data were taken from:

[1] Takeuchi and Komamine 1978, 1982

[2] Asamizu and Nishi 1980

[3] Blaschek et al. 1981

[4] Pilet et al. 1984

(Fulcher et al. 1976). The formation of β-1.3-glucans in regenerating protoplasts may be either a wound response or a de novo synthesis effect like in cell plate formation during cell division.

As already mentioned, culture conditions may influence cell-wall regeneration in protoplasts. 2.4-dichlorophenoxyacetic acid (2.4-D) was reported to stimulate cell-wall regeneration in tobacco mesophyll protoplasts (Takebe and Nagata 1973), whereas in Convolvulus protoplasts it had no effect on cell-wall regeneration (Horine and Ruesink 1972). In tobacco leaf cells, plasmolysis changed the sugar composition of the cell-wall (Boffey and Northcote 1975). The cell-wall composition in cultured tobacco cells was influenced by the phytohormone composition of the medium and by osmolarity (Blaschek and Franz 1983). Mannitol, 2.4-D and 6-benzylamino-purine at various concentrations did not significantly affect the sugar composition of regenerated walls from *Vinca rosea* protoplasts (Takeuchi and Komamine 1982). These in part contradictary results require more detailed investigation on the significance of culture conditions for cell-wall formation.

The cell-wall of regenerating tobacco mesophyll protoplasts was composed mainly of non-cellulosic polysaccharides with glucose predominating (65%) and a content of only about 5% cellulose (Blaschek et al. 1981a). Compared to mesophyll tissue and suspension or callus cultured cells, a quite different cell-wall was formed (Fig. 3). Methylation analysis and enzymatic degradation showed the presence of about 65% 1.3- and 35% 1.4- linkages in the non-cellulosic glucans (Blaschek et al. 1981b). A fractionation of the cell-wall material similar to that mentioned for *Vinca rosea* (Takeuchi and Komamine 1981) demonstrated that a decreasing solubility of glucans was accompanied by increasing amounts of 1.4-linkages (Blaschek, unpublished results) indicating the occurrence of mixed linked 1.3-1.4-glucans besides callose and cellulose. Arabinose, galactose and uronic acids were the main constituents of the ECP-fraction, which was similar to that of suspension cultured tobacco cells. This corresponds to the polyuronides and 3.6-arabinogalactan reported for the ECP of *Vinca rosea* protoplasts (Takeuchi and Komamine 1978).

The possibility of myo-inositol being involved in the formation of uronides and polyoses was studied in tobacco protoplasts by feeding labeled myo-inositol or unlabeled myo-inositol together with labeled glucose (Blaschek et al. 1981). It could be demonstrated, that the myo-inositol pathway (Loewus et al. 1973) was operative in cell-wall regenerating tobacco protoplasts.

The cell-wall composition of maize root cells from the elongation zone was compared with regenerating protoplasts isolated from these cells (Pilet et al. 1984) (Fig. 3). Again, both wall materials proved to be different. The protoplast wall had a low cellulose content, increased levels of mannose, xylose and galactose and lower amounts of uronic acids and rhamnose. Most likely pectic material was lost into the medium. Only small amounts of non-cellulosic glucans, but larger quantities of glucomannans were inserted into the cell walls of the regenerating protoplasts. It remains uncertain whether this is a common and remarkable difference between cell walls formed by dicot and monocot protoplasts.

Conclusion

All these results give the impression that for some reason the cell-wall of regenerating protoplasts does not resemble the genuine primary wall of naturally growing plant tissue. In a plant already the cells of very young tissue are programmed for a specific differentiation, which might be reflected in the cell-wall composition, as it is well known that differentiation causes changes in the cell-wall composition (Thornber and Northcote 1962, Nevins et al. 1968, Franz 1972, Huwyler et al. 1979, Nishitani and Masuda 1979, Van Holst et al. 1980). Protoplasts, however, are certainly in a state of dedifferentiation. Moreover, protoplast walls might show the composition of a very young cell-wall, which is perhaps most easily comparable with the cell-wall formed in the cell plate during cytokinesis. Under this aspect even cell walls isolated from very young tissue represent a mixture of very young and older primary cell walls. Unfortunately, however, some wall material in the case of protoplasts is lost to the medium. If the type, amount, and sequence of polysaccharide synthesis for a wall is controlled by a sort of feedback mechanism by the cell, the permanent loss of wall material to the medium may disturb this control and result in an altered, unusual cell-wall composition.

Acknowledgment. This work was supported by the Deutsche Forschungsgemeinschaft SFB 43, A 1

References

Asamizu T, Nishi A (1980) Regenerated cell-wall of carrot protoplasts isolated from suspension cultured cells. Physiol Plant 48:207–212

Asmizu T, Tanaka K, Takebe J, Nishi A (1977) Change in molecular size of cellulose during regeneration of cell-wall on carrot protoplasts. Physiol Plant 40:215–218

Benziman M, Haigler CH, Brown RM, White AR, Cooper KM (1980) Cellulose biogenesis: polymerization and crystallization are coupled processes in *Acetobacter xylinum.* Proc Nat Acad Sci USA 77: 6678–6682

Blaschek W, Franz G (1983) Influence of growth conditions on the composition of cell-wall polysaccharides from cultured tobacco cells. Plant Cell Rep 2:257–260

Blaschek W, Haaß D, Koehler H, Franz G (1981a) Cell-wall regeneration by *Nicotiana tabacum* protoplasts: chemical and biochemical aspects. Plant Sci Lett 22:47–57

Blaschek W, Haaß D, Koehler H, Franz G (1981b) Cell-wall regeneration by *Nicotiana tabacum* protoplasts: chemical and biochmical aspects. In: Robinson DG, Quader H (eds) Cell walls 81. Wissenschaftliche Verlagsgesellschaft, Stuttgart, pp 109–118

Blaschek W, Koehler H, Semler U, Franz G (1982) Molecular weight distribution of cellulose in primary cell walls. Planta (Berl) 154:550–555

Boffey SA, Northcote DH (1975) Pectin synthesis during the wall regeneration of plasmolyzed tobacco leaf cells. Biochem J 150:433–440

Bonnet HT, Newcomb EH (1966) Coated vesicles and other cytoplasmic components of growing root hairs of radish. Protoplasma 62:59–75

Brown RM, Montezinos D (1976) Cellulose microfibrils: visualization of biosynthetic and orienting complexes in assoziation with the plasma membrane. Proc Natl Acad Sci USA 73:143–147

Burgess J, Linstead PJ (1979) Structure and association of wall fibrils produced by regenerating tobacco protoplasts. Planta (Berl) 146:203–210

Burgess J, Linstead PJ, Bonsall VE (1978) Observation on the time course of wall development at the surface of isolated protoplasts. Planta (Berl) 139:85–91
Clarke AE, Stone BA (1963) Chemistry and biochemistry of β-1.3-Glucans. Rev Pure Appl Chem 13:134–156
Cocking EC (1970) Virus uptake, cell-wall regeneration, and virus multiplication in isolated plant protoplasts. Int Rev Cytol 28:89–124
Cocking EC (1972) Plant cell protoplasts – isolation and development. Annu Rev Plant Physiol 23:29–50
Currier HB (1957) Callose substance in plant cell. Am J Bot 44:478–488
Davey MR, Mathias RJ (1979) Close-packing of plasma membrane particles during wall regeneration by isolated higher plant protoplasts – fact or artefact. Protoplasma 100:85–99
Delmer DP (1977) The biosynthesis of cellulose and other plant cell-wall polysaccharides. In: Loewus FA, Runeckles VC (eds) Recent Adv Phytochem, vol 11. Plenum, New York, pp 45–77
Doohan ME, Palevitz BA (1980) Microtubules and coated vesicles in guard cell protoplasts of *Allium cepa* L. Planta (Berl) 149:389–401
Fowke LC (1978) Ultrastructure of isolated and cultured protoplasts. In: Thorpe TA (ed) Frontiers of plant tissue culture. Proc 4th Int Congr Plant Tissue Cell Culture, Calgary, Alberta, pp 223–233
Fowke LC, Gamborg OL (1980) Applications of protoplasts to the study of plant cells. Int Rev Cytol 68:9–51
Fowke LC, Rennie PJ, Constabel F (1983) Organelles associated with the plasma membrane of tobacco leaf protoplasts. Plant Cell Rep 2:292–295
Franke WW, Herth W (1974) Morphological evidence for de novo formation of plasma membrane from coated vesicles in exponentially growing cultured plant cells. Exp Cell Res 89:447–451
Franz G. (1972) Polysaccharid-Metabolismus in den Zellwänden wachsender Keimlinge von *Phaseolus aureus.* Planta (Berl) 102:334–347
Fulcher RG, McCully ME, Setterfield G, Sutherland J (1976) β-1.3-Glucan may be associated with cell plate formation during cytokinesis. Can J Bot 54:539–542
Galbraith DW (1981) Microfluorimetric quantitation of cellulose biosynthesis by plant protoplasts using Calcofluor White. Physiol Plant 53:111–116
Galun E (1981) Plant protoplasts as physiological tools. Ann Rev Plant Physiol 32:237–266
Giddings TH Jr, Brower DL, Staehelin LA (1980) Visualization of particle complexes in the plasma membrane of *Micrasterias denticulata* associated with the formation of cellulose fibrils in primary and secondary walls. J Cell Biol 84:327–339
Gregory DW, Cocking EC (1965) The large-scale isolation of protoplasts from immature tomato fruit. J Ce.. Biol 24:143–146
Haigler CH, Brown RM Jr, Benziman M (1980) Calcofluor White ST alters the in vivo assembly of cellulose microfibrils. Science (Wash DC) 210:903–905
Hanke DE, Northcote DH (1974) Cell-wall formation by soybean callus protoplasts. J Cell Sci 14:29–50
Herth W (1983) Arrays of plasma-membrane "rosettes" involved in cellulose microfibril formation of Spirogyra. Planta (Berl) 159:347–356
Herth W, Meyer Y (1977) Ultrastructural and chemical analysis of the wall fibrils synthesized by tobacco mesophyll protoplasts. Biol Cell 30:33–40
Herth W, Franke WW, Bittiger H, Kuppel A, Keilich G (1974) Alkali resistant fibrils of β-1.3- and β-1.4-glucans: structural polysaccharides in the pollen tube wall of *Lilium longiflorum.* Cytobiologie 9:33–40
Horine RK, Ruesink AW (1972) Cell-wall regeneration around protoplasts isolated from Convolulus tissue culture. Plant Physiol 50:438–445
Huwyler HR, Franz G, Meier H (1978) β-1.3-glucans in the cell walls of cotton fibres (*gossypium arboreum* L.) Plant Sci Lett 12:55–62
Huwyler HR, Franz G, Meier H (1979) Changes in the composition of cotton fibre cell walls during development. Planta (Berl) 146:635–642

Klein AS, Delmer DP (1979) Characterization of the regenerating wall of soybean protoplasts. Plant Physiol Suppl. (Bethesda) 63:51
Klein AS, Montezions D, Delmer DP (1981) Cellulose and 1.3-glucan synthesis during the early stages of wall regeneration in soybean protoplasts. Planta (Berl) 152:105–114
Lloyd CW, Slabas AR, Powell AJ, Lowe SB (1980) Microtubules, protoplasts and plant cell shape. Planta (Berl) 147:500–506
Loewus FA, Chen M, Loewus MW (1973) The myo-inosital oxidation pathway to cell-wall polysaccharides. In: Loewus FA (ed) Biogenesis of plant cell-wall polysaccharides. Academic Press, New York, pp 1–27
Maeda H, Ishida N (1967) Specificity of binding of hexapyranosyl polysaccharides with fluorescent brightener. J Biochem (Tokyo) 62:276–278
Maltby D, Carpita NC, Montezinos D, Kulow C, Delmer DP (1979) β-1.3-glucan in developing cotton fibers. Structure, localization, and relationship of synthesis to that of secondary wall cellulose. Plant Physiol (Bethesda) 63:1158–1164
Mankarios AT, Jones CFG, Jarvis MC, Threllfall DR, Friend J (1979) Hydrolysis of plant polysaccharides and GLC analysis of their constituent neutral sugars. Phytochemistry (Oxf) 18:419–422
McNeil M, Darvill AG, Albersheim P (1979) The structural polymers of the primary cell walls of dicots. Fortschr Chem Org Naturst 37:191–249
Meyer Y, Herth W (1978) Chemical inhibition of cell-wall formation and cytokinesis, but not of nuclear division in protoplasts of *Nicotiana tabacum* L. cultivated in vitro. Planta (Berl) 142:253–262
Meyer Y, Herth W. (1982) Interaction of cell-wall formation and cell division in higher plant cells. In: Brown RM Jr (ed) Cellulose and other natural polymer systems; biogenesis, structure and degradation. Plenum, New York, pp 149–165
Mühlethaler K. (1967) Ultrastructure and formation of plant cell walls. Annu Rev Plant Physiol 18:1–23
Mueller S, Brown RM Jr (1980) Evidence for an intramembrane component associated with a cellulose microfibril synthesizing complex in higher plants. J Cell Biol 84:351–376
Mueller S, Brown RM Jr (1982) The control of cellulose microfibril deposition in the cell-wall of higher plants
Mueller S, Brown RM, Scott TK (1976) Cellulosic microfibrils: nascent stages of synthesis in a higher plant cell. Science (Wash DC) 194:949–951
Nevins DJ, English PD, Albersheim P (1968) Changes in cell-wall polysaccharides associated with growth. Plant Physiol (Bethesda) 43:914–922
Nishitani K, Masuda Y (1979) Growth and cell-wall changes in azukibean epicotyls I. Changes in wall polysaccharides during intact growth. Plant Cell Physiol 20:63–74
Pilet PE, Blaschek W, Senn A, Franz G (1984) Comparison between maize root cells and respective regenerating protoplasts: wall polysaccharides. Planta (Berl) 161:465–469
Prat R, Roland J-C (1971) Etude ultrastructurale des premiers stades de néoformation d'une enveloppe par les protoplastes végétaux séparés mécaniquement de leur paroi. CR Hebd Seanc Acad Sci Ser D Sci Nat 273:165–168
Prat R, Williamson FA (1976) Chronologie de la sécretion de parois par les protoplastes végétaux. Soc Bot Fr (Coll Secret Veget) 123:33–45
Preston RD (1974) The physical biology of plant cell walls. Chapman and Hall, London
Quader H (1981) Interruption of cellulose microfibril crystallization. Naturwissenschaften 68:428–429
Reinert J, Bajaj YRS (eds) (1977) Applied and fundamental aspects of plant cell tissue and organ culture. Springer, Berlin Heidelberg New York
Ring SG, Selvendran RR (1978) Purification and methylation analysis of cell-wall material from *solanum tuberosum*. Phytochemistry (Oxf) 17:745–752
Robenek H, Peveling E (1977) Ultrastructure of the cell-wall regeneration of isolated protoplasts of *Skimmia japonica* thunb. Planta (Berl) 136:135–145
Roberts E, Seagull RW, Haigler CH, Brown RM Jr (1982) Alteration of cellulose microfibril formation in eukaryotic cells: Calcofluor White interferes with microfibril assembly and orientation in *Oocystis apiculata*. Protoplasma 113:1–9

Ryser U (1979) Cotton fibre differentiation: occurrence and distribution of coated and smooth vesicles during primary and secondary wall formation. Protoplasma 98:223–239

Schulz GV, Marx M (1954) Über Molekulargewichte und Molekulargewichtsverteilungen nativer Cellulosen. Chem 14:52–95

Spencer FS, MacLachlan GA (1972) Changes in molecular weight of cellulose in the pea epicotyl during growth. Plant Physiol (Bethdesda) 49:58–63

Staehelin LA, Giddings TH (1982) Membrane-mediated control of cell-wall microfibrillar order. In: Developmental order: its origin and regulation. Liss, New York, pp 133–147

Takebe I, Nagata T (1973) Culture of isolated tobacco mesophyll protoplasts. Colloq Int Cent Natl Rech Sci 212:175–187

Takeuchi Y, Komamine A (1978) Composition of the cell-wall formed by protoplasts isolated from cell suspension of *Vinca rosea.* Planta (Berl) 140:227–232

Takeuchi Y, Komamine A (1981) Glucans in the cell walls regenerated from *Vinca rosea* protoplasts. Plant Cell Physiol 22:1585–1594

Takeuchi Y, Komamine A (1982) Effects of culture conditions on cell division and composition of regenerated cell walls in *Vinca rosea* protoplasts. Plant Cell Physiol 23:249–255

Thornber JP, Northcote DH (1962) Changes in the chemical composition of a cambial cell during its differentiation into xylem and phloem tissue in trees. Biochem J 82:340–346

Updegraff DM (1969) Semimicro determination of cellulose in biological materials. Anal Biochem 32:420–424

Van Holst GJ, Klis FM, Bonman F, Stegwee D (1980) Changing cell-wall compositions in hypocotyls of dark-grown bean seedlings. Planta (Berl) 149:209–212

Van der Valk P, Fowke LC (1981) Ultrastructural aspects of coated vesicles in tobacco protoplasts. Can J Bot 59: 1307–1313

Van der Valk P, Rennie PJ, Connolly JA, Fowke LC (1980) Distribution of cortical microtubules in tobacco protoplasts. An immunofluorescense microscopic and ultrastructural study. Protoplasma 105:27–43

Williamson FA, Fowke LC, Constabel FC, Gamborg OL (1976) Labelling of concanavalin A sites on the plasma membrane of soybean protoplasts. Protoplasma 89:305–316

Williamson FA, Fowke LC, Weber G, Constable FC, Gamborg O, (1977) Mikrofibril deposition on cultured protoplasts of vicia hajastana. Protoplasma 91:213–219

Willison JHM (1976) Synthesis of cell walls by higher plant protoplasts. In: Peberdy JF, Rose AH, Rogers HJ, Cocking EC (eds) Microbial and plant protoplasts. Academic, New York, pp 283–298

Willison JHM (1981) Secretion of cell-wall material in higher plants. In: Tanner W, Loewus FA (eds) Encyclopedia of plant physiology, new series vol 13B. Plant carbohydrates II. Extracellular carbohydrates. Springer, Berlin, Heidelberg, New York, pp 513–541

Willison JHM, Cocking EC (1972) The production of microfibrils at the surface of isolated tomato-fruit protoplasts. Protoplasma 75:397–403

Willison JHM, Cocking EC (1975) Microfibril synthesis at the surface of isolated tobacco mesophyll protoplasts, a freeze-etch study. Protoplasma 84:147–159

Willison JHM, Grout BWW (1978) Further observations on cell-wall formation around isolated protoplasts of tobacco and tomato. Planta (Berl) 140:53–58

Willison JHM, Klein AS (1982) Cell-wall regeneration by protoplasts from higher plants. In: Brown RM Jr (ed) Cellulose and other natural polymer systems; biogenesis, structure and degradation. Plenum, New York, pp 61–85

Glucan Synthases and Cell-Wall Regeneration in Fungal Protoplasts

M. FÈVRE and V. GIRARD

Introduction

Protoplasts from filamentous fungi exhibit a very efficient capacity to regenerate a new cell-wall and to revert to normal colonies. The entire process takes place in about 24 h. These properties led us to use the protoplasts as a model for studying the processes involved in cell-wall synthesis. Moreover, the apparent simplicity of the fungal cell-wall structure, compared to higher plant cell walls, is an advantage which permits to elucidate the biosynthesis of 1–3 β glucans and cellulose or chitin. Recent reviews have described patterns of fungal protoplast regeneration and reversion (Peberdy 1979, Wessels et al. 1976). In our studies, protoplasts from *Saprolegnia monoica,* a cellulosic cell-wall fungus, were used to study the activities of the plasma membrane bound glucan synthases and the biosynthesis of wall components.

Material and Methods

Protoplast Production

Mycelia 48 h old were converted to protoplasts during incubation in lytic medium containing Drisetease (5 mg ml^{-1}) cellulase (1 mg ml^{-1}) and 0.5 M sorbitol as stabilizer. Protoplasts were separated by filtration through bolting cloth and washed by centrifugation (Gaugy and Fèvre 1982, Girard and Fèvre 1984a).

Isolation of the Plasma Membrane

Protoplasts were labelled with 250 μg ml^{-1} Con A as described previously (Girard and Fèvre 1984a). When radioactive Con A was used, labelling was performed with 1 ml protoplast suspension and (^{3}H)-acetyl Con A. Non-fixed Con A eliminated by centrifugations. The suspension was homogenized using as glass homogenizer with a Teflon pestle. Plasma membranes were collected by centrifugation of the lysate at 500 g for 10 min. Particulate enzymes were layered over a linear density gradient (32 ml) of 1.04 to 1.25 g.ml^{-1} Telebrix and centrifuged at 25,000 r.p.m. for 16 h.

The Physiological Properties of Plant Protoplasts
Edited by P. E. Pilet

Enzyme Assays

1–4 β glucan synthases were assayed in the presence of 0.3 nmol UDP (^{14}C) glucose and 5.5 μmol $MgCl_2$. When enzymes were assayed for 1–3 β glucan synthesis, $MgCl_2$ was omitted and a higher substrate level was provided [400 nmol UDP (^{12}C) glucose]. The incorporation of radioactivity into glucans was measured after precipitation with ethanol and filtration through glass fibre filters (Fèvre and Rougier 1981). ATP ase was assayed at pH 6.0 in the presence or absence of Na-vanadate (Girard and Fèvre 1984a).

Protoplast Regeneration

Protoplasts were suspended in a liquid culture medium containing sorbitol and 1 μ Ci ml^{-1} of (^{14}C) glucose. The suspension (10^6 protoplasts ml^{-1}) was incubated at 24°C in a shaking incubator. Samples were taken at intervals and protoplasts were collected by centrifugation after precipitation overnight at –20°C by 2 vol. ethanol (Girard and Fèvre 1984b). The cellulose fraction was collected as the non-hydrolysable material in acetic-nitric acid reagent (Updegraff 1969).

Results and Discussion

β-Glucan Synthases Have a Transmembrane Orientation in the Plasma Membrane

Particulate glucan synthases isolated from mycelium or protoplasts and separated by isopycnic centrifugation were mainly associated with membrane equilibrating at the density of 1.16 g cm^{-1} and exhibiting ATPase activity (Girard and Fèvre 1984a). This ATPase activity, vanadate sensitive and oligomycin insensitive, is probably the best single marker for fungal plasma membrane (Scarborough 1975; Bowman et al. 1981). Glucan synthases are also associated with internal membranes, endoplasmic reticulum (Girard and Fèvre 1984a), dictyosomes and Golgi visicles (Fèvre 1979, Fèvre and Rougier 1981). Protoplasts provide the advantage of labelling the plasma membrane with exogenous markers (Scarborough 1975). Concanavalin A which binds to the cell surface of *Saprolegnia* protoplasts, and causes them to agglutinate, was used as a specific label. Con A-labelled plasma membranes could be isolated by low speed centrifugation then fractionated by isopycnic centrifugation (Girard and Fèvre 1984a). 1–3 β and 1–4 β glucan synthases sedimented as a single peak corresponding to those of vanadate sensitive ATPase and (^{3}H) Con A, indicating that both synthases are associated with the plasma membrane (Fig. 1). Electron microscopy showed that the membranes, stabilized by the lectin against fragmentation and vesiculisation, were isolated as sheets (Girard and Fèvre 1984a).

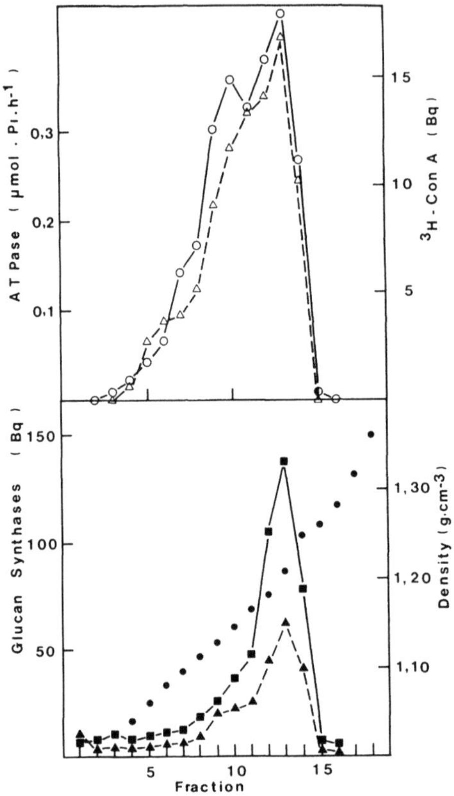

Fig. 1. Distribution of glucan synthase activities, ATPase and (^{3}H)-Con A in a continuous Telebrix density gradient. *Saprolegnia* protoplasts were labelled with acetyl-Con A. Plasma membranes were isolated by low speed centrifugation, layered on the density gradient and centrifuged 16 h at 25,000 r.p.m. (SW 27 Beckmann rotor). Fractions were collected and each enzyme activity was assayed in duplicate. 1–4 β glucan synthase (■), 1–3 β glucan synthases (▲) (^{3}H) Con a (△), ATP ase (○). The value of the ATP ase represents the difference in activity of the assays conducted in the absence and presence of Na vanadate

The accessibility of the synthase to the action of external inactivating agents was studied by incubating intact protoplasts in proteases or glutaraldehyde. When treatments were applied prior the homogenization, 1–3 β and 1–4 β glucan synthases activities were greatly reduced, indicating that the enzymes bound to the plasma membrane faced the outside of the cell (Girard and Fèvre 1984a). Regenerating protoplasts were unable to use external UDP-glucose as a substrate, showing that the substrate-binding site of the plasma membrane glucan synthases faces the inside of the cell (Girard et al. 1984). These results thus indicate the transmembrane orientation of the synthases in the plasma membrane; this location agrees with the model (terminal complexes-rosettes) described by cytologists for higher plants (Giddings et al. 1980, Mueller and Brown 1980).

Patterns of Cell-Wall Regeneration

Light microscopy autoradiography of protoplasts incubated for 5 min in (^{3}H)-glucose nutrivitve medium, just after isolation or after 5h regeneration, showed that cell-wall synthesis occurred over the whole surface (Girard et al. 1984). Eighty to 90% of the protoplasts were capable of producing a new cell-wall. Cell-wall regeneration was characterized by the deposition of a network of cellulose microfibrils

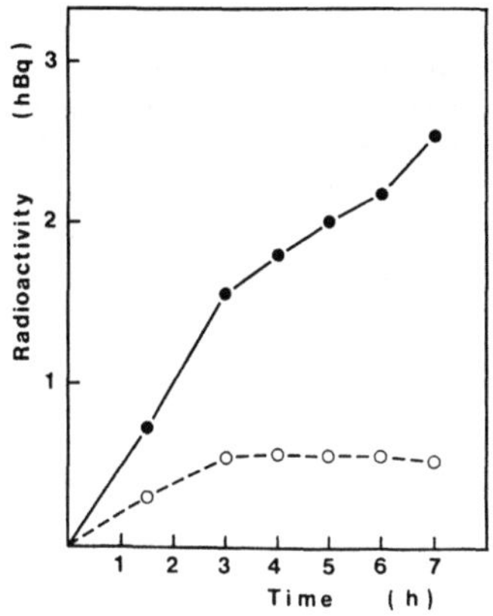

Fig. 2. Incorporation of (^{14}C)-glucose into cell-wall polysaccharides during regeneration of protoplasts in liquid medium containing 1μ Ci ml^{-1} of (^{14}C)-glucose (•) total cell-wall polysaccharides (○) cellulosic fraction insoluble in acetic/nitric acid

insoluble in acetic-nitric acid (Updegraff 1969) and clearly visible after 2 h of regeneration (Girard et al. 1984). Electron microscopy observations suggested also that the early stages of regeneration (up to 6 h) occurred independently of the participation of the Golgi apparatus. Cell-wall synthesis is probably due to the enzymes present at the cell surface and remained intact during protoplast isolation.

β glucan synthases activity of protoplasts increased during the course of cell-wall formation. 1–4 β glucan synthases, which exhibit a higher activity in plasma membrane than internal membranes, increased rapidly since the beginning of regeneration. They were three to four times more active after 4 h cultivation.

β 1–3 glucan synthases of the plasma membrane increased more slowly (x 2 after 4 h regeneration), while enzymes associated with the internal membranes showed a lag phase of 3 h (Girard et al. 1984).

This difference of the in vitro activity of the enzymes was also correlated with the different rates of in vivo production of the cell-wall polymers. Within minutes, (^{14}C)-glucose was transferred to cell-wall glucans when protoplasts were regenerated in a radioactive glucose medium. The production of cellulose was rapid at the beginning of incubation, then declined after 3 h, while other cell-wall polysaccharide deposition continued (Girard et al. 1984; Girard and Fèvre 1984b) (Fig. 2).

Localization of β Glucan Synthases Along the Hyphae

The use of protoplasts provides a means to study the distribution of glucan synthases along the yphae in relation to the apical growth. Lytic digestion of fungal mycelium represents a sequential release of cytoplasm as protoplasts, from the apical region and subsequently from the subapical and distal regions (Isaac et al. 1978; Gaugy and Fèvre 1982). Protoplasts collected at different times of production (i.e. arising from different zones of the hyphae) exhibited differences in their glucan synthase equipment. 1–4 β glucan synthase activities were highest in "early" protoplasts (i.e. apical zones), then decreased in the protoplasts produced from the non-elongating zones. The distribution of 1–3 β glucan synthases was the opposite; activities were four times higher in "late" protoplasts (i.e. subapical zones), than in protoplasts from growing zones (Girard and Fèvre 1984b). The successive populations of proto-

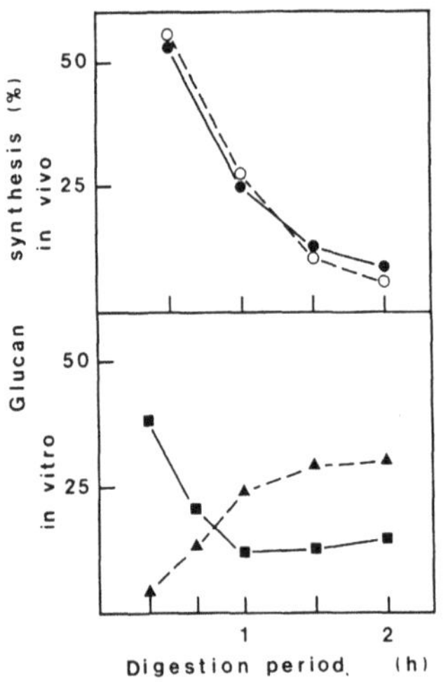

Fig. 3. Glucan synthase activity and glucan synthesis of protoplasts released at different times of incubation. Glucan synthase activity corresponds to the incorporation of glucose from UDP-glucose into polysaccharides by the plasma membrane fraction of protoplasts extracts (i.e. glucan synthesis in vitro). Glucan synthesis corresponds to the incorporation of radioactive glucose into cell-wall glucans of intact protoplasts after 6h regeneration (i.e. glucan synthesis in vivo). Results are expressed as the percentage of the total of the activity exhibited by the different samples of protoplasts
1–4 glucan synthase activity (■)
1–3 glucan synthase activity (▲)

plasts showed differences in the capacity to regenerate a new cell-wall. Early protoplasts exhibited the highest rate of glucose incorporation. It was interesting to note that protoplasts which have a low 1–4 β glucan synthases activity, also exhibited a low rate of cell-wall regeneration even if their potential 1–3 β glucan synthase activity was very high (Girard and Fèvre 1984b) (Fig. 3).

Conclusions

Protoplasts provide a simple experimental system to study the relationship between glucan synthases, glucan synthesis and morphogenesis. Protoplasts used as a gentle procedure for producing cell-free systems, showed that 1–4 β and 1–3 β glucan synthases have a transmembrane location in the plasma membrane of the cellulosic cell-wall fungus *Saprolegnia.*

Wall biogenesis by regenerating protoplasts started by the deposition of a microfibrillar net of cellulose with the subsequent deposition of other material and thus was very similar to wall regeneration in higher plants (Willison and Klein 1982). In *Saprolegnia*, this sequential production of the polymers reflected the variation of the activity of plasma membrane-bound synthases, which occurred after the onset of cultivation.

Another advantage of the protoplast system was the recovery of the cytoplasm of the different zones of hyphae which exhibit a highly polarized growth. The capacity of protoplasts to regenerate cell-walls which decreased from early protoplasts (growing zone cytoplasm) to late protoplasts (non-elongating zone cytoplasm), corresponds to the characteristics of hyphal apical growth. Autoradiographic studies have shown that, in *Saprolegnia* hyphae, polysaccharide deposition is located in the first 10 μm of the apex (Fèvre and Rougier 1982).

All these results obtained through the different advantages offered by protoplasts, indicate that a sequential synthesis of cell-wall polymers must exist in *Saprolegnia* hyphae. Cellulose will first be produced, then 1–3 β glucan deposition will occur during the later stages of cell-wall formation.

References

Bowman EJ, Bowman BJ, Slayman CW (1981) Isolation and characterization of plasma membranes from wild-type *Neurospora crassa.* J Biol Chem 256:12336–12342

Fèvre M (1979) Glucanases, glucan synthases and wall growth in *Saprolegnia monoica.* In: Burnet JH, Trinci APG (ed) Fungal wall and hyphal growth. Cambridge Uni Press, Cambridge, pp 225–263

Fèvre M, Rougier M (1981) β 1–3 and β 1–4 glucan synthesis by membrane fractions from the fungus *Saprolegnia.* Planta (Berl) 151:232–241

Fèvre M, Rougier M (1982) Autoradiographic study of hyphal cell-wall synthesis of *Saprolegnia.* Arch Microbiol 131:212–215

Gaugy D, Fèvre M (1982) Protoplast production from *Saprolegnia monoica.* Microbios 34: 89–98

Giddings TH, Brower DL, Staehelin LA (1980) Visualization of particles complexes in the plasma membrane of *Micrasterias denticulata* associated with the formation of cellulose fibrils in primary and secondary cell-walls. J Cell Biol 84:327–339

Girard V, Fèvre M (1984a) β 1–4 and β 1–3 glucan synthases are associated with the plasma membrane of the fungus *Saprolegnia.* Planta (Berl) 160:400–406

Girard V, Fèvre M (1984b) Distribution of (1–3) β and (1–4) β glucan synthases along the hyphae of *Saprolegnia.* J Gen Microbiol 130:1557–1562

Girard V, Capellano A, Fèvre M (1984) β glucan synthesis and glucan synthase activities during early stages of cell-wall regeneration by protoplasts from *Saprolegnia monoica.* J Gen Microbiol 130:2367–2374

Isaac S, Ryder WS, Peberdy JF (1978) Distribution and activation of chitin synthase in protoplast fractions released during the lytic digestion of *Aspergillus nidulans* hyphae. J Gen Microbiol 105:45–50

Mueller SC, Brown RM Jr (1980) Evidence for an intramembrane component associated with a cellulose microfibril-synthesizing complex in higher plants. J Cell Biol 84:315–326

Peberdy JF (1979) Wall biogenesis by protoplasts. In: Burnet JH, Trinci APJ (ed). Fungal walls and hyphal growth. Cambridge Univ Press, Cambridge, pp 49–70

Scarborough GA (1975) Isolation and characterization of *Neurospora crassa* plasma membrane. J Biol Chem 250:1106–1114

Updegraff DM (1969) Semi-micro determination of cellulose in biological materials. Anal Biochem 32:420–424

Willison JHM, Klein AS (1982) Cell-wall regeneration by protoplasts from higher plants. In: Brown RMJ (ed) Cellulose and other natural polymer systems. Plenum, New York, pp 61–85

Wessels JGH, Valk P, van der Vries OMH de (1976) Wall synthesis by fungal protoplasts. In: Peberdy JF, Rose AH, Rogers HJ, Cocking CC (ed) Microbial and plant protoplasts. Academic, London, pp 267–281

Fatty Acids in Protoplasts

E. HARTMANN and K. HOCK

Introduction

Protoplast isolation and regeneration of protoplasts is known from many plants (Binding et al. 1981). In spite of great success in producing and handling protoplasts, there are still many plant species which are very resistant against protoplast isolation, or against regeneration. Regengeration of plant protoplasts is more or less a field of testing and trying than understood by known parameters. Therefore great difficulties often accompany the installation of an efficient protoplast regeneration system, even if the isolation provides a high yield of viable protoplasts.

The question arises, whether regeneration is regulated by genetically fixed high or low regeneration efficiency of the species, or depends on stress conditions urgently connected with the isolation methods. The tissue is losing its correlative cellular association under the influence of cell-wall-degrading enzymes in strong hypertonic medium, resulting in a dramatic disturbance of cell metabolism. This stress might be so severe that many cell types cannot overcome the situation.

Fleck et al. (1980, 1982) observed dramatic changes in the proteins synthesized by freshly isolated protoplasts. Protoplasts were missing many proteins synthesized in isotonic cells, such as the small subunits of ribulose bisphosphate carboxylase.

However, some new proteins directly related to the high osmotic strength of the medium were observed, the so-called osmotic shock proteins. These proteins seem to be analogous to other stress proteins reported in plants in response to various environmental stresses (Webster 1980; Key et al. 1981). The function of these proteins seems to be comparable to those shock proteins known from animals, such as the heat shock proteins of *Drosophila* described by Ashburner and Bonner (1979).

In isolated plant protoplasts especially the plasmamembrane is exposed to high concentrations of different substances, but has still to function as a selective barrier, to regulate transport, and to maintain signal processing. We think that this regulation is not possible without a high capability of the surface membrane to adapt to the stress condition. Unfortunately, we do not know much about the connection between membrane constitution and membrane metabolism. Therefore, the danger of discussing unfounded speculations is great.

Nevertheless, the only way to reduce the ignorance about the involvement of membranes in stress physiology are experiments, and data. We investigated changes in the lipid metabolism, and lipid patterns induced by protoplast isolation, and during protoplast regeneration. We report here about our data of fatty acid patterns. Changes

The Physiological Properties of Plant Protoplasts
Edited by P. E. Pilet

in fatty acid patterns of plant tissues are often affected by environmental stresses. It is well established that freezing temperatures lead to an increase in the amount of polyunsaturated fatty acids. This metabolism mainly occurs in the structural lipids as a direct result of which optimum membrane fluidity is maintained (Breidenbach and Warring 1977, Radwan et al. 1978, Mazliak 1981, Willemont 1983, Gawer et al. 1983). Water and osmotic stress can also result in a decrease of polyunsaturated fatty acids and modulation of membrane lipids (Lilijenberg and Kates 1982, Thu et al. 1982, Douglas and Paleg 1981). We did not find any report about the influence of the isolation procedure on the lipid pattern of protoplasts, and also no data about modification of lipid patterns during adaptation of regeneration of protoplasts. We hope this paper will stimulate some work in this field.

Material and Methods

Phaseolus vulgaris L. seedlings were grown in moist vermiculite under red light (660 nm) at 20°C for 7 days. The hypocotyl hook was cut off and sterilized by a commercial bleach (10% v/v Domestos). Protoplasts were isolated in a solution containing 2% (w/v) cellulase Onozuka R-10 (Kinki Yakult), 1% (w/v) macerozyme R-10 (Kinki Yakult) and 1% (w/v) Driselase (Kyowo Hakko Kogyo) in 0.7 M mannitol, 20 mM Mes-Tris pH 5.6 and 10 mM $CaCl_2$. After an incubation of 4–6 h at 30°C with gentle shaking, the protoplasts were sieved through a 100 μm mesh net and purified by centrifugation in a Percoll gradient (50% v/v).

Mesophyll protoplasts of *Nicotiana tabacum* L. were isolated according to Nagy and Maliga (1976) and mesophyll protoplasts of *Petunia hybrida* L. according to Durand et al. (1973).

Axenic protonema cultures of the mosses *Ceratodon purpureus, Physcomitrella patens* and *Leptobryum pyriforme* (2n strain) were grown in liquid medium. Koflers medium was modified according to Saxena and Rashid (1981) with glucose concentration reduced to 1 per cent (w/v, pH 5.8). The plant material for isolation of protoplasts was 3 days old.

The protonemata were soaked for 30 min at room temperature in plasmolyticum in 25-ml glass vials. The solution consisted of 3.05 mM $Ca(NO_3)_2$, 1.24 mM KNO_3, 0.92 mM KH_2PO_4, 0.51 mM $MgSO_4$ x 7 H_2O, 0.025 mM $FeSO_4$ x 7 H_2O, 0.44 M mannitol and 1 ml^{-1} Hoaglands A–Z solution. The pH-value was 5.8. The protonema tissue was digested by an enzyme solution of 2.0 per cent (w/v) Cellulase Onuzuka, 1.0 per cent (w/v) Mazerozyme R 10 (Serva, Heidelberg), and 1.5% Rhoament P (Röhm, Darmstadt) in plasmolyticum pH 5.6. Protonemata were incubated for 3 to 4 h at 28°C. The suspension was filtered through a 100 μm mesh nylon net. The resulting protoplast solution was washed 4 times by repeated centrifugation at 100 g for 3 min and resuspended in plasmolyticum pH 5.8.

Protoplasts were cultured in darkness or in continous white light. The culture was performed on plastic multidishes (Nunc Intermed, Roskilde). Each chamber contained 1 ml plasmolyticum and about 5000 protoplasts. For the lipid extraction about 0.3 g fresh weight, e.g. 500 000–1 000 000 protoplasts were used.

Lipids were extracted according to Folch et al. 1957 using a solution of trichloromethane: methanol: 2-propanol (6:2:1 v/v), and separated by column chromatography on silicic acid (SIL-LC 325 mesh; Sigma) according to Gellerman et al. 1975, slightly modified with respect to the plant species. By this method the total lipid fraction was separated in neutral lipids, glycolipids and phospholipids. Fatty acids were determined as methylesters after methylation with H_2SO_4/methanol. Gasliquidchromatographic (GLC) analysis was performed using a Perkin Elmer 900 GLC apparatus with a glass column (1800 x 2 mm) equipped with a FID-detector. The stationary phase was Chromosorb W-AW coated with 15% DEGS-PS (Supelco), carrier gas was nitrogen with a flow rate of 60 $ml min^{-1}$, the column temperature was 200°C. Quantitative determination was performed with standards and calculation according to Ackman 1969.

Results are expressed as mole % of fatty acids of the different lipid fractions. All experiments were reproduced at least four times, and the data represent mean values of three determinations.

Results and Discussion

Protoplasts with good yields, and high viability could be isolated from hypocotyls of young seedlings, from primary and secondary leaves, from callus, and from roots of *Phaseolus vulgaris.* It was not possible to get regeneration from any of these protoplast populations (Hock 1984). In contrast to beans, regeneration could be induced by using mesophyll protoplasts from leaves of *Nicotiana tabacum, Vicia faba, Pisum sativum,* and *Petunia hybrida.* Besides higher plants, mosses were used for protoplast isolation. Mosses generally show a high regeneration capability, and in most cases are very resistant against environmental stresses. The protoplasts were isolated from the protonemata. Protoplasts from *Physcomitrella patens* (see also Grimsley et al. 1977; Burgess and Lindstead 1981), *Ceratodon purpureus* (Jahnen and Hartmann 1983) regenerated very well (Fig. 1), whereas those from *Leptobryum pyriforme* did not.

The lipids were separated in three fractions: neutral lipids, glycolipids and phospholipids. The amounts of the fatty acids are expressed as percentage of the total amount of fatty acids of a distinct lipid fraction. Besides these fatty acids there were traces of others which were not taken into consideration.

The fatty acid pattern of *Phaseolus vulgaris* is shown in Fig. 2A. Neutral lipids provide no dominant fatty acids as in the case with glycolipids. There linolenic acid represents 64.6% of the total glycerolipid fraction. In the phospholipid fraction the unsaturated fatty acids linoleic acid (36.2%) and linolenic acid (28.6%) also represent two third of the total fatty acids. The comparison between hypocotyl tissue and protoplasts clearly shows the pronounced decrease of polyunsaturated fatty acids and the increase of saturated fatty acids such as palmitic acid, or stearic acid.

The glycolipids and the phospholipids of *Nicotiana tabacum* also show high amounts of unsaturated fatty acids, which are not as dominant in the neutral lipid

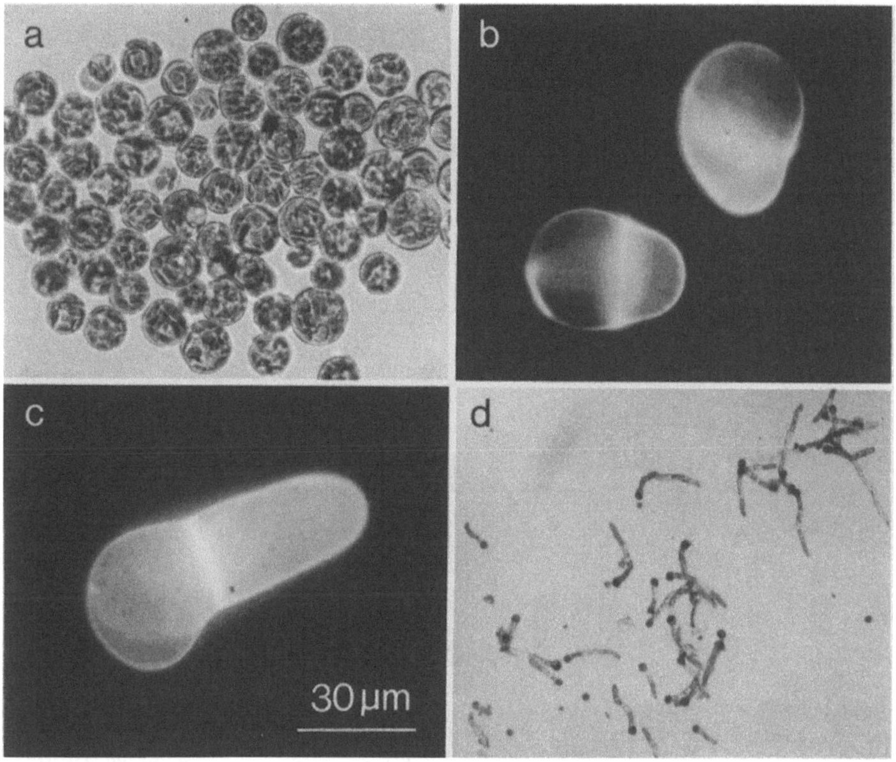

Fig. 1 a–d. Regeneration of protoplasts from protonemata of the moss *Leptobryum purpureus.* **a** Protoplast-preparation; **b** Calcofluor-staining of protoplasts 24 h after isolation; **c** Calcofluor-staining of a 48-h-old young protonema-filament; **d** 10-day-old protonemata regenerated form protoplasts

fraction as in the other two fractions. The relative amounts of fatty acids from both intact tissue and protoplasts show much less fluctuations than the fatty acid patterns from beans (Fig. 2B). In the case of *Petunia hybrida* a more specified analysis was performed. Besides intact tissue and protoplasts, homogenized isotonic tissue and osmotically stressed tissue were investigated (Fig. 2C). The overall fatty acid patterns are comparable with the patterns already described. The glycolipid fraction demonstrates the continous decrease in linoleic acid in a sequence: (1) intact tissue, (2) homogenized tissues, (3) protoplasts. But also in this tissue the changes are not as strong as in *Phaseolus vulgaris.* There is no tremendous shift between unsaturated and saturated fatty acids. Nevertheless, for the understanding of the effects, it might be important to recognize that patterns of plasmolyzed tissues are more comparable with patterns from intact tissues than with those from homogenized isotonically treated ones.

The fatty acid patterns of the analysed mosses are presented in Fig. 3. We separated the discussion of this analysis from that of higher plants because mosses show some characteristics in the fatty acid patterns not common for higher plants. Mosses contain to high amounts long-chained polyunsaturated fatty acids, and even

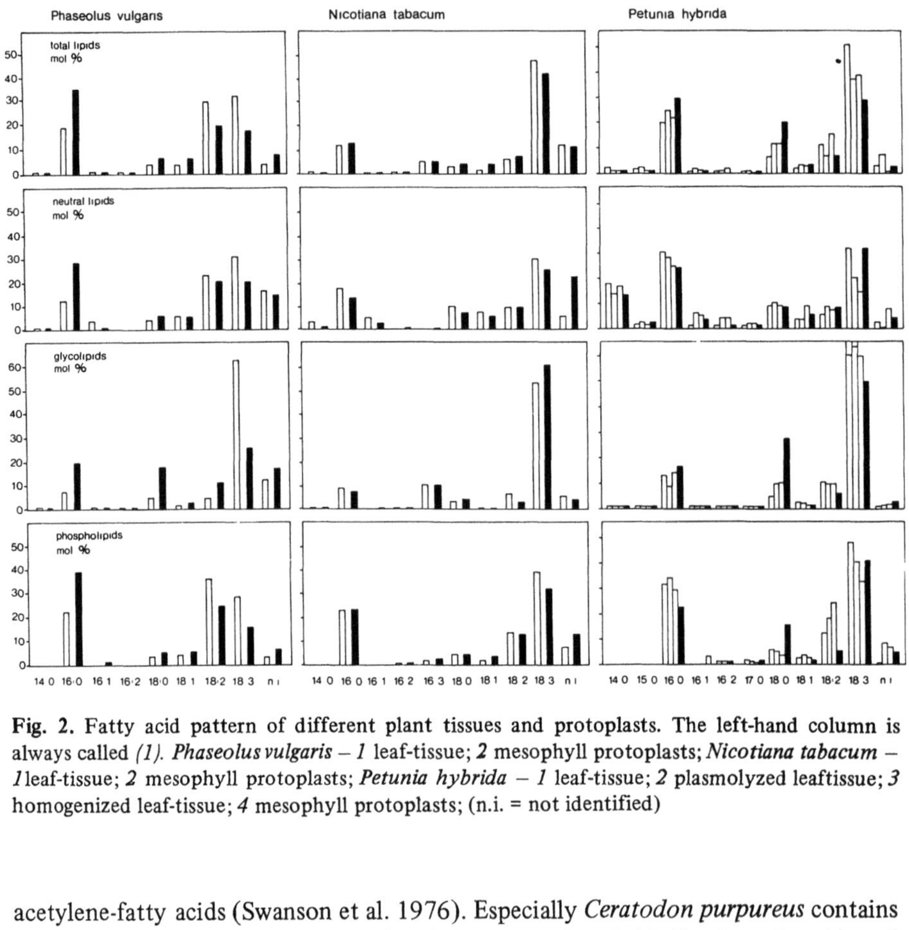

Fig. 2. Fatty acid pattern of different plant tissues and protoplasts. The left-hand column is always called *(1)*. *Phaseolus vulgaris* – *1* leaf-tissue; *2* mesophyll protoplasts; *Nicotiana tabacum* – *1* leaf-tissue; *2* mesophyll protoplasts; *Petunia hybrida* – *1* leaf-tissue; *2* plasmolyzed leaftissue; *3* homogenized leaf-tissue; *4* mesophyll protoplasts; (n.i. = not identified)

acetylene-fatty acids (Swanson et al. 1976). Especially *Ceratodon purpureus* contains the acetylenic long-chain fatty acids 18:2 ac, octadeca-9,12-dien-6-ynoic acid, and 18:3 ac, octadeca-9,12-15-triene-6-ynoic acid. The 18:3 ac was only detectable in the neutral lipid fraction, whereas the distribution of the 18:2 ac acid depended on the age and the treatment of the protonemata (Rauch 1984).

We compared the fatty acid patterns of two mosses *Leptobryum pyriforme* and *Ceratodon purpureus.* The fatty acid patterns from *Physcomitrella patens* were identical with those from *Ceratodon.* We chose these mosses, because they grow excellently in liquid culture, and are well known in different physiological parameters. All provided high yields of viable protoplasts, which were completely different in regeneration efficiency. *Ceratodon* and *Physcomitrella* protoplasts regenerated very well, whereas those from *Leptobryum* died.

The fatty acid pattern of *Leptobryum* protoplasts showed some general similarities with those from higher plants. The amount of polyunsaturated fatty acids was much higher in the structural lipid fractions, represented by glycolipids and phospholipids, than in the neutral lipid fraction. The changes between the fatty acid patterns of intact protonemata and isolated protoplasts of *Leptobryum pyriforme* demonstrate a strong shift from polyunsaturated to saturated fatty acids. This trend con-

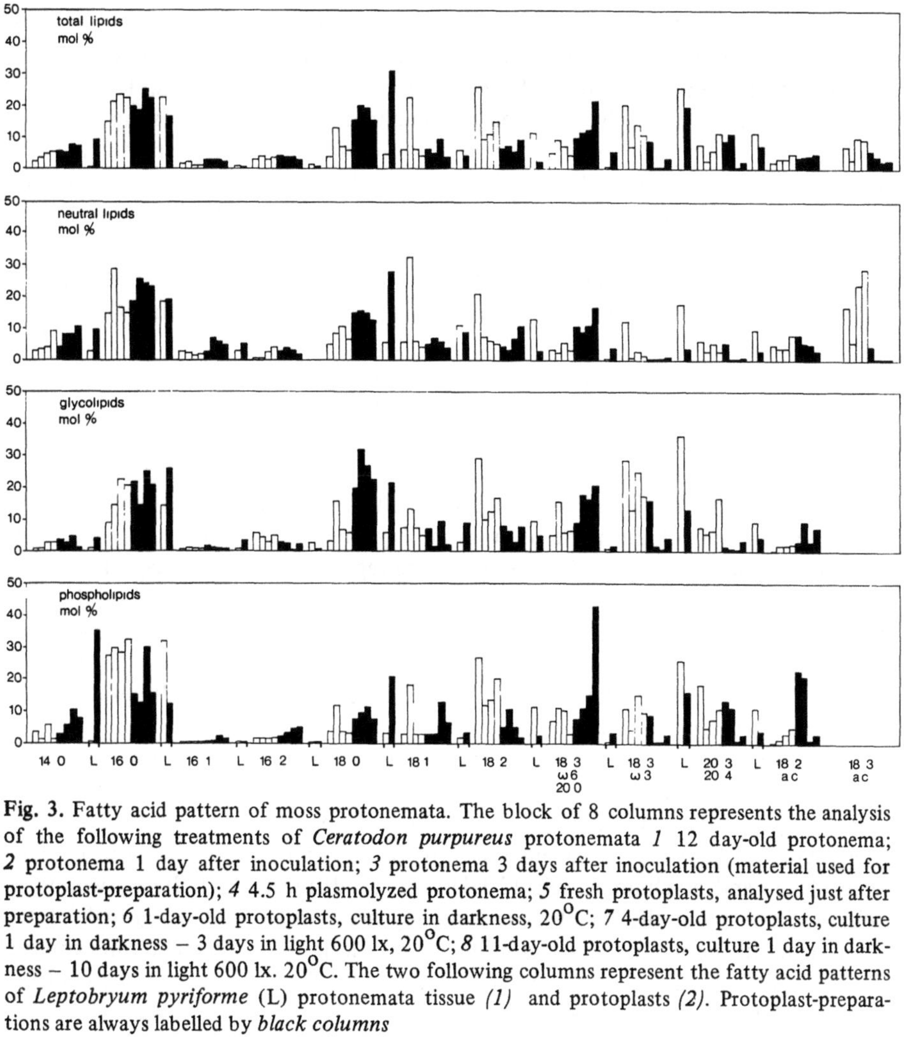

Fig. 3. Fatty acid pattern of moss protonemata. The block of 8 columns represents the analysis of the following treatments of *Ceratodon purpureus* protonemata *1* 12 day-old protonema; *2* protonema 1 day after inoculation; *3* protonema 3 days after inoculation (material used for protoplast-preparation); *4* 4.5 h plasmolyzed protonema; *5* fresh protoplasts, analysed just after preparation; *6* 1-day-old protoplasts, culture in darkness, 20°C; *7* 4-day-old protoplasts, culture 1 day in darkness – 3 days in light 600 lx, 20°C; *8* 11-day-old protoplasts, culture 1 day in darkness – 10 days in light 600 lx. 20°C. The two following columns represent the fatty acid patterns of *Leptobryum pyriforme* (L) protonemata tissue *(1)* and protoplasts *(2)*. Protoplast-preparations are always labelled by *black columns*

cerned all polyunsaturated fatty acids. The same shift in degree of desaturation was already observed in beans.

The phospholipid fraction showed an additional interesting alteration conducted by protoplast isolation. The amount of myristic acid increased very much in protoplasts, whereas this fatty acid was only detectable in small traces in the intact protonemata. This reaction concerns the amount of palmitic acid, which decrease in comparable amounts.

In *Ceratodon* protoplasts was the reduction of the amounts of the typical polyunsaturated fatty acids linoleic acid and linolenic acid very similar to *Leptobryum.* Stearic acid showed a comparable increasing tendency. There was no alteration between palmitic acid and myristic acid, but there were strong changes concerning the acetylenic fatty acids typical for *Ceratodon purpureus.* The amount of 18:3 ac, only present in the neutral lipid fraction, decreased from 25.7% to 11.2% in the protoplasts. The fatty acid 18:2 ac changed extremely in the phospholipid fraction and increased from 3.0% up to 22.2% in the fresh protoplasts. Compared with this increase, there were only small changes in the other lipid fractions. The increase of the arachidonic acid (20:4) in the phospholipid fraction is interesting. Unfortunately, there is no information about a physiological function of arachidonic acid in mosses. We only know that arachidonic acid levels depend very sensitively on altered environmental conditions (Euler 1983). This is also demonstrated by the following data about protoplast regeneration.

The regeneration efficiency of *Ceratodon* protoplasts gave us the possibility to investigate the lipid pattern alterations during regeneration (Rauch 1984). These data are summarized in Fig. 3. The analysis started with 12-day-old material which was used for vegetative subculture of the protonemata. The fatty acid pattern was determined form cultures 1 day after inoculation, and from 3-day-old material, which is used for protoplast isolation. The protonemata were analysed 4.5 h after the application of the isolation medium without enzymes, but with all other compounds, to see the effects of plasmolysis on fatty acid patterns. The next stage was the determination of the patterns from freshly prepared protoplasts, followed by the analysis of the fatty acid patterns during regeneration after 1 day, 4 days and 11 days. Cell-wall regeneration occurred very fast. After about 1 h synthesis of wall material could be detected by calcofluor staining; after 6 h the cell-wall had regenerated, and 12–24 h after isolation of protoplasts the first cell division occurred. Ten days after isolation of protoplast, small protonemata cultures developed (Jahnen and Hartmann 1983).

The comparison of protonemata before and after inoculation show a pronounced shift form unsaturated to saturated fatty acids, as was already recognized as typical for protoplast isolation. During the 3 days' growth of fresh inoculated protonemata the degree of this shift is weakened, clearly demonstrated by the levels of linolenic acid (18:3) and arachidonic acid (20:4). This result means that already wounding, which stimulates in the intact moss tissue regeneration and induces growth of new protonema filaments, has a strong influence on the fatty acid metabolism expressed by an altered desaturation degree. The data would further support the hypothesis that wounding has a stronger effect than osmotic stress. If this were true, fatty acid pattern from plasmolyzed tissue should have closer relationships to unwounded tissue than to isolated protoplasts. Exactly this result was observed. The protonemata were treated with hypertonic solution as long as it lasted to isolated protoplasts. Figure 3 show the results, and it is obvious that the patterns from 3-day-old untreated and 4.5-h osmotically stressed protonemata coincide much more than protoplasts of the same age. There was no strong reduction in unsaturated fatty acids, and the level of stearic acid did not increase. The only remarkable effect oncerned the amount of the arachidonic acid fraction, but in contrast to protoplast not the phospholipid fraction was mediated, but the glycolipid fraction.

Regeneration of protoplasts has many resemblances with the growth induction of protonemata by wounding connected with the inoculation. The recovery of the normal protonemal condition, expressed as fatty acid patterns, was shorter after wounding than with protonemata derived from protoplasts. The regenerated cells showed the most different fatty acid pattern 4 days after protoplast isolation. The usually very important fatty acid linolenic acid reached very low concentrations. The low levels increased very slowly during the regeneration period of 11 days. It might be possible that the acids 18:3w6/20:0, which could not be separated by our gaschromotographic procedure, became functional like linolenic acid. Nevertheless, this fraction showed the strongest increase, and after 11 days reached in the phospholipid fraction 43.9% of the total fatty acid content, and even 22.1% of the total lipid fraction, compared with 4.3% in normal protonemata.

Conclusion

There is no question about the fact that the isolation of protoplasts is a strong stress condition for the cells. The lipid metabolism may be an important indicator to recognize, and even to elucidate plant stress. But the question arises, which factor is the important one, which one affects regeneration efficiency?

We consider that there are at least two important factors involved in the modulation of membrane constitution, the wounding effect, and osmotic stress. The question about the influence of osmotic stress is still open, because there are not enough data to be discussed more intensively. But it seems to be possible that this stress has no very effects in intact tissues, but is more effective in isolated protoplasts.

In contrast, much more is known about wounding effects. It is well documented by many papers, that wounding of a tissue leads to drastic changes in the cell metabolism (Kahl 1982). There is a fast reduction of endoplasmic reticulum (Jackman and Van Stevenick 1967). Laties (1976), and Theologis and Laties (1981) reported strong influences of wounding on lipid pattern of potato tubers. They described a rapid and strong degradation of phospholipids and glycolipids, and a decrease in the unsaturated fatty acids linoleic acid and linolenic acid in potato disks during the first hours after wounding.

The answer of the wounding stimulus is different, but in most cases increased proteins synthesis of organells was observed (Nakano and Asahi 1970, Hackett et al. 1960). Included in this stimulated protein synthesis are enzymes of the fatty acid synthesis (Ben Abdelkader 1969). Phospholipids, which are most generally degraded after wounding, are mainly synthesized in the endoplasmic reticulum (Moore et al. 1973, Hock and Hartmann 1981). The recovery and increased formation of rough endoplasmic reticulum would explain the restitution of the phospholipid pattern 24 h after the wounding (Laties 1976), induced through de novo synthesis of lipid synthesizing enzymes (Ben Abdelkader and Auderset 1972). Wounding also stimulates ethylene production (Burg 1962), which may function as an effector for membrane degradation or as an effector to induce regeneration (woundhealing). Such effects of ethylene should be considered to be very important for the regulation of the still very obscure process of protoplast regeneration.

Acknowledgments. The authors thank the Deutsche Forschungsgemeinschaft (DFG) for supporting these studies by a grant to Ha; Renate Müller for capable technical assistance, and Doris Bertleff for her help and suggestions on the preparation of the manuscript.

References

Ackman RG (1969) Gas-liquid chromatography of fatty acids and esters. In: Lowenstein JM (ed) Methods of enzymology, Vol 14 Lipids, Academic Press, London, pp 329–381

Ashburner M, Bonner JJ (1979) The induction of gene activity in *Drosophila* by heat shock. Cell 17:241–254

Ben Abdelkader A (1969) Influence de la survie sur la biosynthese des phospholipids dans les microsomes de tubercules de pomme de terre. C R Acad Sci Paris 268:2406–2409

Ben Abdelkader A, Auderset G (1972) Formation de polyribosomes dans des tranches de tubercules de pomme de terre maintenues en survie. C R Acad Sci Paris 274:1311–1314

Binding H, Nehls H, Kock R, Finger J, Mordhorst G (1981) Comparative studies on protoplast regeneration in herbaceous species of the dicotyledoneae class. Z Pflanzenphysiol 101: 119–130

Breidenbach RW, Warring AJ (1977) Response to chilling of tomato seedlings and cells in suspension cultures. Plant Physiol 60: 190–192

Burg S (1962) The physiology of ethylene formation. Ann Rev Plant Physiol 13:265–303

Burgess J, Linstead PJ (1981) Studies on the growth and development of protoplasts of the moss, *Physcomitrella patens,* and its control by light. Planta (Berl) 151:331–338

Douglas TJ, Paleg LG (1981) Lipid composition of *Zea mays* seedlings and water stress-induced changes. J Exp Bot 32:499–508

Durand J, Potrykus I, Donn G (1973) Plantes issues de protoplastes de *Petunia.* Z Pflanzenphysiol 69:26–34

Euler R (1983) Untersuchungen zum Lipidstoffwechsel eines Laubmoosprotonemas unter ververschiedenen Kulturbedingungen. Thesis, Mainz

Fleck J, Durr A, Fritsch C, Lett MC, Hirth L (1980) Comparison of proteins synthesized in vivo and in vitro by mRNA from isolated protoplasts. Planta 148:453–454

Fleck J, Durr A, Fritsch C, Vernet T, Hirth L (1982) Osmotic shock "stress proteins" in protoplasts of *Nicotiana silvestris.* Plant Sci Lett 26:159–165

Folch J, Lees M, Szanley HS (1957) A simple method for the isolation and purification of total lipids from animal tissue. J Biol Chem 226:497–509

Gawer M, Sansonetti A, Mazliak P (1983) Lipid composition of tobacco cells cultivated at various temperatures. Phytochemistry 22:855–859

Gellerman JL, Anderson WH, Richardson DG, Schlenk H (1975) Distribution of arachidonic and eicosapentaenoic acid in the lipids of mosses. Biochim Biophys Acta 388:277–290

Grimsley NH, Ashton NW, Cove DJ (1977) The production of somatic hybrids by protoplast fusion in the moss *Physcomitrella patens.* Molec gen Genet 154:97–100

Hackett DP, Haas DW, Griffiths SK, Niederpruem DJ (1960) Studies on the development of cyanide resistant respiration in potato tuber slices. Plant Physiol 37:8–19

Hock K (1984) Untersuchungen zur Kompartimentierung der Enzyme des pflanzlichen Acetylcholin-Metabolismus mit Hilfe pflanzlicher Protoplasten. PhD Thesis, Mainz

Hock K, Hartmann E (1981) Localization of phosphatidylcholine synthesizing enzymes in etiolated bean seedlings *(Phaseolus vulgaris* L.). Plant Sci Lett 21:389–396

Jackman ME, Van Steveninck RFM (1967) Changes in the neoplasmic reticulum of beetroot slices during aging. Austr J Biol Sci 20:1063–1068

Jahnen W, Hartmann E (1983) Microfluorometric determination of early regeneration of moss protoplasts of *Ceratodon purpureus* (Hedw.). In: Potrykus I, Harms CT, Hinnen A, Hütter R, King PJ, Shillito RD (eds) Protoplasts 1983, 6th INternational Protoplast Symposium, Basel, Birkhäuser Verlag, pp 182–184

Kahl G (1982) Molecular biology of wound healing: The conditioning phenomenon. In: Kahl G, Schell JS (eds) Molecular Biology of plant tumors, Academic Press, London New York, pp 211–267

Key JL, Lin CY, Chen YM (1981) Heat schock proteins in higher plants. Proc Natl Acad Sci USA 78:3526–3530

Laties GG (1976) Membrane destruction and resynthesis in salt absorbing tissue. In: Proceedings of the workshop on salt effects on plant structures and processes, Riverside Cal, Agricultural Res Service US

Lilijenberg C, Kates M (1982) Effect of waterstress on lipidcomposition of oat seedling root cell membranes. In: Wintermans FJGM, Kuiper PJC (eds), Biochemistry and metabolism of plant lipids, Elsevier Biomedical Press, Amsterdam, pp 441–444

Mazliak P (1981) Regulation acourt terme et a long terme de l'activite des enzymes membranaires par la temperature. Physiol Veg 19:543–563

Moore TS, Lord JM, Kagawa T, Beevers H (1973) Enzymes of phospholipid metabolism in the endoplasmic reticulum of castor bean endosperm. Plant Physiol 52:50–53

Nagy JI, Maliga P (1976) Callus induction and plantlet regenration from mesophyll protoplasts of *Nicotiana sylvestris*. Z Pflanzenphysiol 78:453–455

Nakano M, Asahi T (1970) Biochemical studies on mitochondrial formed during aging of sliced potato tissue. Plant & Cell Physiol 11:499–502

Radwan SS, Grosse-Detinghaus S, Mangold HK (1978) Lipids in plant tissue cultures. VI. Effect of temperature on the lipids of *Brassica napus* and *Tropaeolum majus* cultures. Chem Phys Lipids 22:177–184

Rauch V (1984) Beeinflussung des Lipidmusters durch Isolierung und Regeneration von Protoplasten aus Moosprotonemen. Thesis, Mainz

Saxena PK, Rashid A (1981) High frequency regeneration of *Funaria hygrometrica* protoplasts isolated from low calcium protonemal suspension. Plant Sci Lett 23, 117–122

Swanson ES, Anderson WH, Gellerman JL, Schlenk H (1976) Ultrastructure and lipid composition of mosses. Bryologist 79, 337–349

Theologis A, Laties GG (1981) Wound-induced membrane lipid breakdown in potato tuber. Plant Physiol 68:53–58

Thu PTA, Flood C, Dasilva JV (1982) Effects of water stress on lipid and fatty acid compositions of cotton leaves. In: Wintermans FJGM, Kuiper PJC (eds), Biochemistry and metabolism of plant lipids, Elsevier Biomedical Press, Amsterdam, pp 451–454

Webster PL (1980) "Stress" protein synthesis in pea root meristem cells? Plant Sci Lett 20: 141–145

Willemot C (1983) Rapid degradation of polar lipids in forst damaged winter wheat crown and root tissue. Phytochemistry 22:861–863

Proline in Protoplasts: The Chemical Potential of Proline and Stress Sensitivity of Cells

E. PAHLICH

Introduction

Proline and some other compatible solutes accumulate in plants in response to environmental stress. The stress factors (high and low temperature, salinity, water deficit), physiological responses towards stress and the functional aspect of proline accumulation have been reviewed by Paleg and Aspinall (1981) and Miflin (1980).

Little information exists explaining the accumulation of proline as a consequential effect of stress. The stress-borne trigger mechanisms are not yet known. Also the "master reactions" (if there are any) which are perturbed by stress factors remain to be elucidated. Therefore the proline accumulation system still remains an interesting model for the study of environmentally induced metabolic acclimatizations (Harborne 1977) and/or adaptions.

The stress-related proline pool should be regarded as the sum of various processes. These include proline producing and consuming reaction sequences and diffusion and/or transport processes. As judged from recent concepts of metabolic network dynamics (Heinrich and Rapoport 1974, Kacser and Burns 1973, Reich and Sel'kov 1981, Heinrich and Rapoport 1983, Kacser 1983), neither the reaction sequences nor the transport steps acting on the proline pool have been analyzed in an appropriate way. Measurements of single enzyme activities or determinations of single enzyme properties before and after stress treatment are fragmentary and hence insufficient to characterize complex dynamic systems. Taking this into account, very much research still has to be done in order to explain proline accumulation from a mechanistic point of view.

Recently it has been shown that plant cell cultures can also respond to water and salt stress by proline accumulation (see for instance Handa et al. 1983, Watad et al. 1983 and preceding chapters in this volume). In this respect, cell cultures behave very much like "intact" plants and might provide adequate model systems for the study of the mechanism of stress-induced metabolic perturbations.

In this chapter we report on milieu-dependent changes of the chemical potential of proline. From in vitro experiments it is seen that the chemical potential (the thermodynamic activity) of proline can be changed on a broad scale by alterations of the solute milieu (water potential, solute composition, pH). Thus changes of the solute milieu result in "concentration changes" of proline and hence in changes of respective reaction and transport rates. Based on these observations, we present a mechanism which might explain "primary stress effects" on mainly thermodynamic

The Physiological Properties of Plant Protoplasts
Edited by P. E. Pilet

grounds. These in vitro experiments help to explain the findings on the proline pool dynamics in potato cell cultures which are performed in parallel.

Methods

The experimental conditions for the determination of the thermodynamic activity "a" of proline in different solutions have been described elsewhere (Pahlich and Stadermann 1984). Diffusion experiments were carried out with special Disc micro chambers (500 μl) (Schütt, Göttingen). Proline determinations were performed following the procedure of Bates et al. (1973). The potato cell cultures were grown as described by Mühlbach and Sänger 1981. We are indepted to Prof. Sänger and Dr. Mühlbach for providing the cell cultures.

The Cellular Proline Pool: a Model

Figure 1 represents an extremely simplified model of the input and output reactions constituting the cytoplasmic proline pool. The results on proline distribution in isolated tobacco protoplasts (Pahlich et al. 1983) caused us to include a storage pool (vacuole) in the model. For the sake of clearness, stoicheiometric relations, details of the reaction sequences and diffusion-reaction interactions are not included in the model.

The variation with time of the proline pool can be described by Eq. (1). As can be seen from this equation, there are at least seven reaction systems which can be effected individually or in concert by stress treatment and thus produce proline pool fluctuations. The reaction rates are represented by $v_1 - v_6$.

$$\left[\frac{d[\mathrm{Pro}]_{\mathrm{Cyt}}}{dt} = \right] \begin{array}{l} V\ \mathrm{input} - V\ \mathrm{output} \\ = v_1 + v_4 + v_{5'} - (v_2 + v_3 + v_{5'} + v_6) \end{array} \qquad (1)$$

Equation (2) shows the respective reaction rate expressions in more detail. It is obvious that "concentrations" and rate and transport constants are the respective targets of stress effects.

$$\frac{d[\mathrm{Pro}]_{\mathrm{Cyt}}}{dt} = k_1[X \cdot \mathrm{NADH} \cdot \mathrm{ATP}] + k_4 K_{\mathrm{Vac}}[\mathrm{Pro}]_{\mathrm{Cyt}} + k_{5'}[Y] \qquad (2)$$

$$- k_2[\mathrm{Pro}]_{\mathrm{Cyt}} \cdot [\mathrm{NAD}^+] + k_3[\mathrm{Pro}]_{\mathrm{Cyt}} + k_5[\mathrm{Pro}]_{\mathrm{Cyt}} + k_6[\mathrm{Pro}]_{\mathrm{Cyt}}$$

(An introduction and discussion of basic interactions in rate processes is given by Pahlich 1984).

From Eq. (2) it can be seen that proline links the reaction systems. Any changes of the proline "concentration" brought about by water potential changes will produce a motion of the whole reaction system. An appropriate measure of the susceptibility of respective fluxes toward concentration changes is the "elasticity" of the reations of a sequence (Kacser and Burns (1973). No such analysis is available at present for reaction processes depicted in Fig. 1.

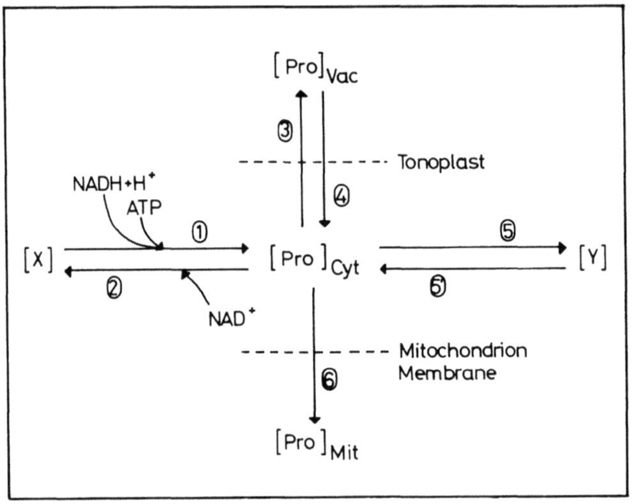

Fig. 1. The cytoplasmic pool of proline: a simplified flux scheme of input and output processes. [X] = precursor pool, $[Pro]_{Cyt}$, $[Pro]_{Vac}$ and $[Pro]_{Mit}$ are the respective proline pools in the cytoplasm, the vacuole and the mitochondria. Numbers symbolize the following rates

$$(v = \frac{d[Pro]_{Cyt}}{dt})$$

1 Biossynthesis: $v_1 = k_1 \, [X \cdot NADH \cdot ATP]$;

2 Degradation: $v_2 = k_2 \, [Pro]_{Cyt} \cdot [NAD]$;

3 Deposition: $v_3 = k_3 \, [Pro]_{Cyt}$;

4 Storage mobilization: $v_4 = k_4 \cdot K_{Vac} \cdot [Pro]_{Cyt}$, $[Pro]_{Vac} \; [Pro]_{Cyt} = K_{Vac}$

5 Protein synthesis and hydrolysis: $v_5 = k_5 \, [Pro]_{Cyt}$,

$$v_{5'} = k_{5'} \cdot [Y]$$

6 Export into Mitochondria (and/or into free space)

$$v_6 = k_6 \, [Pro]_{Cyt}$$

k's represent rate constants of the respective reaction sequences and transport processes

Furthermore Eq. (2) shows that stress treatment of cells might act on rate constants (catalytic and transport constants). In order to gain insight into whether this happens or not, the "sensitivities" of reaction steps must be determined (Kacser and Burns 1973). This information is also not available at present.

In order to analyze the phenomenon of proline accumulations as outlined, we have startet to investigate the effect of water potential changes on the chemical potential of proline. The question was whether milieu-dependent alterations of the chemical potential of proline (and other cell constituents) are large enough to evoke reaction and diffusion rate changes. The results suggest an additional and as yet unnoticed aspect in the field of stress research.

Results and Discussion

The Thermodynamic Activity of Proline

One of the peculiar colligative properties of proline is its solubility in water (Greenstein and Winitz 1961) and in solutions (Krishna Sastry et al. 1982). Intuitively, this solubility can be regarded as important for maintaining dissolved proline even under unfavorable solute conditions.

An appropriate description of the properties of a solute in solution is its chemical potential. The chemical potential μ is defined as (Nobel 1974)

$$\mu = \mu_0 + RTlna + \overline{V}P + zFE + mgh$$

μ_0 is a reference level, a the thermodynamic activity, $\overline{V}$ the partial molal volume, zFE an electrochemical and mgh a gravitanional term. The other symbols have the usual meaning. In our investigation we made the approximation that the chemical potential is adequately described by the thermodynamic activity a, thus neglecting the additional terms. The thermodynamic activity (the "corrected" concentration) is defined as

$$a = \gamma \cdot m$$

a= thermodynamic activity, γ= activity coefficient (which changes with the environment), m= molality. The acitivty is a measure of the "thermodynamically active part of the concentration" (Nobel 1974) and hence describes the biochemical availability of a reactant.

The activity of proline changes with respect to water potential, solute composition and solute pH to various degrees, as seen in Fig. 2 (also Pahlich and Stadermann 1984). The most pronounced changes of the thermodynamic activity of proline are measured in PEG solutions. In KCl and saccharose solutions the activity varies to a lesser extent. The activity changes are pH-dependent. Under all conditions measured, the activity of proline is higher then the weighed portion of the moeality, which indicates that the activity coefficient exeeds unity.

If water potential changes are applied to cells, a similar behaviour of the activity of proline would also be expected. The proline activity in different cell compartments changes differentially with stress because of the distinct "solute composition" of single compartments. Regarding the activity changes in PEG as "typical" for the situation in the free cytoplasm and in the KCl and the saccharose solution as "typical" for the vacuole shows that identical water potential variations lead to new intercompartmental gradients, of proline because of activity changes.

Similar effects can be expected also for other amino acids and metabolites. Schrier and Robinson (1974) measured marked differential effects of salt on the activity coefficients of alanine and glycin and of di- and tripeptides. Even tissue-specific activites of ions have been described (Wiggins 1975).

The few results cited show that single amino acids, peptides and even ions respond specifically, and to different extents, towards milieu changes. This differential response towards milieu changes is a prerequisite to disturb, even to destabilize a previously homoeostatic metabolic system. Stress brought about by water loss or

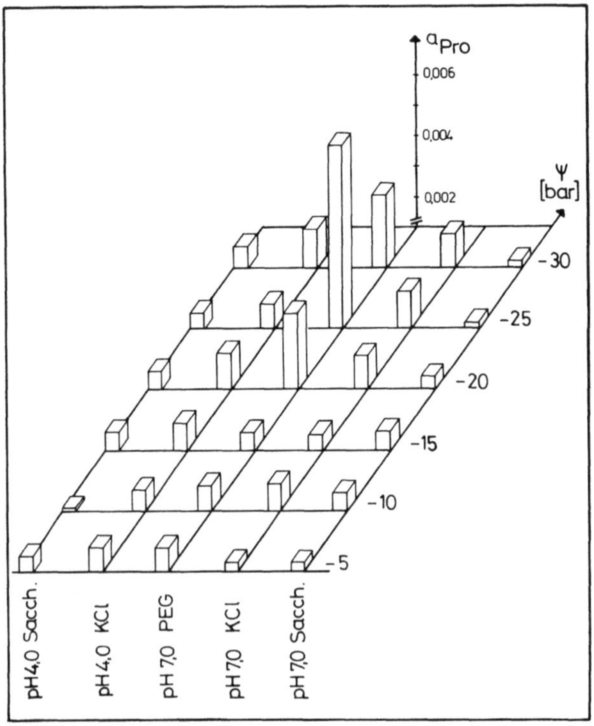

Fig. 2. Thermodynamic activity changes of proline in ternary solutions. PEG (polyethyleneglycol), KCl and Saccharose solutions of different water potential and pH containing proline at 10^{-3} molal concentrations. *Bars* represent the respective thermodynamic activities of proline

extremes of salt, temperature- or pH-changes can thus alter "effective concentrations" by purely thermodynamic reasons. This type of signal-response effect might be part of the stress trigger. The resulting perturbations of concentrations "induce" pool movements which finally end up in new states of metabolic stability. (For a more extended discussion of stability problems see Reich and Sel'kov 1981, Boiteux et al. 1980, Schiffman 1980.)

Thermodynamic Activity and Diffusion of Proline

"The difference in chemical potential of a substance in two phases determines the direction in which the substance will diffuse spontaneously" (Slatyer 1967). In order to show whether the determined proline activities fulfill this requirement we are performing diffusion experiments. A typical result is shown in Table 1.

Both chambers of the diffusion cell, separated by a dialysis membrane, contained initially 10^{-3} molale solutions of proline (t=O) in PEG-water (–30 bar) and in water (0 bar). From the results shown in Fig. 2 it is seen that proline in PEG-water has a higher activity then in water. In fact, proline diffuses spontaneously into the water phase thus producing an "up hill" gradient as judged from the resulting molalities shown in Table 1.

Table 1. Thermodynamic activity and diffusion of proline

t(h)	Molality A (PEG, ψ=−30 bar)	B (H_2O dest.)
0	1.00×10^{-3}	1.000×10^{-3}
6	0.95×10^{-3}	1.003×10^{-3}
12	0.67×10^{-3}	1.027×10^{-3}
18	0.60×10^{-3}	1.034×10^{-3}

A: Inner chamber of the dialysis cell with a fixed volume of 0.5 ml. Proline is dissolved in a PEG solution (ψ=−30 bar) at a 10^{-3} molal concentration
B: Outer dialysis chamber with a volume of 10 ml. Proline is dissolved in dest. water also at a 10^{-3} molal concentration
Numbers represent the final molal concentrations of proline within the chambers after 6, 12 and 18 h of diffusion at 25 °C. The proline diffuses from the PEG solution into the water phase which is in accordance with the measured thermodynamic activity difference of proline within these solutions

We expect this to happen also in cells. Changing the water potential means that the thermodynamic activity of proline (and perhaps the other compatible solutes) also changes, but differentially in the cytoplasm and other cell compartments (vacuole, mitochondria).

Thus water stress applied to cells forms new gradients across cell membranes, thus being the "signal" for the establishment of new intracompartimental pools. From the results shown in Fig. 2 it may be judged that stress changes the activity of proline in the cytoplasm (here PEG solutions) to a higher extent then in the vacuole (here KCl or saccharose). Therefore an increased influx of proline into the vacuole during increasing stress can occur. This is in accordance with results on the vacuole/extravacuole distribution of proline in tobacco protoplasts (Pahlich et al. 1983). The proline withdraws from the active metabolic pool and accumulates in the vacuole.

It would be interesting to know whether the masured vacuole/extravacuole pools of proline (Leigh et al. 1981, Pahlich et al. 1983) represent thermodynamic equilibria or not. This would enable us to recalculate energy profiles of the biosynthetic reaction sequence (Pahlich et al. 1981, 1982) and help to understand the changes of the proline pool system during stress in terms of thermodynamic concepts.

Time Course of Proline Accumulation with Respect to Growth in Cell Cultures

The time course of proline formation in potato suspension cultures (not shown here) bears some interesting details. The proline pool peaks in the late growth phase. This indicates a diphasic pattern of the input-output relations of the proline pool. The proline pool-forming sequences outweigh the consuming reaction sequences in the early growth phase. Later, this reverses. Generally the proline pool in stressed cells is increased as compared to the respective control. The shape of the time-dependent proline pool profile, however, is very similar in both control and stressed samples, through the absolute amount of proline in stressed cells is increased. Irrespective of an externally unchanged stress effect, the proline pool first increases and later decreases.

A similar behavior of proline accumulation was observed with water-stress-adapted tomato cells (Handa et al. 1983). From this it is clear that stress treatment of cells produces a quantitative change of the proline pool. But there is a decline of the proline pool in the late exponential phase of growth in stressed and unstressed cells. Despite a possible protection function of proline at some time during growth phase this profile shows: the proline pool changes have to be regarded primarily as indications of metabolic transitions between functionally different bi- or multistable steady states (Schiffman 1980). The role of activity changes of proline in the course of these changes is not yet clear. But these metabolic transitions are necessarily accompanied by concentration changes in other cell constituents. Indeed concentration changes in other cell constituents have been shown in stress-treated cells (Heyer and Nabors 1981, Handa et al. 1983). The effect of stress specifically on the concentration of proline and other compatible solutes coincides with the fact that these molecules possess activity coefficients greater than one in concentrated solutions, and – as hitherto shown for proline – that the activity coefficient also at low concentrations responds very flexibly towards environmental changes. We see a close connection between these physical properties and the physiological effect.

Conclusions

Treating the accumulation of proline in cells as a consequential effect of stress requires analysis of the system parameters and variables, i.e. the thermodynamic and kinetic properties of the respective reations and transport processes. The outstanding effect of stress application on the proline pool results at least in part from the outlined thermodynamic properties of this amino acid. Changes of the respective solute environment in the cytoplasm and the other compartments of the cell results in differential alterations of the chemical potential of proline. Though not proved in vivo this effect can be deduced from in vitro experiments.

This concept can be extended to a general hypothesis. The changes of the chemical potentials of single ions, metabolites and other cell constituents by increasing or decreasing stress are not proportional. This differential behavior initiates a metabolic perturbation because of altered diffusion gradients, changed reaction rates, disturbed feedback and feedforward regulations, altered allosteric interactions, etc.

Furthermore also proteins and membranes are molded in their structure by stress-induced solute changes, giving rise to altered rate and transport constants (Pahlich, in progress).

Most of the pronounced stress effects on reaction rates would be expected in those reaction sequences which contain metabolites and catalysts capable of responding with large chemical potential changes. The classical key-lock concept of E. Fischer for signal recognition and signal verification has to be extended. Stress signals like changes of water content, temperature, pressure etc. are realized in metabolism primarily via solute perturbations as predicted by the Gibb's-Duhem relation (Moore and Hummel 1976).

Acknowledgments. The author expresses his graditude to Mr. N. Crompton (M.Sc.) for editing the manuscript. I thank Mrs. M. Reeh for preparing the drawings and typing the manuscript.

References

Bates LS, Wildren RP, Teary JD (1973) Rapid determination of free proline for waterstress studies. Plant Soil 39:205–207

Boiteux A, Hess B, Sel'kov EE (1980) Creative functions of instability and oscillations in metabolic systems. Curr Top Cell Regul 17:171–201

Greenstein IP, Winitz M (1961) Chemistry of the amino acids, Vol 1. Wiley, New York

Handa S, Bressan RA, Handa AK, Carpita NC, Hasegawa PM (1983) Solutes contributing to osmotic adjustment in cultured plant cells adapted to water stress. Plant Physiol (Bethesda) 73:834–843

Harborne JB, (1977) Introduction to ecological biochemistry. Academic London

Heinrich R, Rapoport TA (1974) A linear steady-state treatment of enzymic chains. General properties, control and effector strenght. Eur J Biochem 42:89–95

Heinrich R, Rapoport TA (1983) The utility of methematical models for the understanding of metabolic systems. Biochem Soc Trans 11:29–35

Heyser JW, Nabors MW (1981) Growth, water content and solute accumulation of two tobacco cell lines cultured on sodium chloride, Dextran and Polyethylene glycol. Plant Physiol (Bethesda) 68:1454–1459

Kacser H (1983) The control of enzyme systems in vivo: elasticity analysis of the steady-state. Biochem Soc Trans 11:35–40

Kacser H, Burns JA (1973) The control of flux. In: Rate control of biological processes. Sym Soc Biol Vol 27. Cambridge Press, Cambridge, pp 65–104

Krishna Sastry KS, Udayakumar M, Devendra R, Mekkri AA (1982) Solubility of proline and its biological significance. Curr Sci (Bangalore) 51:485–486

Leigh RA, Ahmad N, Wyn RG (1981) Assessment of glycinebetaine and proline compartmentation by analysis of isolated beet vacuoles. Planta (Berl) 153:34–41

Miflin BJ (ed) (1980) In: Stumpf PK, Conn EE (ed inchief) The Biochemistry of plants. A comprehensive treatise, Vol 5. Amino acids and derivates. Academic, New York

Moore WJ, Hummel DO (1976) Physikalische Chemie. Walter de Gruyter, Berlin

Mühlbach HP, Sänger HL (1981) Continous replication of potato spindle tuber viroid (PSTV) in permanent cell culture of potato and tomato. Bio Sci Rep 1:79–87

Nobel PS (1974) Introduction to biophysical plant physiology. Freeman, San Francisco

Pahlich E (1984) Prinzipien der Geschwindigkeitskontrolle von Stoffwechselabläufen. Orientierungshilfe für die praktische Arbeit. Vieweg, Braunschweig

Pahlich E (to be published) Water stress induced metabolic perturbations: primary mechanisms and theoretical background

Pahlich E, Stadermann T (1984) The thermodynamic activity of proline in ternary solutions of different water potentials. Z Pflanzenphysiol 115:91–96

Pahlich E, Jäger H-J, Kaschel E (1981) Thermodynamische Betrachtungen über die reversible Reaktionssequenz Glutaminsäure Prolin. Z Pflanzenphysiol 101:137–144

Pahlich E, Jäger H-J, Horz M (1982) Weitere Uuntersuchungen zur thermodynamischen Struktur der Biosynthesesequenz Glutaminsäure Prolin in wassergestressten Buschbohnen. Z Pflanzenphysiol 105:475–478

Pahlich E, Kerres R, Jäger H-J (1983) Influence of water stress on the vacuole/extravacuole distribution of proline in protoplasts of Nicotiana rustica. Plant Physiol (Bethesda) 72: 590–591

Paleg LG, Aspinall D (eds) (1981). The physiology and biochemistry of drought resistence in plants. Academic, New York

Reich JG, Sel'kov EE (1981) Energy metabolism of the cell. A theoretical treatise. Academic, London

Schiffman Y (1980) Bifurcation in the privileged two-dimensional reaction-diffusion system as the ligand-induced redistribution, and biochemical control as its functional significance. Prog Biophys Mol Biol 36:88–122

Schrier EE, Robinson RA (1974) Free energy relations in aqueous amino acid and peptide solutions containing sodium chloride. J Solution Chem 3:493–501

Slatyer RO (1967) Plant-water relations. Academic Press, London

Watad AA, Reinhold L, Lerner HR (1983) Comparison between a stable NaCl-selected Nicotiana cell line and the wild type. K^+, Na^+ and proline pools as a function of salinity. Plant Physiol (Bethesda) 73:624–629

Wiggins PM (1975) Cellular functions of a cell in a metastable equilibrium state. J Theor Biol 52:99–11

The Biosynthesis and Catabolism of Indole-3-Acetic Acid in Protoplasts

G. SANDBERG and A. CROZIER

Introduction

There is much interest in the involvement of the auxin, indole-3-acetic acid (IAA), in various aspects of higher plant growth and development (see Letham et al. 1978). Information on IAA pool sizes and how they are controlled by biosynthetic and catabolic processes is an integral part of an understanding of the mechanisms by which IAA influences growth. In recent years, investigations of endogenous IAA levels have made use of procedures such as combined gas chromatography-mass spectrometry (GC-MS) in the single and multiple ion monitoring modes (Rivier and Pilet 1974), high performance liquid chromatography (HPLC) with fluorescence detection (Crozier et al. 1980) and immunoassays (Weiler 1982), in order to obtain reliable quantitative data. While these techniques have been utilized in elegant studies on the production and hydrolysis of IAA conjugates in *Zea mays* kernels (see Bandurski 1984, Cohen and Bialek 1984) they have not as yet found widespread application in investigations on IAA biosynthesis and catabolism. As a consequence, progress in this aspect of auxin biochemistry compares most unfavourably with equivalent research on abscisic acid (see Neill et al. 1984), cytokinins (see McGaw et al. 1984), and gibberellins (see Crozier 1983a,b).

There are many reports in the literature on the production and degradation of IAA by both in vivo and in vitro test systems which give the impression of a thoroughly researched field in which the enzymes involved have been identified and the intermediates in the metabolic pathways carefully evaluated (see Schneider and Wightman 1978). However, a review by Bearder (1980) reveals a quite different situation and it is evident that few of the presumed intermediates of IAA metabolism (see Fig. 1) have been identified on the basis of strict physico-chemical evidence. There are also shortcomings with the generally accepted picture of IAA catabolism that is summarised in Fig. 2. Although many of the components in the non-decarboxylative pathway have been detected in extracts from higher plants, experimental evidence on the exact catabolic routes operating in individual species is very limited indeed (Bandurski and Nonhebel 1984). Decarboxylative catabolism of IAA to products such as 3-hydroxymethyloxindole and indole-3-aldehyde has been extensively investigated in vitro using assorted IAA oxidase preparations and commercially available enzymes such as horseradish peroxidase. Unfortunately equivalent information from in vitro higher plant test systems is scant and very few of the compounds potentially involved have been satisfactorily identified as either endogenous constituents or catabolites of IAA (Sandberg et al. 1984).

The Physiological Properties of Plant Protoplasts
Edited by P. E. Pilet

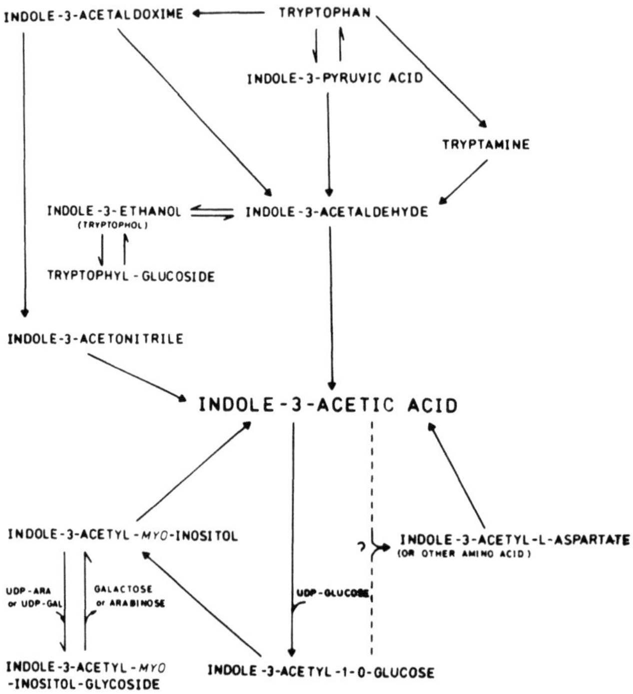

Fig. 1. IAA biosynthesis pathways. (For structures see Cohen and Bialek 1984)

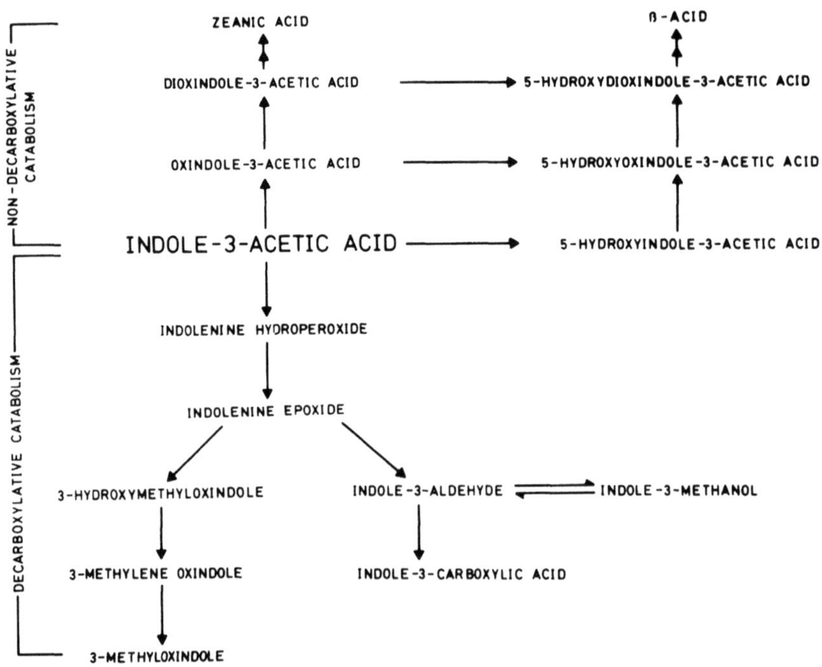

Fig. 2. IAA catabolism pathways. (For structures see Sembdner et al. 1980)

Protoplasts have been widely used in photosynthesis research and are now being utilized with increasing frequency in studies on various aspects of the compartmentation of plant growth regulators (see Garcia-Martinez et al. 1981, Hartung et al. 1981a, b, Kramers 1981, Weiler et al. 1982). Work with auxins includes studies on the production and/or catabolism of IAA by protoplasts from spinach *(Spinacia oleracea)* (Heilmann et al. 1982), barley *(Hordeum vulgare)* (Sandberg et al. 1982, 1983), Scots pine *(Pinus sylvestris)* (Sandberg and Hällgren 1985; Sundberg et al. 1985b) and pea *(Pisum sativum)* (Sandberg and Crozier, unpublished data).

Spinacia oleracea

Heilmann et al. (1982) used leaf mesophyll protoplasts from spinach to investigate intracellular localization of IAA biosynthesis. It was observed that the uptake of [^{14}C]tryptophan by chloroplasts was more extensive than that of intact protoplasts. Protoplasts, as well as chloroplasts and a cytoplasmic fraction derived from protoplasts all converted [^{14}C]tryptophan to [^{14}C]IAA. The overall yield of [^{14}C]IAA produced by the chloroplasts and the cytoplasmic fraction was greater than that synthesized by intact protoplasts. It was suggested that this could be a consequence of the movement of [^{14}C]tryptophan across the plasmalemma acting as the rate limiting step in [^{14}C]IAA production in protoplasts. However, one should not read too much into the findings of Heilmann et al. (1982) as sample purification was minimal and the analysis of [^{14}C]IAA in [^{14}C]tryptophan incubates was based on thin layer chromatography (TLC), which does not provide sufficient resolution for reliable identifications (Reeve and Crozier 1980).

Hordeum vulgare

Schneider et al. (1972) have reported that light-grown barley shoots contain IAA, indole-3-aldehyde, tryptamine, 5-hydroxytryptamine, indole-3-ethanol and indole-3-aldehyde, and indole-3-lactic acid; while Gibson et al. (1972) have published information on the following conversions in the same tissue (1) [3-^{14}C]tryptophan to tryptamine, indole-3-pyruvic acid, indole-3-acetaldehye, IAA and indole-3-aldehyde and (2) [2-^{14}C]tryptamine to indole-3-acetaldehyde, IAA and indole-3-aldehyde. Although these papers provide interesting background information, they should be viewed with caution as no mass spectrometric data were obtained and the identifications were based on paper chromatography and TLC retention properties. Furthermore, in the metabolism studies the precursors were fed in amounts well in excess of physiological levels, i.e. 80 mg tryptamine to 700 g of shoots.

The biosynthesis and catabolism of IAA by protoplasts, chloroplasts and a cytoplasmic fraction from barley leaves have been investigated by Sandberg et al. (1982, 1983). Combined gas chromatography-mass spetrometry (GC-MS) was used to demonstrate that IAA is an endogenous constituent of *Hordeum* protoplasts while HPLC was utilized to show (1) protoplasts converted enzymatically [5-^{3}H]tryptophan to [^{3}H]IAA, and (2) both chloroplasts and a cytoplasmic fraction, isolated from protoplasts that had previously been fed [5-^{3}H]tryptophan, contained [^{3}H] IAA. Chloroplasts and cytoplasmic preparations isolated from protoplasts and thereafter incubated with [5-^{3}H]tryptophan also synthesized [^{3}H]IAA, although in both instances the pool size was less than half of that detected in in vivo feeds. This clearly differs from data obtained by Heilmann et al. (1982) in equivalent experiments with spinach mesophyll protoplasts.

There was no significant difference in the amount of [5-^{3}H]tryptophan-derived [^{3}H]IAA that accumulated in barley protoplasts and chloroplasts incubated in light and darkness. Similarly, light had no effect on the [^{3}H]IAA pool produced by the cytoplasmic fraction. Protoplasts did, however, catabolise [2-^{14}C]IAA faster in light than in darkness and, although the rate of degradation was slower than in protoplasts, chloroplasts also catabolised [2-^{14}C]IAA more rapidly when incubated in light. Light did not alter the catabolic activity of the cytoplasmic fraction. The data, thus, indicate that the influence of light on IAA catabolism in protoplasts is mediated primarily via effects on chloroplasts rather than the cytoplasm. However, it should be noted that the combined rate of degradation of [2-^{14}C]IAA by chloroplast and cytoplasmic fractions in the light was much lower than that of intact protoplasts. It seems therefore, that the breakdown of intracellular compartmentation reduces the ability of chloroplasts to respond to light and/or adversely affects the activity of the catabolic enzymes because of a mixing of proteins, co-factors and ions that are normally strictly segregated.

As light enhances the rate of [2-^{14}C]IAA catabolism by barley protoplasts yet does not alter significiantly the amount of [^{3}H]IAA formed from [5-^{3}H]tryptophan, the size of the [^{3}H]IAA pool in the light must be maintained by an increased rate of synthesis from [5-^{3}H]tryptophan. Light, thus, appears to enhance the rate of IAA turnover in protoplasts. Similar inferences can be drawn regarding the effect of light on IAA turnover in chloroplast preparations.

Pinus sylvestris

In a series of investigations, GC-MS has been used to identify and quantifiy IAA (24.5 ± 6.5 ng g^{-1} f.wt.), indole-3-ethanol (46.2 ± 4.1 ng g^{-1}), indole-3-methanol (19.5 ± 1.4 ng g^{-1}) and indole-3-carboxylic acid (2.3 ± 0.4 ng g^{-1}) in extracts from needles of Scots pine (Sandberg et al. 1981, 1984, Sandberg 1984, Sundberg et al. 1985a). In addition, the presence of the conjugate, indole-3-acetyl-L-aspartate has also been established (Andersson and Sandberg 1982). Metabolism studies with *Pinus* needles have demonstrated the conversion of [2-^{14}C] and [3-^{14}C]tryptophan to [^{14}C]indole-3-ethanol and in turn the conversion of [^{14}C]indole-3-ethanol to [^{14}C]IAA. Surprisingly, [2-^{14}C]IAA was not catabolised to [^{14}C]indole-3-carboxylic acid in detectable quantities indicating that at best only a minor portion of the endogenous indole-3-carboxylic acid pool originates from IAA (Sandberg 1984). In addition, a crude enzyme preparation from *Pinus* needles has been shown to convert [$^2H_5$]IAA to [$^2H_5$]indole-3-methanol (Sundberg et al. 1985b).

Sandberg and Hällgren (1985) have recently developed procedures for isolating protoplasts from shoots of dark-grown *Pinus sylvestris* seedlings. Studies on the metabolism of [1-^{14}C]- and [2-^{14}C]IAA by these protoplasts have been initiated and preliminary information is available on two of three catabolites that have been detected (Sandberg and Hällgren 1985; Sundberg et al. 1985b). One catabolite co-chromatographs with indole-3-methanol in both reversed- and normal-phase HPLC systems. The other has not been identified but it is very polar when analysed by ion pair HPLC, cannot be hydrolysed to yield IAA and appears to be a product of the non-decarboxylative IAA catabolism pathway (see Fig. 2) as it retains the C-l label.

Pisum sativum

The chlorinated indoles, 4-chloroindole-3-acetic acid, methyl 4-chloroindole-3-acetic acid, monomethyl 4-chloroindole-3-acetyl-L-aspartate, α-N-carbomethoxy-acetyl-D-4-chlorotryptophan and α-N-carboethyloxyacetyl-D-4-chlorotryptophan have been identified in extracts from immature seed of *Pisum sativum* (Marumo et al. 1986, a, b, Marumo and Hattori 1971, Hattori and Marumo 1972). HPLC and GC-MS analyses have shown that shoots of light-grown *Pisum sativum* c v. Meteor seedlings contain IAA and indole-3-ethanol (Brown, Crozier and Sandberg, unpublished data). IAA has also been identified as an endogenous constituent of *Pisum* chloroplast preparations. Preliminary studies indicate that the chloroplasts convert [5-^{3}H]tryptophan to [^{3}H]IAA, while investigations with [1-^{14}C]- and [2-^{14}C] IAA suggest that the preparations decarboxylate IAA. Although the chloroplast may catabolites have not yet been identified conclusively, a crude enzyme system capable of converting [2-^{14}C]IAA to [^{14}C]indole-3-methanol has been isolated from *Pisum* chloroplast preparations (Brown, Crozier and Sandberg, unpublished data). It has also been reported that indole-3-aldehyde, indole-3-methanol and indole-3-carboxylic acid are IAA decarboxylation products in etiolated pea stem sections (Magnus et al. 1971). There are, however, certain limitations with this study (1) none of the substrates were isotopically labelled (2) they were fed in very large amounts (3–4 mg per 10 g of tissue) and (3) the putative products were identified only on the basis of the TLC data and colour reactions.

Sandberg and Crozier (unpublished data) have used protoplasts from shoots of *Pisum sativum* cv. Meteor to survey the effects of incubation time in digestion medium, pH and osmotic environment on the metabolism of [2-^{14}C]IAA. Protoplasts subjected to 2 h in digestion medium catabolised ca 30% of the applied substrate during a subsequent 5 h incubation with [2-^{14}C]IAA. HPLC revealed the presence of five catabolites. Protoplasts prepared after a 4-h incubation in digestion medium catabolised up to 35% of the added [2-^{14}C]IAA and there were small but distinct differences in the spectrum of catabolites detected by HPLC (Fig. 3).

In order to investigate the effect of the osmotic environment, protoplasts were incubated with [2-^{14}C]IAA in media containing 450, 475, 500, 525 and 550 mM l^{-1} sorbitol for 5 h. The data obtained are presented in Table 1. The amount of radioactivity taken up by the protoplasts was broadly similar in all instances while the amount of catabolism observed was also similar except for the highest sorbitol concentration where it was reduced. Thus, the osmotic environment does not affect markedly the uptake and catabolism of [2-^{14}C]IAA by *Pisum sativum* protoplasts.

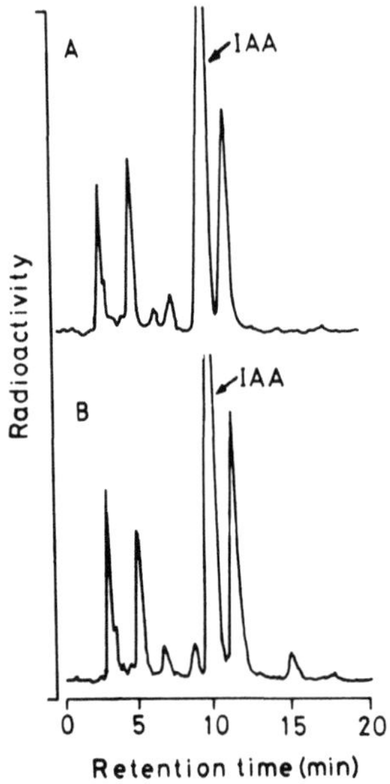

Fig. 3. Effect of time in digestion medium on subsequent metabolism of [2-^{14}C]IAA by protoplasts from leaves of *Pisum sativum*. Methanolic extracts from protoplasts analysed by isocratic, reverse-phase HPLC with a radioactivity monitor operating in the heterogeneous mode. **A** 2-h digestion period; **B** 4-h digestion period

Table 1. Uptake and catabolism of [2-^{14}C]IAA by protoplasts from *Pisum sativum* leaves incubated in assay medium containing different concentrations of sorbitol. Total radioactivity in the protoplasts expressed as percent of [2-^{14}C]IAA initially added. Radioactivity associated with IAA catabolites determined by HPLC and expressed as percent of [2-^{14}C]IAA initially added. Protoplasts incubated at 20°C for 5 h under ambient light

Concentration of sorbitol	Total radioactivity recovered in the protoplasts (%)	Radioactivity associated with IAA catabolites in the protoplasts (%)
450 mM l^{-1} (1)	24.1	12.1
(2)	19.1	8.7
475 mM l^{-1} (1)	21.0	11.6
(2)	22.7	12.3
500 mM l^{-1} (1)	17.3	15.0
(2)	19.0	11.9
525 mM l^{-1} (1)	28.8	13.0
(2)	21.9	7.4
550 mM l^{-1} (1)	20.2	2.3
(2)	18.1	0.4

Table 2. Effects of pH on the uptake and catabolism of [2-^{14}C]IAA by protoplast preparations from *Pisum sativum* leaves. Total radioactivity in the protoplasts and the incubation medium expressed as percent of [2-^{14}C]IAA initially added. Radioactivity associated with IAA in the protoplasts and the incubation medium, determined by HPLC and also expressed as percent of [2-^{14}C]IAA initially added. Protoplasts incubated at 20°C for 5 h under ambient light

Incubation conditions	Total radioactivity (%) Protoplasts	Medium	Radioactivity associated with IAA(%) Protoplasts	Medium
pH 3.0 (1)	36.1	61.3	33.9	57.8
(2)	34.4	60.8	30.1	58.2
pH 4.0 (1)	31.0	64.0	28.7	59.6
(2)	33.0	66.9	24.3	61.0
pH 5.0 (1)	19.6	80.2	14.8	61.1
(2)	22.5	75.6	9.0	63.5
pH 6.0 (1)	14.7	80.1	4.9	59.3
(2)	12.4	87.0	8.6	71.0
pH 7.0 (1)	13.4	83.8	3.5	68.8
(2)	15.5	82.4	7.2	65.6

The effect of pH on the uptake of [2-^{14}C]IAA by *Pisum* protoplasts was investigated by incubating preparations at pH 3.0–8.0 for a 5-h period, after which the amount and nature of the radioactivity associated with both the protoplasts and the incubation medium were determined. The data obtained are presented in Table 2. Uptake of radioactivity was greatest at pH 3.0 and 4.0 while [2-^{14}C]IAA catabolism was most rapid at pH 7.0. It would seem that the plasmalemma is more permeable to IAA under acidic conditions when the molecule is not charged and in the undissociated form. This observation is in keeping with the findings of Rubery and Sheldrake (1973, 1974) obtained with suspension-cultured crown gall-infected cells of *Parthenocissus tricuspidata*. In contrast to uptake, catabolism of [2-^{14}C]IAA by *Pisum* protoplasts was more extensive at higher pH's (Table 2).

Conclusions

Protoplasts represent an interesting test system with which to study the biosynthesis and catabolism of IAA and other plant growth regulators. It is, however, necessary to bear in mind that protoplasts are exposed to artificial conditions and that they are not merely cells without cell walls. They are prepared, stored and incubated in concentrated solutions of sugar or plastic polymers to which the plasmalemma is more or less impermeable. This artificial environment inevitably disturbs the water balance of the cell, as it is difficult to obtain identical water potentials in the protoplasts and the medium. As a consequence, protoplast preparations are almost always under either positive or negative water stress. In addition, removal of the cell-wall, which contains important proteins for cell metabolism, results in protoplasts being exposed to digestive enzymes, preparations of which frequently contain an array of contaminants. It is perhaps not surprising, in the circumstances, that several studies on different aspects of metabolism have shown that protoplasts can behave quite differently to intact cells (see Galun 1981).

Studies with *Pisum* protoplasts discussed in the previous section have shown that the pH of the incubating medium can have a major bearing on the uptake and metabolism of IAA. It is evident that before detailed studies are initiated with plant hormones, the effects of not only pH but also incubation time in digestion medium, and osmotic environment should be investigated.

There are special problems to be overcome when studying IAA and related compounds. The use of adequate analytical procedures is, of course, essential otherwise inaccurate data of very dubious value will be obtained. It is also necessary to recognise that IAA itself and some of its precursors and catabolites are unstable under certain conditions. For instance, TCL and drying can results in a 30% conversion of tryptophan to IAA (Epstein et al. 1980). Under acidic conditions indole-3-acetonitrile breaks down to yield a number of indoles, including IAA. IAA can be degraded to indole-3-aldehyde and indole-3-methanol and, in turn, indole-3-methanol breaks down very rapidly in an acidic environment (Sundberg et al. 1985a). These problems are not insurmountable if (1) purification procedures are chosen with care, (2) an antioxidant such as sodium diethyldithiocarbamate or 2,6-di-tert-butyl-4-methylphenol is added to all extraction, partitioning and chromatography solvents and (3) appropriate controls are run wherever possible.

Acknowledgments. The authors would like to thank the British Council for financial assistance.

References

Anderson B, Sandberg G (1982) Identification of endogenous N-(3-indoleacetyl) aspartic acid Scots pine (*Pinus sylvestris* L) by combined gas chromatographmass spectrometry, using high-performance liquid chromatography for quantification. J Chromatogr 238:151–156

Bandurski RS (1984) Metabolism of indole-3-acetic acid. In: Crozier A, Hillman JR (eds). The biosynthesis and metabolism of plant hormones. Soc Exp Biol Seminar Series No 23. Cambridge Univ Press, Cambridge, pp 183–200

Bandurski RS, Nonnebel H (1984) Auxins. In: Wilkins MB (ed) Advanced plant physiology. Pitman, London, pp 1–20

Bearder JR (1980) Plant hormones and other growth substances – their background, structures and occurrence. In: MacMillan J (ed) Encyclopedia of plant physiology, new series, vol 9 Hormonal regulation of development I. Molecular aspects of plant hormones. Springer, Berlin Heidelberg New York, pp 9–112

Cohen JD, Bialek K (1984) The biosynthesis of indole-3-acetic acid in higher plants. In: Crozier A, Hillman JR (eds) The biosynthesis and metabolism of plant hormones. Soc Exp Biol Seminar Series No 23. Cambridge Univ Press, Cambridge, pp 165–181

Crozier A (1983a) The biochemistry and physiology of gibberellins, vol 1. Praeger, New York

Crozier A (1983b) The biochemistry and physiology of gibberellins, vol 2. Praeger, New York

Crozier A, Loferski K, Zaerr JB, Morris RO (1980) Analysis of picogram quantities of indole-3-acetic acid by high performance liquid chromatography-fluorescence procedures. Planta (Berl) 150:366–370

Epstein E, Cohen JD, Bandurski RS (1980) Concentration and metabolic turnover of indole in germinating kernels of *Zea mays* L. Plant Physiol (Bethesda) 65:415–421

Galun E (1981) Plant protoplasts as physiological tools. Annu Rev Plant Physiol 32:237–266

Garcia-Martinez JL, Ohlrogge JB, Rappaport L (1981) Differential compartmentation of gibberellin A_1 and its metabolites in vacuoles of cowpea and barley leaves. Plant Physiol (Bethesda) 68:865–867

Gibson RA, Schneider EA, Wightman F (1972) Biosynthesis and metabolism of indoyl-3-acetic acid II. In vivo experiments with ^{14}C-labelled precursors of IAA in tomato and barley shoots. J Exp Bot 23:381–399

Hartung W, Gimmler B, Heilmann B, Kaiser G (1981a) The site of abscisic acid metabolism in mesophyll cells of *Spinacia oleracea*. Plant Sci Lett 18:359–364

Hartung W, Heilmann B, Gimmler H (1981b) Do chloroplasts play a role abscisic acid synthesis. Plant Sci Lett 22:235–242

Hattori H, Marumo S (1972) Monomethyl-4- chloroindolyl-3-acetyl-L-aspartate and absence of indolyl-3-acetic acid in immature seeds of *Pisum sativum*. Planta (Berl) 102:85–90

Heilmann B, Hartung W, Gimmler H (1982) The site of indole-3-acetic acid synthesis in mesophyll cells of *Spinacia oleracea*. Z Naturforsch 37:174–178

Kramers RMJ (1981) Cytokinin metabolism. Ph D thesis, Univ College of Wales, Aberystwyth

Letham DS, Goodwin PG, Higgins TJV (1978) Phytohormones and related compounds – a comprehensive treatise, vol 2. Elsevier, Amsterdam

Magnus V, Iskric S, Kveder S (1971) Indole-3-methanol- a metabolite of indole-3-acetic acid in pea seedlings. Planta (Berl) 97:116–125

Marumo S, Hattori H (1971) Isolation of D-4-chlorotyptophan derivatives as auxin related metabolites from immature seeds of *Pisum sativum*. Planta (Berl) 90:208–211

Marumo S, Hattori H, Abe H, Munakata K (1968a) Isolation of 4-chloroindoyl-3-acetic acid from immature seeds of *Pisum sativum*. Nature (Lond) 219:959–960

Marumo S, Hattori H, Abe H, Munakata K (1968b) Isolation of a novel auxin, methyl 4-chloroindolecetate from immature seeds of *Pisum sativum*. Agric Biol Chem 32:117–118

McGaw BS, Scott IM, Horgan R (1984) Cytokinin biosynthesis and metabolism. In: Crozier A, Hillman JR (eds) The biosynthesis and metabolism of plant hormones. Soc Exp Biol seminar series No 23. Cambridge Univ Press, Cambridge, pp 105–133

Neill SJ, Horgan R, Walton DC (1984) Biosynthesis of abscisic acid. In: Crozier A, Hillman JR (eds) The biosynthesis and metabolism of plant hormones. Soc Exp Biol seminar series No 23. Cambridge Univ Press, Cambridge, pp 43–70

Reeve DR, Crozier A (1980) Quantitative analysis of plant hormones. In: MacMillan J (ed) Encyclopedia of plant physiology, new series, vol 9: Hormonal regulation of development I. Molecular aspects of plant hormones. Springer, Berlin Heidelberg New York, pp 203–280

Rivier L, Pilet PE (1974) Indolyl-3-acetic acid in cap and apex of maize roots: identification and quantification by mass spectrometry. Planta (Berl) 120:107–112

Rubery PH, Sheldrake AR (1973) Effect of pH and surface charge on cell uptake of auxin. Nature (Lond) 244:285–288

Rubery PH, Sheldrake AR (1974) Carrier-mediated auxin transport. Planta (Berl) 118:101–121

Sandberg G (1984) Biosynthesis and metabolism of indole-3-ethanol and indole-3-acetic acid by *Pinus sylvestris* L needles. Plant (Berl) 161:398–403

Sandberg G, Anderson B, Dunberg A (1981) Identification of 3-indole-acetic acid in *Pinus sylvestris* L by gas chromatography-mass spectometry, and quantitative analysis by ion pair reversed phase liquid chromatography with spectroflurimetric detection. J Chromatogr 205, 125–137

Sandberg G, Hällgren J-E (to be published 1985) Catabolism of 3-indole acetic acid in protoplasts from etiolated seedlings of Scots pine (*Pinus sylvestris* L.). Plant Cell Reports

Sandberg G, Jensen E, Crozier A (1982) Biosynthesis of indole-3-acetic acid in protoplasts, chloroplasts and a cytoplasmic fraction from barley (*Hordeum vulgare* L). Planta (Berl) 156:541–545

Sandberg G, Jensen E, Crozier A (1983) Effects of light on the catabolism of [2-^{14}C]-3-indole acetic acid in protoplasts, a chloroplast-rich fraction and a crude cytoplasmic preparation from barley (*Hordeum vulgare* L). Plant Cell Environ 6:111–115

Sandberg G, Jensen E, Crozier A (1984) Analyses of 3-indole carboxylic acid in Scots pine (*Pinus sylvestris* L) needles. Phytochemistry (Oxf) 23:99–102

Sundberg B, Sandberg G, Jensen E (to be published 1985a) Identification and quantification of indole-3-methanol in etiolated seedlings of Scots pine (*Pinus sylvestris* L) Plant Physiol (Bethesda) 11

Sundberg B, Sandberg G, Jensen E (1985b) Catabolism of 3-indole acetic acid to 3-indole methanol in a crude enzyme extract and in protoplasts from Scots pine (*Pinus sylvestris* L.) (submitted for publication)

Sembdner G, Gross D, Leibisch HW, Schneider G (1980) Biosynthesis and metabolism of plant hormones. In: MacMillan J (ed) Encyclopedia of plant physiology, news series, vol 9. Hormonal regulation of development. I. Molecular aspects of plant hormones. Springer, Berlin Heidelberg New York, pp 281–344

Schneider EA, Wightman F (1978) Auxins. In: Letham DS, Goodwin PB, Higgins TJV (eds) Phytohormones and related compounds – a comprehensive treatise, vol 1. Elsevier, Amsterdam, pp 29–105

Schneider EA, Gibson RA, Wightman F (1972) Biosynthesis and metabolism of indole-3-ylacetic acid I. The native indoles of barley and tomato shoots. J Exp Bot 23:152–170

Weiler EW (1982) Plant hormone immunoassay. Physiol Plant 54:230–234

Weiler EW, Schnabel H, Hornberg C (1982) Stress related levels of abscisic acid in guard cells protoplasts of *Vicia faba*. Planta (Berl) 154:24–28

Auxin Receptors in Tobacco Leaf Protoplasts

K.R. LIBBENGA, A.C. MAAN, P.C.G. VAN DER LINDE, A.M. MENNES
and P.A.A. HARKES

Introduction

We have used different plant cell and tissue cultures to study the interactions between hormones and target cells, thereby restricting ourselves to auxins.

We assume that auxin and other plant hormones, like animal hormones, initiate their physiological effects by interaction with receptor proteins.

To verify this hypothesis we have demonstrated the presence of specific, high-affinity auxin-binding proteins in auxin-sensitive cultured cells, derived from tobacco tissues.

The characteristics of these binding proteins have been worked out in detail using tobacco callus cultures (see Table 1). We were able to distinguish two different auxin-binding proteins: a binding protein that is present in the cytoplasm and/or the nucleus (Oostrom et al. 1975, 1980; Van der Linde et al. 1984) and a membrane-bound binding protein (Vreugdenhil et al. 1979, Maan et al. 1983).

Studies with cell-free systems have shown that the cytoplasmic/nuclear binding proteins are probably involved in the regulation of transcription (Van der Linde et al. 1984), while the membrane-bound binding proteins might be involved in the phosphorylation of membrane proteins (Maan et al., in prep.).

Besides these auxin-binding proteins we detected a membrane-bound binding protein for N-1-naphthylphthalamic acid (NPA), a potent inhibitor of polar auxin transport (Morgan and Soding 1958). Recent studies have shown that auxin binds to these proteins too, but with lower affinity (Maan et al., submitted).

One of the systems we use for our investigations consists of isolated tobacco mesophyll protoplasts, which require auxin and cytokinin for cell division. Such a system is very well suited for the study of the role these binding proteins might play in cell division and/or cell elongation.

Our aim is to determine the concentrations of these auxin-binding proteins in the various cellular components during the first cell cycle. This may shed some light upon their physiological function. Although this investigation is still in progress, some of the results are worth mentioning.

The Physiological Properties of Plant Protoplasts
Edited by P. E. Pilet

Table 1. Properties of different binding proteins from *Nicotiana tabacum*

	Cytoplasmic receptor	Membrane-bound auxin receptor	Membrane-bound NPA receptor
K_d (M)	$6x10^{-9}$ (IAA)	$2x10^{-7}$ (NAA)	$8x10^{-10}$ (NPA)
Maximal conc. pmol/mg^{-1} prot.	0.2	50	1
Temp (°C) of binding assay	25	25	0
pH-Optimum	7–7.5	5	4
Specificity (K_d compet.)			
– IAA	–	$2x10^{-5}$	10^{-4}-10^{-5}
– NAA	10^{-8}	–	10^{-4}-10^{-5}
– 2,4-D	$3x10^{-8}$	$3x10^{-6}$	10^{-5}
– TRP	$> 10^{-5}$	$> 10^{-3}$	$> 10^{-2}$
Occurrence in cell	Cytoplasm/ nucleus	Plasmalemma?	Plasmalemma
Presence in tissues:			
– Callus	+	+	+
– Cell-susp.	+	+[b]	+
– Leaves	n.t.	+	+
– Protopl.	–	+[a]	+[a]
– Plant tips	+	n.t.	n.t.
– Stem pith	–[c]	+	+

n.t. = not tested
[a] Only after 2–3 days
[b] Only in NAA-kinetin requiring cell-suspensions, not in 2,4-D requiring cell suspensions
[c] Detectable after 1 day of culture.
Table compiled from Bogers et al. (1980), Maan et al. (subm.), Oostrom et al. (1975, 1980), Van der Linde et al. (1984 and unpubl.), Vreugdenhil et al. (1979, 1980)

Materials and Methods

Plants

Tobacco plants (*Nicotiana tabacum* var. White Burley) were grown from surface-sterilized seeds in a greenhouse at 22°C and 70–80% relative humidity under continuous illumination.

Culture of Protoplasts

The medium for the isolation and culture of protoplasts consisted of (concentrations as mg l^{-1}, unless otherwise indicated): $NaH_2PO_4 \cdot H_2O$: 150; $Ca(H_2PO_4)_2 \cdot H_2O$: 50; $CaCl_2 \cdot 2H_2O$: 900; KNO_3: 2500; NH_4NO_3: 250; $(NH_4)_2SO_4$: 134; $MgSO_4 \cdot 7H_2O$: 250; $FeNa(EDTA)_2$: 37.5; micronutrients and vitamins as in B5 (Gamborg et al. 1968); 6-benzylaminopurine: 1; NAA: 3; xylose: 250; sucrose: 1%; mannitol: 0.7 M; pH 5.6.

Isolation

Protoplasts were isolated from surface-sterilized leaves. The cells were plasmolysed by floating peeled leaf pieces on the medium during 1 h at 25°C. The medium was then replaced by isolation medium containing an additional 0.5% (w/v) macerozyme and 2% (w/v) cellulase Onozuka R-10 (Yakult Biochemicals Co. Ltd., Japan). The incubation (under the same conditions) was continued until 90% of the cell walls were digested (2–4 h). The protoplasts were separated from cellular debris by filtration through a 63-μm filter and then collected by centrifuging for 5 min at 60 g. Spherical protoplasts were floated on a hypertonic sucrose solution (medium containing 23% (w/v) sucrose instead of 0.7 M mannitol) and washed twice with the isolation medium. They were cultured at a concentration of 5×10^4 cells ml^{-1} in plastic Petri dishes (diam. 10 cm) containing 10 ml of culture medium for up to 12 days at 25°C in the dark.

Isolation of auxin-binding proteins fractions. After culturing for the desired time the protoplasts were collected by centrifugation (5 min at 60 g). Membrane particles were obtained by homogenization in 2 vol of buffer A (50 mM 2-morpholinoethanesulfonic acid, 10 mM $MgCl_2$, 0.5 M sucrose) at pH 5 for the auxin-binding protein and pH 4 for the NPA-binding protein. For the cytoplasmic auxin-binding protein buffer B (50 mM Tris-HCl, 0.75 M KCl, 2 mM dithiothreitol, 2 mM Na_2EDTA, pH 7.5) was used. Homogenization took place in a Potter-Elvehjem glass homogenizer (10 strokes, 1500 r.p.m.). The homogenate was centrifuged at 145,000 g for 60 min. For membrane-bound binding proteins the pellet was resuspended in buffer A at the appropriate pH, for the cytoplasmic binding protein the supernatant was concentrated and dialysed against 20 mM Tris-HCl, pH 7.5.

Binding Assays

For each binding protein a different binding assay was used.

a) The presence of the cytoplasmic binding protein was determined according to Oostrom et al. (1975, 1980). [^{3}H]-IAA (spec. act. 10^{15} Bq mol $^{-1}$) was used as the radioligand. Using different concentrations of unlabelled IAA in the assay, a concentration range from 2.5×10^{-9} to 2×10^{-7} M was obtained.
b) For membrane-bound binding proteins the procedure described by Vreugdenhil et al. (1979) was followed with [^{14}C]-NAA (spec. act. 2.2×10^{12} Bq mol^{-1}) as the radioligand. Dilution with unlabelled NAA gave a concentration-range from

5×10^{-8} to 10^{-4} M NAA was used because it has a higher affinity for these membrane-bound binding proteins than the natural auxin, IAA.

c) The membrane-bound NPA-binding proteins were detected by a procedure described by Maan et al. (submitted) using [^{3}H]-NPA (spec. act. 1.1×10^{15} Bq mol^{-1}). Dilution with unlabelled NPA gave a concentration range of 5.10^{-10} to 10^{-5} M.

Results and Discussion

Membrane-Bound Auxin-Binding Proteins

The properties of the membrane-bound auxin-binding proteins are listed in Table 1. These binding proteins occur among others in microsome preparations of fully expanded leaves. Vreugdenhil et al. (1980) found that these binding proteins are missing, or at least cannot be detected, in freshly isolated protoplasts derived from these leaves (detection level ca. 2 pmol mg^{-1} protein). However, upon culturing the protoplasts, the concentration of binding proteins, expressed per mg of membrane protein, increases rapidly after the onset of cell division (Fig. 1).

The fact that these binding proteins do not appear before the onset of cell division might be an indication that they have a regulatory role that is not directly connected with cell division.

To further support this assumption we tried to select cells that do not contain this binding protein but nevertheless divide normally. We were able to derive two cell suspension lines, one grown on NAA and kinetin, the other grown on 2,4-dichlorophenoxyacetic acid (2,4-D). Both lines have similar growth rates, but the latter lacks the membrane-bound auxin-binding protein. Thus cell division is apparently not correlated with the presence of this binding protein.

Vreugdenhil et al. (1980) found that the binding proteins from protoplasts are not destroyed or inactivated by osmotic stress, but most probably by proteolytic enzymes that are present in the cellulase and macerozyme preparations used for protoplast isolation.

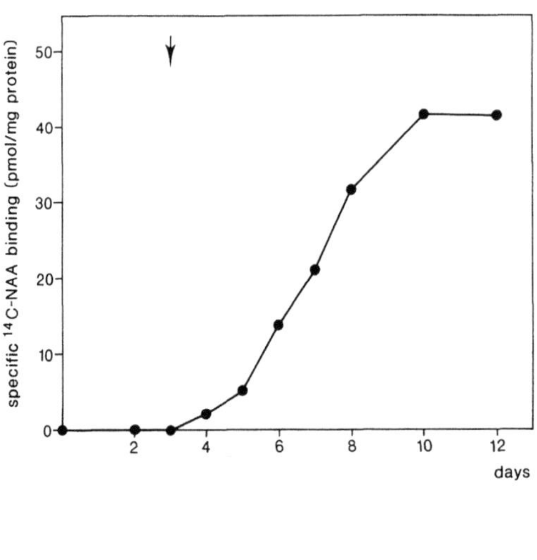

Fig. 1. Amount of auxin-binding protein after protoplast isolation. Start of cell division is indicated by an *arrow*

If these proteins do not penetrate the plasmalemma of the protoplasts – and up to now there are no reports describing endocytosis of proteins into protoplasts – then only the proteins in the cell wall and at the external face of the plasmalemma will be degraded. Furthermore, it is known from animal systems that proteins taken up by endocytosis are usually transferred to lysosomes and destroyed (Silverstein et al. 1977).

We can exclude a possible localization in the cell wall, because we have found that cell wall fragments from tobacco callus are pelleted at low speed centrifugation, while binding proteins are not. Moreover, the density in linear sucrose gradients of particles containing the binding proteins is much lower than that of cell wall fragments (Vreugdenhil et al. 1981).

An obvious explanation for the disappearance of the auxin-binding proteins is that they are localized at the external face of the plasmalemma and are destroyed by proteolytic enzymes during protoplast isolation. A localization of the binding proteins at the external face of the plasmalemma is in agreement with the pH optimum of 5 for auxin binding.

The microsome preparations of tobacco callus, cell suspension cultures and leaves also contain a particular class of NPA-binding proteins (Maan et al., submitted). The properties of these binding proteins are also listed in Table 1. Since NPA-binding proteins are generally considered as plasmalemma markers (Lembi et al. 1971, Ray 1977) this gives us a good probe to test for the destruction of plasmalemma proteins during the preparation of protoplasts. Figure 2 shows that, indeed, the NPA-binding proteins cannot be detected (detection level ca. 0.1 pmol mg^{-1} protein) in freshly isolated protoplasts. However, like the auxin-binding proteins, they reapear after culturing the protoplasts for 2–3 days.

These NPA-binding proteins are generally considered to constitute part of an auxin-carrier system, see e.g. Goldsmith (1982) and Hertel (1983). The striking resemblance in the disappearance and reappearance of the NPA-binding proteins and the auxin-binding proteins might suggest that they are, in fact, identical. However, this is not the case because in microsome preparations from the 2,4-D-cell line the NPA-binding proteins are still present while the auxin-binding proteins have disappeared (see Fig. 3). This implies that the plasmalemma of the cells contain both auxin carriers and auxin receptors, discussed by Maan et al. (submitted).

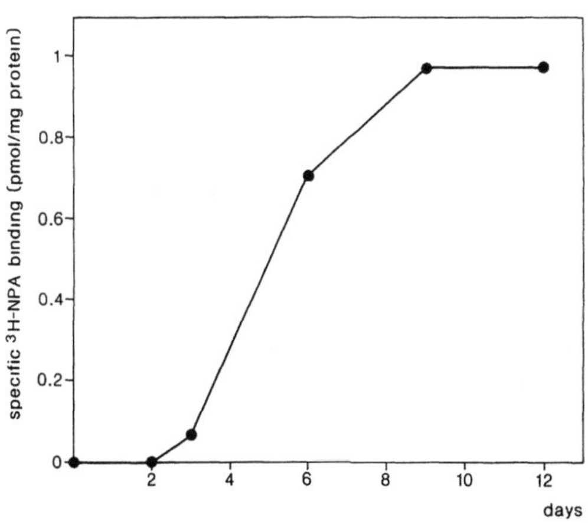

Fig. 2. Amount of NPA-binding protein after protoplast isolation

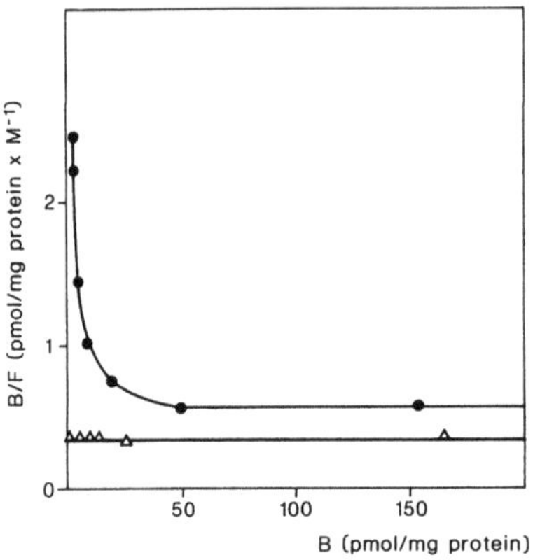

Fig. 3. Binding of NAA (△–△) and NPA (●–●) plotted according to Scatchard (1949). *B* bound ligand; *F* free ligand

Cytoplasmic/Nuclear Auxin-Binding Proteins

It is likely that regenerating protoplasts contain an auxin-receptor system that is involved in triggering cell division. One candidate, the membrane-bound auxin receptor, has already been ruled out due to reasons discussed above.

There is another auxin-binding protein, already mentioned in the introduction, that can be extracted from the cytoplasm and/or the nucleus of cultured tobacco cells. The main properties of this binding protein are given in Table 1. From this table it is obvious that the membrane-bound and the cytoplasmic binding protein differ substantially. The cytoplasmic/nuclear binding proteins are present in all kinds of proliferating cells and tissues examined thus far, including the 2,4-D-requiring cell line.

However, although we expected this binding protein to be present in freshly isolated protoplasts, we were not able to demonstrate its presence in 1- or 2-day-old protoplasts. On second thoughts this is not strange, since both the amount of binding sites (pmol mg^{-1} fresh weight, see Table 1) in other tissues and the amount of protein that is isolated from protoplasts are probably too low to detect specific auxin-binding.

References

Bogers RJ, Kulescha Z, Quint A, Van Vliet ThB, Libbenga KR (1980) The presence of a soluble auxin receptor and the metabolism of 3-indoleacetic acid in tobacco-pith explants. Plant Sci Lett 19:311–317

Gamborg OL, Miller RA, Ojima K (1968) Nutrient requirements of suspension cultures of soybean root cells. Exp Cell Res 50:151–158

Goldsmith MHM (1982) A saturable site responsible for polar transport of indole-3-acetic acid in sections of maize coleoptiles. Planta (Berl) 155:68–75

Hertel R (1983) The mechanism of auxin transport as a model for auxin action. Z Pflanzenphysiol 112:53–67

Lembi CA, Morré DJ, Thomson K-S, Hertel R (1971) N-1-naphthylphthalamic acid binding of a plasma membrane-rich fraction from maize coleoptiles. Planta (Berl) 99:37–45
Maan AC, Vreugdenhil D, Bogers RJ, Libbenga KR (1983) The complex kinetics of auxin-binding to a particulate fraction from tobacco pith callus. Planta (Berl) 158:10–15
Maan AC, Kühnel B, Beukers J, Libbenga KR (1984) Naphthylphtalamic acid binding sites in cultured cells from *Nicotiana tabacum*. Planta (submitted)
Morgan DG, Soding H (1958) Über die Wirkungsweise von Phthalsäuremononaphthylamid (NPA) auf das Wachstum der Haferkoleoptile. Planta (Berl) 52:235–249
Oostrom H, Van Loopik-Detmers MA, Libbenga KR (1975) A high affinity receptor for indoleacetic acid in cultured tobacco pith explants. FEBS Lett 59:194–197
Ostrom H, Kulescha Z, Van Vliet ThB, Libbenga KR (1980) Characterization of a cytoplasmic auxin receptor from tobacco pith callus. Planta (Berl) 149:44–47
Ray PM (1977) Auxin binding sites are localized on membranes of the endoplasmic reticulum. Plant Physiol (Bethesda) 59:594–599
Scatchard G (1949) The attraction of proteins for small molecules and ions. Ann NY Acad Sci 57:660–672
Silverstein SC, Steinman RM, Cohn ZA (1977) Endocytosis. Annu Rev Biochem 46:669–722
Van der Linde PCG, Bouman H, Mennes AM, Libbenga KR (1984) A soluble auxin-binding protein from cultured tobacco tissues stimulates RNA synthesis in vitro. Planta (Berl) 160: 102–108
Vreugdenhil D, Burgers A, Libbenga KR (1979) A particle-bound auxin receptor from tobacco pith callus. Plant Sci Lett 16:115–121
Vreugdenhil D, Harkes PAA, Libbenga KR (1980) Auxin-binding by particulate fractions from tobacco leaf protoplasts. Planta (Berl) 150:9–12
Vreugdenhil D, Burgers A, Harkes PAA, Libbenga KR (1981) Modulation of the number of membrane-bound auxin-binding sites during the growth of batch-cultured tobacco cells. Planta (Berl) 152:415–419

Some Physiological Properties of Protoplasts from Gravireacting Maize Roots

P.E. PILET

Introduction

Protoplasts, with their exposed plasmalemma, have been used in studies of membrane integrity (Fowke et al. 1983). It has been clearly shown that they respond to auxin (such as indole-3-acetic acid: IAA) by increasing their rate of vacuolation until they ultimately burst (Cocking 1962, Pilet 1971b, 1981). Such plasmalemma disruption, induced by IAA, may depend on the endogenous auxin level of the cells from which protoplasts have been obtained. If onion roots were treated with IAA before the preparation of protoplasts which were then incubated in medium containing IAA, the protoplasts burst at a higher IAA concentration (Pilet 1971b, 1981). Similarly, when using mesophyll protoplasts obtained from two species of tobacco and two of their tumor-prone interspecific hybrids (having an auxin level about 10 to 30 times higher than the two initial species) the bursting response of protoplasts showed that protoplasts from the tumorous hybrid tolerated auxin levels up to 30 times higher than did protoplasts from parent plants (Bayer 1973). It seems clear that protoplasts burst only after exogenous IAA reaches a level comparable with that of endogenous IAA present in the cells from which they were prepared (Pilet 1972, Hall and Cocking 1974, Pilet 1981).

When roots were kept in a horizontal position, they bent downwards. Such positive gravitropism may be the consequence of a redistribution of endogenous hormones (Jackson and Barlow 1981) such as auxin and abscisic acid (Pilet and Elliott 1981) which could induce a differential growth between the upper and lower parts of the root (Pilet and Ney 1981). According to the classical Cholodny-Went theory (Audus 1975), a supra-optimal auxin level in the lower side of the extending part of the gravireacting roots may induce the inhibition of the elongation of this side and thus a positive root curvature.

There is very little information on the fine regulation of the extension of cells from the upper and lower parts of the growing and gravireacting root zone. IAA has been shown to be bound to specific plasmalemma proteins (Hertel et al. 1972). This was recently demonstrated for maize roots membrane fractions (Moloney and Pilet 1981). It is now believed that some binding sites located on the plasmalemma play a key role in the recognition of the informational signal due to some hormones as IAA, ABA etc. (Rubery 1981) and could control their polar transport from cell to cell (Kende and Gardner 1976). Consequently it was of interest to analyse – precisely at the plasmalemma level – the possible changes occurring in elongating cells from the two sides of gravistimulated roots. It was decided to use

The Physiological Properties of Plant Protoplasts
Edited by P. E. Pilet

freshly prepared protoplasts, having not yet reformed their cell walls (Pilet 1973), and for which the plasmalemma was directly accessible for quantitative experiments. Protoplasts are known to keep most of the characteristics of the cells from which they were obtained (Cocking 1972, Pilet 1981, Seen and Pilet 1982). It has been reported that, at least in roots, it was not so difficult to prepare protoplasts from several distinct tissues (Pilet et al. 1972), i.e. cells from the extension zone of maize roots (Senn and Pilet 1980).

One of the parameters used for characterizing the plasmalemma properties is the zeta potential (Grout et al. 1972, Pilet and Senn 1974, Pilet 1981, Senn and Pilet 1982). Such potential directly depends on the surface charges of the plasmalemma (Haydon 1961, Pilet and Senn 1974, Nagata and Melchers 1978, Senn and Pilet 1981, 1982) and can easily by calculated from the electrophoretic mobility of the protoplasts (Pilet and Senn 1974, Senn and Pilet 1981). These charges are mostly due to a combined ionization of surface group as proteins, amino acids, lipids and polysaccharides (Brinton and Lauffer 1959) and can be modified by ion adsorption, surface active agents, enzyme treatments etc. (Ambrose 1966). Any changes in the fine structure of plasmalemma will be detected by some variations in the zeta potential of the protoplasts (Haydon 1961, Weiss 1969, Tenforde 1970, Pilet and Senn 1974).

In the present report, two problems will be discussed. At first, the purpose of this study was (1) to test the sensitivity to IAA (analysis of the bursting response) of the maize root protoplasts and (2) to ascertain whether the "survival" of protoplasts treated by IAA was different when they were obtained from the lower or the upper sides of the elongating part of gravireacting roots. In a second part, variations of the zeta potential – related to several presentation times and light conditions – will be analysed for protoplasts prepared from cells of the extension zone (upper and lower sides) of horizontal and vertical maize roots. The results will be compared to some data on the rate of gravireaction and of the differential growth of maize roots also given for several times of the gravipresentation.

Material and Methods

Preparation of Protoplasts

Primary maize (*Zea mays* cv. LG 11) roots were obtained as previously described (Pilet 1977). Selected caryopses were placed in the dark (22°C) between moist paper towels on plastic frames. Roots elongated vertically; when they reached 15 ± 1 mm in lenght, they were used for preparing protoplasts. Fragments from the elongating part (1 to 5 mm counted from the tip) were rapidly dipped into Petri dishes containing 2 ml of 2.5 mM cystein + 3% sorbitol solution (Senn and Pilet 1980) and then incubated (60 min) in a giratory shaker (75 r.p.m.) in the dark. The medium was then pipetted out and replaced by an enzyme mixture (cellulysin Calibiochem 1%, pectinase Serva 0.5%, hemicellulase Sigma 0.5% in 2 ml of 2 mM 2 (N-morpholino) ethane sulphonic acid [MES]-NaOH buffer: pH 5.7). The Petri dishes where then replaced in darkness on the shaker (75 RPM, 25°C). After a 3-h incubation, protoplasts were filtered and washed by three successive centrifugations (100 g, 5 min) in buffer (Senn and Pilet 1982).

IAA Treatment and Bursting Analyses

Protoplasts were kept in MES solution (pH 5.7) containing 3% sorbitol (Senn and Pilet 1981). The protoplast suspensions (10 μl) were mixed, at zero time, with the MES buffer (10 μl containing 0, 10 or 50 μM IAA) and kept for 30 or 60 min (dark, 25°C) to be viewed under the light microscope. The number of undamaged protoplasts, was measured with Thoma's counting chamber. Protoplasts incubated without IAA were used as controls. Standard deviations of the mean are reported for each result and significant differences have been assessed by the *t*-test.

Zeta Potential Measurement

As the ionic strength and consequently the pH plays an essential role in electrophoretic analyses (Heard and Seaman 1960, Tenforde 1970, Senn and Pilet 1981), an ionic strength of 0.002 was chosen. Electrophoretic buffers (pH 6.0 and 7.0) were prepared by mixing suitable amounts of NaH_2PO_4/Na_2HPO_4. After the last centrifugation, protoplasts were resuspended in EB, containing sorbitol 3% and lactose 5%. Measurements were performed using a MK II microelectrophoresis apparatus (Rank brothers, UK) connected to a microcomputer (32 K CBM, Commodore USA). Zeta potential was calculated by using the Smoluchowski equation (Shaw 1969, Senn and Pilet 1981, 1982). All these experiments were repeated at least five times; for each assay 25 ± 5 protoplasts have been used for zeta potential determination.

Root Growth and Gravireaction

Each 30 min, *lengths* of upper and lower sides of gravireacting roots were measured by using magnified (10 times) shadowgraphs as previously done for the curvature measurements (Pilet and Ney 1981). Each data reported represents the mean of at least 70 ± 8 roots. As for the curvature data, the standard error of the mean has been calculated and significant differences assessed by the *t*-test. Each hour, macrophotographs (camera Nikon FE, objective macro Nikon 105 mm / f 4.0) were taken and *curvature* of each root measured at 10 times magnification (Beffa and Pilet 1982). Each determination, for a given time, was repeated at least four times (total roots measured: 45 ± 5 per assay). The standard error of the mean was reported for each mean value and significant differences have been assessed by the *t*-test.

Results and Discussion

Two problems, as indicated before, will be successively discussed: variations in the bursting, caused by IAA, of maize root protoplasts and the changes in their zeta potential.

Protoplast Bursting

In a first series of assays, intact roots were immersed (60 min, dark, 25°C, pH 5.7) with 100 μM or without IAA before protoplast isolation. Protoplasts were then incubated for bursting analyses. As can be seen (Table 1) the number of protoplasts surviving decreases with increasing IAA concentration as previously reported (Cocking 1962, Pilet 1971b, 1981, Bayer 1973). However, for protoplasts derived from cells pretreated with IAA, the bursting is significantly less. This is in agreement with data obtained when using onion root protoplasts (Pilet 1971b) and indicates that the IAA "sensitivity" of the plasmalemma strongly decreases when the cells (from which protoplasts were prepared) have been pretreated with IAA. The present results confirm those obtained by Bayer (1973), who showed that protoplasts respond by bursting only when the exogenous IAA concentration exceeds the IAA content of the cells (Pilet, 1984).

In a second set of experiments, protoplasts from the upper and the lower sides of the extending parts of gravireacting roots – kept 120 min in a horizontal position (white light, 21°C) – were prepared. Then, as before, they were incubated for 60 min (dark, 25°C, pH 5.7) with 3% sorbitol and IAA at different concentrations (0–100 μM). Figure 1 shows that the bursting increases with increasing IAA levels, and this for both kinds of protoplasts tested. These data also clearly indicate that the percentage of protoplasts surviving, for a given IAA concentration, was significantly higher (except for 5 μM IAA) for protoplasts obtained from the cells of the lower part of the root extension zone. From previous results and those presented in Table 1,

Table 1. Relative number [a] (% ± standard deviation) of unburst surviving protoplasts prepared from the extending zone of maize roots and incubated (30 or 60 min; dark 25°C) in a MES buffer (pH 5.7) + 3% sorbitol containing IAA (10 or 50 μM IAA) or not containing IAA. Intact roots pretreated (complete immersion 60 min; dark; 25°C) in MES buffer (pH 5.7) with (100 μM) or without IAA before protoplast preparation

	0 μM IAA		100 μM IAA	
Incubation [b]	30 min	60 min	30 min	60 min
0 μM IAA	91 ± 5	86 ± 7	94 ± 4	90 ± 9
10 μM IAA	53 ± 6	27 ± 8	72 ± 6	61 ± 10
50 μM IAA	32 ± 4	9 ± 3	49 ± 7	38 ± 8

[a] Each value is the average of 500 ± 60 counted protoplasts (4 independent experiments)

[b] IAA given in the incubation medium containing the protoplast suspension

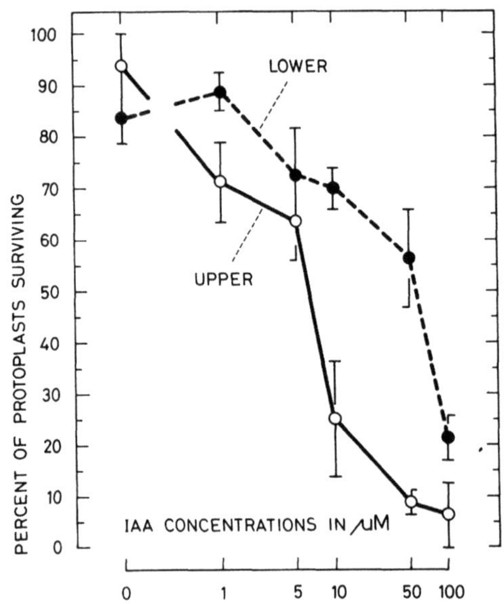

Fig. 1. Relative number (% ± standard deviation) of unburst protoplasts prepared from the upper and the lower parts of elongating zone of gravistimulated (white light, 21 °C, 120 min) maize roots. Protoplasts were incubated (60 min; dark, 25 °C) in a MES buffer (pH 5.7) ± 3% sorbitol containing IAA (1 to 100 μM) or not containing IAA. Each value is the average of 400 ± 50 counted protoplasts (3 independent experiments)

it is difficult not to conclude that the endogenous IAA content in the lower cells of the elongating side of gravireacting roots is higher than that of the upper cells. This is in agreement with the endogenous IAA redistribution often reported (Pilet 1971a, Audus 1975, Thimann 1977) in positively gravireacting roots.

Zeta Potential of Protoplasts

After 3 h of root gravipresentation, protoplasts were prepared from cells of the upper and lower sides of the growing part of the roots. Measurements were done in buffered solution (pH 6.0 and 7.0). Data given in Tables 2 and 3 show that the standard errors of zeta potential were small, then the protoplast population was very homogenous. This could be explained by the fact that they were prepared from quite similar cells: only cortical cells of a small part (1 mm in length) of the elongation zone have been used. Furthermore, an other point should be briefly discussed. As can be seen for some data in Fig. 2A, the standard errors at 90 min are much smaller than these at 0 and 180 min. Such difference could be the result at least of a balance of two processes. First, a stabilisation of the root cells after the roots are placed on the horizontal position could explain the decline in the standard devivation between 0 and 90 min. Thereafter, during root gravireaction, an increasing heterogeneity between root cells could be reflected in an increasing standard error of the zeta potential between 90 and 180 min.

As can be seen in Table 2, the negative zeta potentials were lower at pH 6.0 than at pH 7.0. These results are in agreement with those previously obtained using a similar material but when comparing protoplasts of different maize root zones (Seen and Pilet 1981). The mobility has been shown – at least for animal cells – to depend strongly on the pH of the electrophoresis medium (Shaw 1969), the isoelectric point

Table 2. Zeta potential [a] (in mV ± standard deviation) of protoplasts obtained from cortical cells prepared from maize roots kept 0 h (control) and 3 h in horizontal position (white light). Protoplasts were isolated from cells of the upper and lower sides of the elongating and gravireacting root zone. Variations of zeta potential measured at pH 6.0 and 7.0

pH	0 h	3 h Upper side	Lower side
6.0	-24.0 ± 0.5	-29.3 ± 1.2	-26.3 ± 1.4
7.0	-32.8 ± 0.6	-33.0 ± 0.5	-33.1 ± 0.6

[a] Each value represents the mean of data of at least five separate experiments using 25 ± 5 protoplasts each

Table 3. Zeta potential [a] (in mV ± standard deviation) of protoplasts obtained from cortical cells prepared from maize roots respectively kept 0, 1.5, 3 and 6 h in horizontal position (white light). Protoplasts were isolated from cells of the upper and lower sides of the elongating and gravireacting root zone. Variations of zeta potential measured at pH 7.0

Hours in horizontal position	Upper side	Lower side
0.0		-32.8 ± 0.6
1.5	-31.4 ± 1.1	-30.9 ± 1.1
3.0	-33.0 ± 0.5	-33.1 ± 0.6
6.0	-31.6 ± 0.8	-31.0 ± 1.0

[a] Each value represents the mean of data of at least five separate experiments using 25 ± 5 protoplasts

generally being found between pH 3 and 4 (Seaman and Heard 1960, Heard and Seaman 1960, Cook et al. 1962, Röder et al 1973). Present data indicate that the pH effect is related to some pH-dependent ionogenic surface groups in plasmalemma and it cannot be excluded that these reactive groups are proteins. In these experiments, no significant differences can be reported, after 3 h, between the different kinds of protoplasts tested, but the duration of gravistimulation could present some importance. This will be tested in a second series of experiments.

Protoplasts were prepared from cells of the elongating and gravireacting part of roots (gravipresentation of 0, 1.5, 3 and 6 h). Their negative zeta potential was measured at pH 6.0 and 7.0 (see Table 2). At pH 7.0 and for all the gravistimulation times tested, no difference can be detected between protoplasts prepared from cells of the upper and lower sides of roots (see Table 3). In contrast, at pH 6.0 (Fig. 2A), the variations of negative zeta potential – at least for the first 3 h of gravipresentation – are not similar. It is not surprising that values of zeta potential

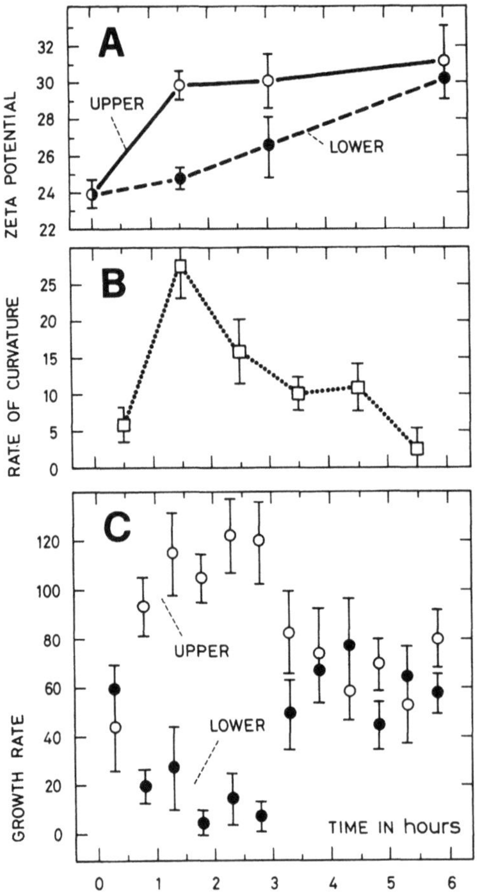

Fig. 2A. Time courses of the zeta potential (in mV ± standard deviation) of maize root protoplasts, suspended in an osmotically stabilized phosphate buffer (pH 6.0:0.002 ionic strength). Protoplasts were isolated, after several gravipresentation times (white light; temperature: 21 ± 1 ^{o}C; relative humidity: 90 ± 5%), from the elongating zone (3 to 4 mm counted from the root tip) the two sides (upper and lower parts) being separated: the upper part corresponding to the caryopse part. Each value is the mean of 5 experiments for which zeta potential of 25 ± 5 protoplasts has been measured for each separate assay (see Table 2). **B** Rate of curvature (in degrees ± standard deviation per h) of gravireacting maize roots for conditions, see **A**). Measurements were performed using macrophotographs. Experiments were repeated at least 4 times (45 ± 5 roots per assay). **C** Growth rate (in μm ± standard deviation per 30 min) of the upper (caryopse side) and lower part or gravireacting maize roots (same conditions as above). Measurements were performed using shadowphotographs; the rates being calculated on the basis of length changes. Each data is the mean of at least 70 ± 8 roots

were different for pH 6 and 7, as a pH 6 is that most inclined to cause differences in the zeta potential under other experimental treatments (Seen and Pilet 1982). This phenomenon is related directly to the influence of the pH environment on the charges of the plasmalemma (Pilet 1981) and not necessarily to the effect of pH on gravireaction. When obtained from the "upper cells", protoplasts react by increasing rapidly their negative zeta potential during the first 90 min of horizontal presentation of roots. Afterwards no significant changes can be obtained. On the contrary, the negative zeta potential of protoplasts prepared from the "lower cells" was, slowly increased. Consequently after 90 and 180 min, zeta potential was significantly higher for the protoplasts from cells of the upper part of gravireacting roots; the maximum differences being found after 90 min of root gravistimulation. From these data it is difficult not to conclude that, during the first 3 h of gravistimulation, some significant changes occur at the plasmalemma level. They are qualitatively similar for plasmalemma of both "upper" and "lower cells"; it was observed that negative zeta potential for the corresponding protoplasts was increasing with increasing time of gravipresentation. Consequently gravtiy seem to have a similar effect on the plasmalemma for both the cortical cells of the upper and lower part of gravistimulated roots. But some quantitative differences have to be noted. The amplitude in the changes of zeta potential is not similar during the first 3 h of gravistimulation. The increase rate was found to be higher for plasmalemma from the "upper cells" than that from the "lower cells" (Pilet et al. 1984). This clearly indicates that the "gravisensitivity" of the plasmalemma depends on the location of the cells. From the present data it can be assumed that the sensitivity is higher for plasmalemma of the cells located in the upper part than of those in the lower part of the elongating zone of gravireactive roots.

It was of interest to report, in the same figure (Fig. 2B), the time-dependent change of the rate of the *downward curvature* (in degree per hour) of roots which bent under the action of the gravity. The maximum rate of gravireaction was obtained after 90 min, that is to say exactly after the time of gravipresentation, when the maximum difference in the zeta potential was obtained between protoplasts prepared from the upper and the lower cortical cells. Now what about the change in the *differential growth* for the identical roots? In the same figure (Fig. 2C) again the growth rate (in μm per 30 min) is reported. As can be seen, significant difference between the elongation rate of upper and lower sides of the extending zone already occurs after 30 min of the horizontal presentation, indicating an increase in the elongation of the upper part and a simultaneous decrease of the lower one (Pilet and Ney 1981). It has to be noticed that such growth distribution remains significant during pratically the same time (150 min) as that obtained for the difference in the zeta potential for protoplasts prepared from the upper and the lower cells.

A last general question has to be discussed. The properties of protoplasts, in particular the zeta potential, permit the analyses of certain characteristics of the plasmalemma of the cells from which the protoplasts were derived. The cell can undergo various treatment, for instance different orientations relative to gravity, and these treatments will cause some changes in the cell which may be detectable already after short periods of time (say 90 min or less, as shown in Fig. 2A). The preparation of maize root protoplasts takes 4 h, and this parameter was kept constant in the present experiments no matter to what treatment the roots had previously been exposed. Since the differences in the properties of the protoplasts are measurable (even when the treatment of the intact roots was short), it follows that at least certain properties of the protoplasts are conserved despite the time required for protoplast preparation (Pilet 1981).

References

Ambrose EJ (1966) Electrophoretic behaviour of plant cells. Progr Biophys Mol Biol 16:243–265

Audus LJ (1975) Geotropism in roots. In: Torrey JG, Clarkson DT (eds) The development and Function of roots. Academic, New York, pp 327–363

Bayer MH (1973) Response of Nicotiana mesophyll protoplasts of normal and tumorous origin to indoleactic acid in vitro. Plant Physiol (Bethesda) 51:898–901

Beffa R, Pilet PE (1982) Elongation and gravireaction of intact and segment roots: light effects. Physiol Plant 54:1–6

Brinton CC, Lauffer MA (1959) The electrophoresis of viruses, bacteria and cells and the microscope method of electrophoresis. In: Bier M (ed) Electrophoresis: theory, methods and applications I. Academic, London, pp 427–492
Cocking EC (1962) Action of growth substances, chelating agents and antibiotics on isolated root protoplasts. Nature (Lond) 193:998–999
Cocking EC (1972) Plant cell protoplasts. Isolation and development. Annu Rev Plant Physiol 23:29–50
Cook GMW, Heard DH, Seaman GVF (1962) The electrokinetic characterization of the Ehrlich carcinoma cells. Exp Cell Res 28:27–39
Fowke LC, Griffing LR, Mersey BG, Van der Valk P (1983) Protoplasts for studies of the plasma membrane and associated cell organelles. In: Potrikus I et al. (eds) Protoplasts 1983, Poster Proc 6th Int Protoplast Symp Basel. Birkhäuser Basel, pp 101–118
Grout BWW, Willison JHM, Cocking EC (1972) Interaction at the surface of plant cell protoplasts. Bioenerg 4:585–602
Hall MD, Cocking EC (1974) The response of isolated *Avena* coleoptile protoplasts to indol-3-acetic acid. Protoplasma 79:225–234
Haydon DA (1961) The surface charge of cells and some others small particles as indicated by electrophoresis. The Zeta potential surface charge relationships. Biochim Biophys Acta 50: 450–457
Heard DH, Seaman GVF (1960) The influence of pH and ionic strength on the electrokinetic stability of the human erythrocyte membrane. J Gen Physiol 43:635–654
Hertel R, Thomson KS, Russo VEA (1972) In vitro auxin binding to particulate cell fractions from corn coleoptiles. Planta (Berl) 107:325–340
Jackson MB, Barlow PW (1981) Root geotropism and the role of growth regulators from the cap: a re-examination. Plant Cell Environ 4:107–123
Kende H, Gardner G (1976) Hormone binding in plants. Annu Rev Plant Physiol 27:267–290
Moloney MM, Pilet PE (1981) Auxin binding in roots: a comparison between maize roots and coleoptiles. Planta (Berl) 153:447–452
Nagata T, Melchers G (1978) Surface charge of protoplasts and their significance on cell interaction. Planta (Berl) 142:235–238
Pilet PE (1971a) Rôle de l'apex radiculaire dans la croissance, le géotropisme et le transport des auxines. Bull Soc Bot Suisse 81:52–65
Pilet PE (1971b) Effets de quelques auxines sur les protoplastes racinaires. CR Acad Sci Paris 273:2253–2256
Pilet PE (1972) Propriétés osmotiques de protoplastes racinaires, mécaniquement et enzymatiquement préparés. Action de traitements auxiniques. CR Acad Sci Paris 275:43–46
Pilet PE (1973) Effect of cell-wall degrading enzymes on protoplast transaminase activity. Experientia (Basel) 29:29–50
Pilet PE (1977) Growth inhibitors in growing and geostimulated maize roots. In: Pilet PE (ed) Plant growth regulation. Springer, Berlin Heidelberg New York, pp 115–128
Pilet PE (1981) Root protoplasts: some physiological properties. In: Göring H, Hoffmann H (eds) Pflanzliche Gewebekultur. Coll Pflanzenphysiol 4. Humboldt Univ Berlin, pp 35–42
Pilet PE (1984) Auxin effect on protoplasts from gravireacting maize roots. Z Pflanzenphysiol 113:373–376
Pilet PE, Elliott MC (1981) Some aspects of the control of root growth and georeaction: the involvement of indoleacetic acid and abscisic acid. Plant Physiol (Bethesda) 67:1047–1050
Pilet PE, Ney D (1981) Differential growth of georeacting maize roots. Planta (Berl) 151:146-150
Pilet PE, Senn A (1974) Effet du Ca^{2+} et du K^{+} sur la mobilité électrophorétique des protoplastes. CR Acad Sc Paris 278:269–272
Pilet PE, Prat R, Roland JC (1972) Morphology, RNase and transaminase of root protoplasts. Plant Cell Physiol 13:297–309
Pilet PE, Hervé R, Senn A (1984) Zeta potential of protoplasts from gravireacting maize roots. Planta (Berl)
Röder C, Cook GMW, Haemmerli G, Sträuli P (1973) Electrokinetic characterization of two ascites tumours: TA3 and Mel 1. J Cell Sci 12:235–251

Rubery T (1981) Auxin receptors. Annu Rev Plant Physiol 32:569–596

Seaman GVF, Heard DH (1960) The surface of washed human erythrocytes as a polyanion. J Gen Physiol 44:251–268

Senn A, Pilet PE (1980) Isolation and some physiological properties of maize root protoplasts. Z Pflanzenphysiol 100:299–310

Senn A, Pilet PE (1981) Electrophoretic mobility, zeta potential and surface charges of maize root protoplasts. Z Pflanzenphysiol 102:19–32

Senn A, Pilet PE (1982) Indoleacetic acid and abscisic acid effects on the electrokinetic properties of maize root protoplasts. In: Marmé D, Marrè E, Hertel R (eds) Plasmalemma and Tonoplast: their functions in the plant cell. Elsevier Biomedical, Amsterdam, pp 69–76

Shaw DJ (1969) (ed) Electrophoresis. Academic, London

Tenforde MB (1970) Microelectrophoretic studies on the surface chemistry of erythrocytes. Adv Biol Phys 13:34–105

Thimann KV (1977) Hormone action in the whole life of plants. The Univ Massachussetts Press, Amherst: 93–130

Weiss L (1969) The cell periphery. Int Rev Cytol 26:63–105

Protoplasts and Gravireactivity

T.H. IVERSEN

Introduction

The direction of growth of plant organs and the differentiation of higher plants depends to a large extent on gravity. When the various organs are turned away from their normal position with respect to gravity, they perform gravitropic (geotropic) movements which enable them to approach their normal orientation. While main roots grow downward in the gravity direction (positive gravitropism), main stems show a negative response and leaves orientate themselves at a certain angle to the direction of gravity (plagiotropism).

It is generally accepted that gravity is perceived at specific locations in the different organs. In roots the site of perception is in the root cap and the gravitropic movement is a result of a reaction chain consisting of a physical phase (susception) and several physiological phases (perception, transmission, reaction) which eventually leads to curvature as a result of a redistribution of plant growth regulators on the two sides of the horizontally stimulated plant organ.

According to the statolith theory sedimentation of movable cell organelles (statoliths) enables the cell to perceive gravity (Haberlandt 1900, Němec 1900). The most obvious statolith candidates are amyloplasts with starch grains which in the normal vertical position are located in the bottom end of each graviperceptive cell (statocyte). Upon reorientation of the plant organ the amyloplasts start moving. Several attempts have been made over the years to demonstrate the dependence of gravitropic reactions on the presence and movement of the statoliths (see e.g. Iversen and Larsen 1971, 1973; Volkmann and Sievers 1979).

Over the years, there have been many attempts to find the primary mechanism of the statolith action within the statocyte. In one of the most interesting models presented in the last decade, it is suggested that the critical factor is the change in the amount and direction of pressure exerted on the original site of statolith sedimentation i.e. on a sensitive endomembrane structure such as the endoplasmic reticulum (ER) (see e.g. Volkmann and Sievers 1979). It has also been suggested that the interaction which occurs between the statolith membrane and the ER-complex may result from electrostatic forces (Iversen 1982). Sine the ER-complex, as a part of the endomembrane system, is interrelated to the plasma membrane, it is assumed that, for example, changes in charges as a result of the statolith/ER interaction may be reflected on the cell surface. Therefore, in part of the present study attempts have been made to isolate statocyte protoplasts and subject these

The Physiological Properties of Plant Protoplasts
Edited by P. E. Pilet

to different g-conditions. This may eventually give rise to changes in charge on the surface of the statocytes – changes which can be detected by electrophoresis.

Since very little is known about the molecular mechanism of initiation of the differentiation processes, simple biological systems like unicellular organisms or tissue cultures are often used to study differentiation. Protoplasts protected from the environment only by the labile plasmamembrane represent a highly sensitive system with respect to external stimuli. The totipotency of the protoplasts and the possibility of their regeneration to mature plants makes protoplasts a valuable tool, for example in the study of the effect of gravity on differentiation (gravimorphoses). Another task of the present study has therefore been to study regeneration of protoplasts exposed to a simulated microgravity environment on a clinostat.

Material and Methods

Isolation of Root Statocyte Protoplasts

Seeds of cress (*Lepidium sativum* L.) and radish (*Raphanus sativus* L.) were sterilized in 15% Klorin® for 45 min, washed with sterile water, and germinated in 9 cm Petri dishes on moist filter paper in the vertical position in the darkroom at 23°C for 24 h and 48 h, respectively. The extreme root tips (0.5–1.0 mm) were cut and preplasmolyzed for 60 min in CPW 13 M (13% mannitol containing KH_2PO_4 27.2 mg l^{-1}, KNO_3 101.0 mg l^{-1}, $CaCl_2 \cdot 2H_2O$ 1480 mg l^{-1}, $MgSO_4 \cdot 7\ H_2O$ 246.0 mg l^{-1}, KI 0.16 mg l^{-1}, $CuSO_4 \cdot 5\ H_2O$ 0.025 mg l^{-1}) at pH 5.8 prior to incubation in darkness on a rotary shaker (30 r.p.m.) with 4% Meicellase (Meijii Seika Kaisha Ltd., Tokyo, Japan) and 0.5% Macerozyme (Yakult Biochemicals Co. Ltd., Nishinomiya, Japan) for 18–20 h at 26°C. After incubation the protoplasts were washed with CPW 13M twice to remove all traces of enzyme. The protoplasts (8.5×10^4 ml^{-1}) were placed on an iso-osmotic density gradient (Harms and Potrykus 1978) and centrifuged for 5 min at 100 g. Samples of protoplasts were taken from two bands at specific densities of 1.039 and 1.047 g cm^{-3} and resuspended in CPW 13M for electromicroscopical preservation and light microscopy and in 13 % mannitol for electrophoresis.

Electrophoresis

A simple electrophoretic system has been developed for the electrophoresis of the protoplasts (Fig. 1). It consists of a slide, two 1-mm-thick agar plates for contact with the platinum electrodes – in order to avoid gas production directly into the electrolyte – and a narrow (1-mm) slit protected by a cover glass for the protoplast suspension. A Shandon SAE 2761 running at 0–50V (1 μA) was used as a power supply and the protoplast movement was followed in the light microscope.

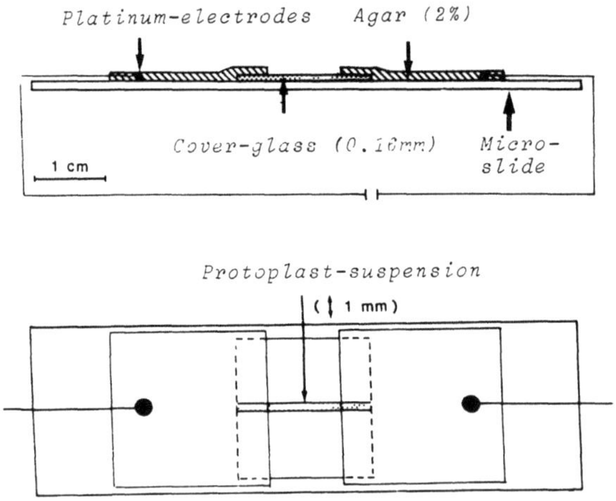

Fig. 1. Equipment for electrophoresis of protoplasts. The unit is placed under the microscope and electrophoresis is followed continuously either manually or by videorecording

Isolation of Mesophyll Protoplasts

For differentiation studies seeds of *Nicotiana sanderae* cv. Crimson Bedder were germinated and grown in soil in a greenhouse at 18°C under fluorescent lamps with an irradiance of 9.0 W m^{-2} at the plant surface. Protoplasts were isolated from leaves taken from 6–8-week-old plants. The leaves were surface sterilized in 7.5% Klorin for 45 min washed (thoroughly) in several changes of sterile distilled water and cut into small strips (25–50 mm^2). The leaf tissue was thereafter plasmolysed in CPW-salts with 13% mannitol (CPW 13M) at 22°C for 1 h.

Cell-wall digestion was performed by incubation of the preplasmolysed tissue in 5% Meicellase and 0.5% Macerozyme R-10 in CPW 13M for 18 h at 26°C on a slow rotation shaker (30 r.p.m.). The protoplast incubation mixture was squeezed gently and the mixture was poured through a nylon filter with a mesh size of 60 μm. After dilution with CPW 13M the protoplast suspension was centrifuged at 80 g for 5 min. The supernatant was poured off, the protoplasts gently resuspended in CPW 13M and centrifuged at 80 g for 7 min.

The pellet was gently resuspended in CPW-salts with 21% sucrose and centrifuged at 120 g for 10 min. After centrifugation, the protoplast layer formed on top of the sucrose solution was removed using a Pasteur pipette. The number of protoplasts was counted using a haemocytometer. Viability of the isolated protoplasts was tested by the FDA-method (Larkin 1976).

Clinostat Experiments

The freshly isolated protoplasts were rotated for different periods of time (0 to 24 h) on a conventional slow rotating clinostat at 2 r.p.m. (Iversen 1974) and a fast rotating clinostat at 50 r.p.m. The fast rotating clinostat (Fig. 2) has been developed at the Deutsche Forschungs- und Versuchsanstalt für Luft- und Raumfahrt (DFVLR), Bonn, West-Germany and the theoretical and physical aspects of the 0-g-simulation have been discussed by Schatz and Briegleb (1982).

Fig. 2 A–C. The fast rotating horizontal clinostat **(A)**. **B** Details showing the plant chamber *(arrow)* and videocamera. **C** A light microscope placed along the horizontal axis of the clinostat. A glass slide **(O)** can be seen where the electrophoretic unit would normally be placed. (Courtsey of Dr. W. Briegleb, Deutsche Forschungs- und Versuchsanstalt für Luft- und Raumfahrt e.V., Cologne, West-Germany)

For clinostat rotation the protoplasts were placed in sterile tubes (inner diameter; 2–3 mm) in two specially constructed plant chambers (Fig. 2). The tubes were placed exactly along the horizontal rotation axis. Since this is most critical on the fast clinostat a laser beam was used here to correct the horizontal axis of the clinostat. The clinostat experiments were performed in darkness.

Cell Culture, Plant Regeneration and Characterization

For cultivation the tobacco protoplasts were dispensed in a liquid Murashige and Skoog (1962) (MS)-medium containing 2.0 mg l^{-1} NAA and 0.5 mg ml^{-1} 6-BAP in 9% mannitol. The suspension was layered on top of 3 mm solidified agar medium (0.6–1% agar) in 5 cm Petri dishes sealed with Parafilm. The suspension was diluted to give a final concentration of 5×10^4 protoplasts ml^{-1}. The cultures were kept in darkness at 23 ± 1°C for 10 days followed by incubation under fluorescent light (25.0 W m^{-2}) with light/dark cycles of 16 h/8 h, respectively. Fresh medium (0.5 ml) containing decreasing concentrations of mannitol was added every 7–10 days. This lowered the osmolarity of the medium and after 4 to 6 weeks the cell colonies were transferred to Gamborg's B5-medium (Gamborg et al. 1968) with 1.0 mg l^{-1} 2.4-D, NAA and 6-BAP in which they were kept for 4 weeks. Root and shoot differentiation were initiated by transferring the callus cultures to B5-D medium containing 0.2 mg l^{-1} IAA and 1.8 mg l^{-1} 6-BAP. After the initiation of roots and shoots, the plantlets were transferred to soil in small pots in the greenhouse. Fourteen weeks after the protoplast isolation the young tobacco plants with 4–6 permanent leaves were transferred to larger pots. Using light microscopy, the initial cell divisons were followed in the clinostat rotated protoplasts and the control group for a period of up to 8 days after isolation. The callus cultures and regenerated plants were characterized morphologically and samples were taken at various developmental stages for cytological analysis.

Light- and Electron Microscopy

Samples of the protoplasts and tissues were taken at different developmental stages and fixed for analysis by light (LM) –, transmission (TEM) – and scanning (SEM) electron microscopy.

For the LM-analysis the protoplasts were either used as fresh squash preparations or fixed, dehydrated and embedded using the same procedure as for TEM. Semithin sections (0.5–1 μm) were cut using glass and diamond knives and an LKB-Ultrotome III. The specimens were examined using phase, interference, and differential interference contrast microscopy.

For TEM, protoplasts were fixed according to the procedure described by Seed (1980). Following dehydration in increasing concentrations of ethanol the protoplasts were embedded in LX 112. Samples of the cell cultures and intact plant organs were cut into small pieces (1 to 2 mm) and fixed in 3% glutaraldehyde in 0.1 M phosphate buffer (pH 7.3) for 2 h at 21°C. After washing in the same buffer the material was postfixed for 1 h at 4°C in 2% OsO_4 in the same buffer. Dehydration was carried out in ethanol of increasing concentrations at room temperature and the tissue was embedded in LX 112. Ultrathin sections were cut on an LKB-Ultrotome III fitted with diamond knives. The sections were stained for 15 min in aqueous 1% uranyl acetate, followed by lead citrade for 5 min. A JEOL 100B electron microscope was used at 60 and 80 kV. For SEM, the protoplasts were fixed in 3% glutaraldehyde for 2 h, washed with buffer and dehydrated with increasing concentrations of acetone. The protoplasts were critical point dried, coated with gold and examined in a JEOL JSM-25 microscope operated at 15 kV (Iversen and Baggerud 1980).

Results and Discussion

Isolation and Characterization of Cress Root Statocytes

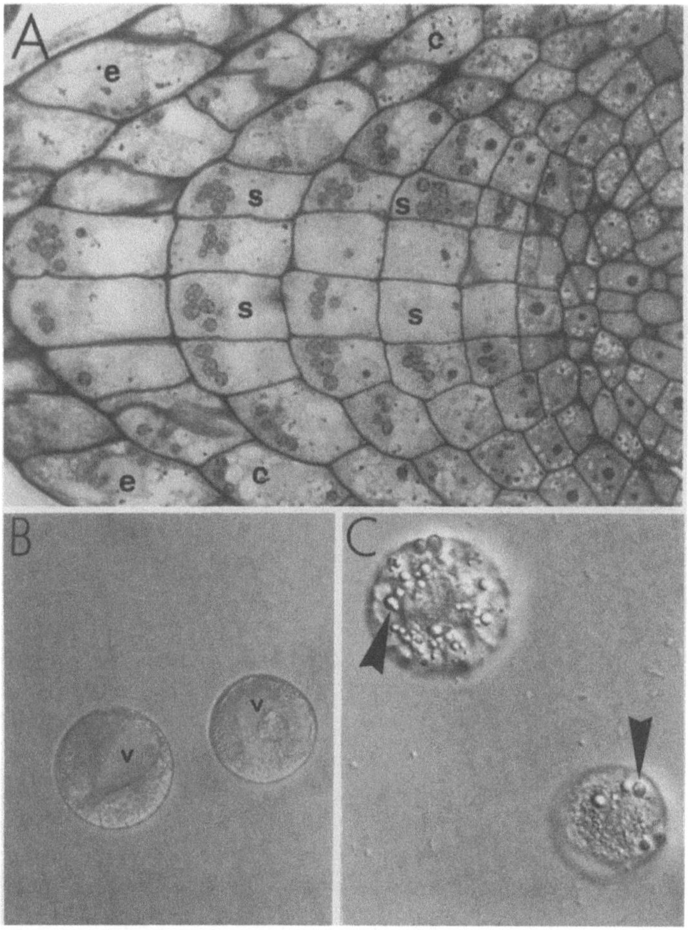

Fig. 3 A–C. Cress root cells and protoplasts. **A** The cress root cap with the columella region with functional statocytes *(s)* cortex *(c)* and epidermal *(e)* cells. **B** protoplasts with large vacuoles *(v)* isolated from cortex cells. C tentatively identified statocytes with amyloplasts *(arrows)*

Most root cap cells contain amyloplasts with starch grains and the main cirteria for distinguishing the peripheral root cap cells from the functional statocytes in the columella region are the extent of vacuolization and the number and size of dilated cisternae (DC) of ER (Iversen 1970).

The yield of root protoplasts of both cress and radish varies depending on the enzymes used, the incubation period, the pretreatment of the plant material etc. as observed, for example by Senn and Pilet (1980). On average 2.5–5x10^4 root protoplasts ml^{-1} were obtained.

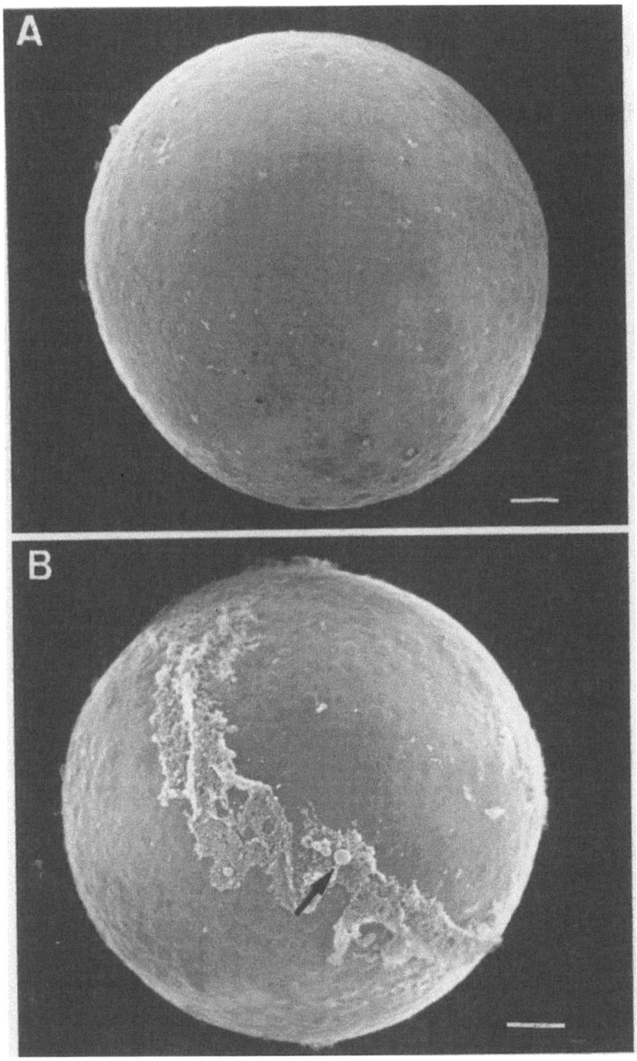

Fig. 4 A, B. Scanning electron micrographs of freshly isolated radish root protoplasts. **A** cell-wall completely removed. **B** residues of cell wall components and plasmodesmata *(arrow)*. *Bars* = 5 μm

Assuming an ideal situation in which only the root cap and the meristematic region have been excised, the lignification of the root cap cells walls is low and the incubation conditions are optimal, such that all cells walls have been totally digested, the percentage of functional statocytes could still only be maximum 10 to 20% of all the root protoplasts. Since these conditions are rarely obtained, the identification and selection of statocyte protoplasts is very complicated.

Figure 3A demonstrates the cress root cap with the columella region and graviperceptive statocytes. Using polarised LM, freshly prepared, highly vacuolized protoplasts from the cortex cells of the cap (Fig. 3B) differ from what we tentatively have identified as root statocyte protoplasts with amyloplasts (Fig. 3C, arrows). Radish root protoplasts appear similar to cress protoplasts and as shown in Fig. 4A the cell-wall has been completely removed. Occasionally residues of cell-wall components and plasmodesmata can be observed (Fig. 4B).

Using TEM, statocytes with starch-free amyloplasts can frequently be identified (Fig. 5) but so far, protoplasts with residues of starch grains have been only occasionally found. The lack of starch might be a result of the incubation procedure used in the protoplast isolation with unpurified cell-wall digesting enzymes and elevated temperature. Such an interpretation is supported by the fact that amylo-

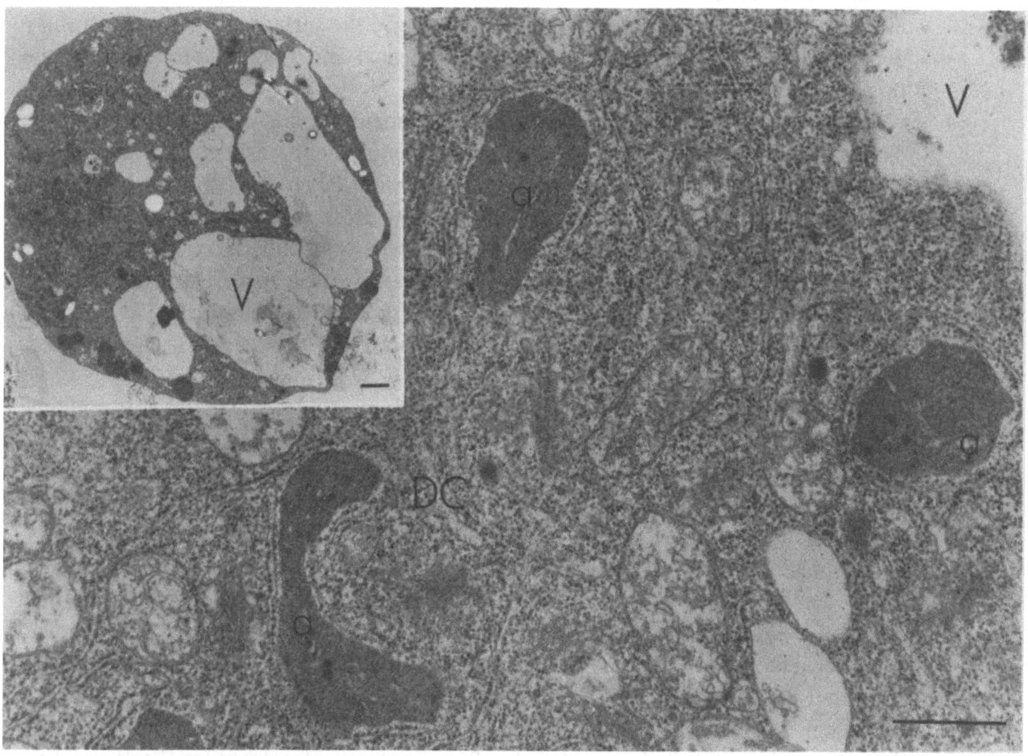

Fig. 5. Transmission electron micrographs of cress statocyte protoplasts. Starch free amyloplasts *(a)* and dilated cisternae of the endoplasmic reticulum *(DC)* can be seen. *Insert* shows a section of the protoplast cut further from the median plane. *V* vacuoles. *Bars* = 1 μm

plasts in root cap cells can be starch-depleted by increasing the temperature (Iversen 1969, 1974). To avoid this problem we have attempted to modify the isolation procedure by using different enzymes, and by decreasing the enzyme concentration, treatment period and temperature, so far without success. However, despite the fact that only starch-depleted statocytes have been found, we feel confident, based on an analysis of the extent of vacuolization and size/distribution of DC, that the protoplasts identified are statocytes.

The starch content in root cells has been found to influence the yield of protoplasts. Crowder et al. (1979) examined 13 cultivars of *Pisum sativum* and found that root cortical protoplast yields were dependent on the cultivar and that the highest yields were obtained from root explants, with no starch or few starchy cells. The authors conclude that some of the variability in protoplasts yields must be explained by other factors than the starch content. Starch-depletion in the protoplasts as a result of the incubation was not found by Crowder et al. (1979).

Electrophoresis of Root Protoplasts

In a detailed study, Seen and Pilet (1981) examined the electrophoretic mobility of protoplasts both on a theoretical basis and in practical experiments. Using a commercially available microelectrophoresis apparatus and protoplasts isolated from different zones of primary maize roots, they found the surface charges were strongly dependent on the cytological origin, physiological and growing properties of the original tissue.

Commercially available electrophoresis equipment was not easily adaptable to the clinostat so a specially modified eletrophoresis equipment was made for the present study (Fig.1). Samples of root cap protoplasts were taken from the upper band on the density gradient and run electrophoretically. Both radish and cress protoplasts have been found to start moving at 2–3 V (1 μA). By gradually increasing the voltage up to 50 V the electrophoretic mobility increases with a speed in the range 5–10 $\mu m\ s^{-1}$. By further increasing the voltage the protoplasts disintegrate. By reversal of the current the movement of the protoplasts is also reversed – this confirms that this a true electrokinetic movement.

With a purified sample of functional statocyte protoplasts, experiments could be designed to help answer the following questions: Will interaction between statoliths and the endomembrane system be reflected as a change in surface charges? Can changes in charges be detected by electrophoresis in protoplasts subjected to a microgravity environment on a clinostat?

At present it is not possible to give definite answers to these questions. It has been demonstrated experimentally that membranes are affected under defined pressure e.g. changes of $2 x 10^6$ dyne cm^{-2} induce action potentials (Zimmermann and Beckers 1978). Enzyme activities are also influenced; acetylcholinesterase is activated by 10 dyne cm^{-2} (Skou 1959), a value comparable to the pressure of 2–4 dyne cm^{-2} exerted by amyloplasts. Surface charges on the amyloplast membrane have been demonstrated in statocytes from rootcaps of coleoptiles (Sack and Leopold 1982). It is therefore possible that interactions between the statolith membrane and the ER-complex may be reflected as changes in the electrostatic forces on the statocyte protoplast surface.

The use of the fast clinostat will also be important in further work in progress in our laboratory. A modified electrophoretic system – based on the same principle as Fig. 1 – will be placed on a microscope on the fast clinostat (Fig. 2C). By video-recording, the electrophoretic movement of statocytes can be followed in a simulated microgravity environment on the rotating clinostat and compared to a control affected by gravity. In the experiments both starch-depleted and starch-containing statocytes will be used in an attempt to elucidate the graviperception mechanism.

Protoplasts, Gravity and Differentiation

In our laboratory we use tobacco as a model plant for differentiation studies since this species can easily be regenerated from protoplasts via callus cultures to flowering plants.

Table 1 shows the effect of clinostat rotation on the cell division of tobacco protoplasts recorded 6 and 8 days after the isolation. Protoplasts rotated on the slow clinostat show a significant decrease in mitotic acitivty compared to the control group. A similar effect has not been found when rotated on the fast clinostat. The rotation period, up to 6 h, does not seem to have any effect on the cell division activity.

Regeneration of protoplasts follows the same pattern for both the rotated and control group. The first cell division in both groups can be observed after 3 days and after 2 months the established callus cultures are ready for differentiation.

Table 1. The effect of rotation on the slow (2 r.p.m., A) and fast (50 r.p.m., B) clinostat on the cell divisions of tobacco protoplasts. The activity was determined 6 and 8 days after the protoplasts were isolated

	Days after isolation			
	6		8	
Rotation period (h)	Control	Rotated	Control	Rotated
A				
2	10.4 ± 1.2	8.4 ± 0.7	11.3 ± 1.0	9.4 ± 0.7
4	9.2 ± 1.5	7.2 ± 1.6	8.0 ± 1.4	6.9 ± 1.1
6	8.1 ± 1.9	7.2 ± 2.1	10.6 ± 1.7	8.4 ± 1.1
B				
2	10.3 ± 1.3	8.9 ± 0.6	11.3 ± 1.0	12.0 ± 0.8
4	11.3 ± 1,4	10.2 ± 1.3	12.0 ± 0.6	9.8 ± 0.9
6	10.7 ± 1.1	9.8 ± 0.9	12.3 ± 0.6	11.9 ± 0.6

± mean error

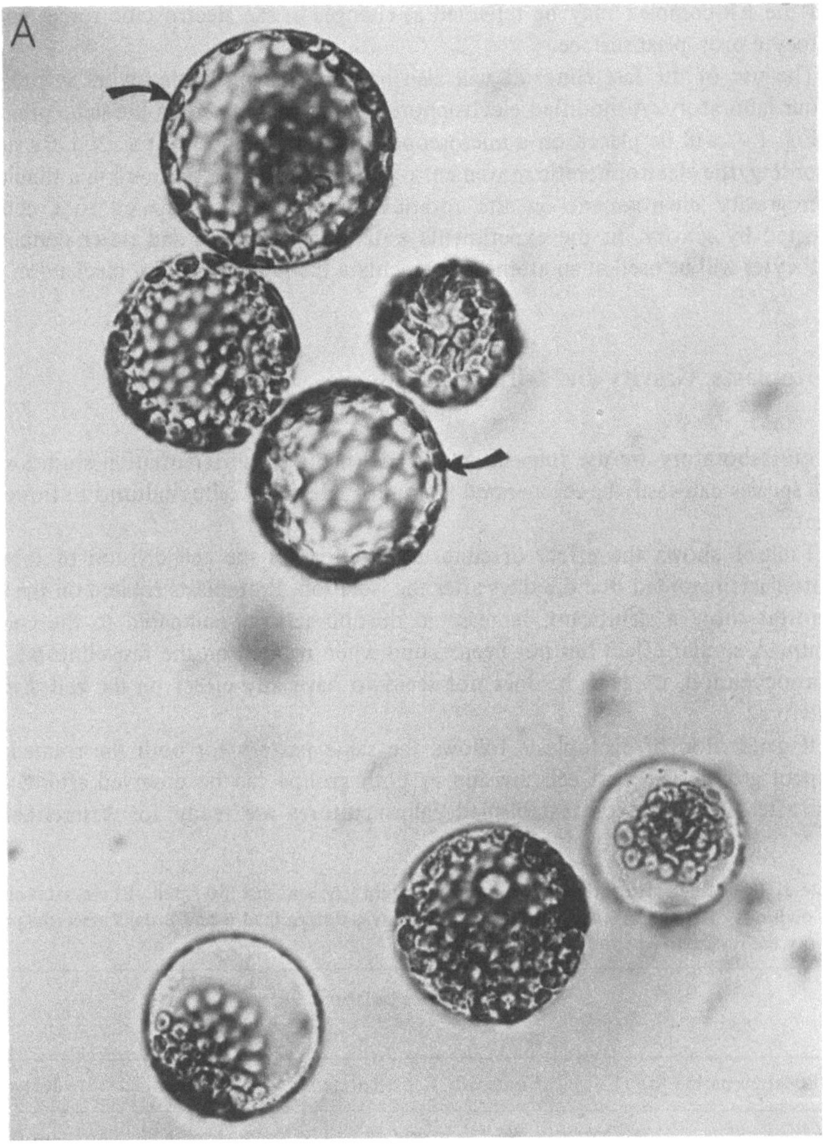

Fig. 6 A–C. Regeneration of tobacco plants after rotation of freshly isolated leaf protoplasts (**A**) with chloroplasts *(arrows)* for different periods of time. **B** 12-week-old plants regenerated from protoplasts rotated on the fast clinostat for 4 h *(FC)* and control group *(C)*. **C** 14-week-old plants transferred to soil after regeneration from 2 and 6 h rotated protoplast. *FC* fast clinostat; *SC* slow clinostat and control groups *(C)*

It has not been possible to find any morphological changes in the plants regenerated from rotated protoplasts (Fig. 6). However, ultrastructurally the content of starch grains in the chloroplasts have been found to decrease after rotation for 24 h on the slow but not on the fast clinostat.

Fig. 6 B, C

It can be concluded that conventional clinostats rotating at 2 r.p.m. interfere with the normal regeneration processes of tobacco protoplasts. It is not possible from such experiments to decide which type of clinostat provides the best simulated microgravity environment. The protoplast culture system seems, however, to be a very sensitive tool for studies of gravity effects and it should therefore be tested in a real microgravity environment during a Space Shuttle flight.

Conclusions

The naked, freshly isolated protoplast, protected from the environment only by the labile plasmamembrane, represents a highly sensitive system with respect to external stimuli, e.g., variations in gravity. In addition the totipotency of the cell makes it possible to regenerate mature plants after gravitational manipulations.

In the search for the primary mechanism of statolith action within graviperceptive root cap cells (statocytes) attempts have been made to isolate such cells as protoplasts. The gravisensitivity of the statocytes seem to be connected with the sensitivity of the endomembrane system (mainly ER) to pressure exerted by the movable statoliths. This may be the basis for the asymmetrical signal which ultimately results in a gravitropic curvature.

If the graviperceptive signal which becomes initiated by the interaction between the statoliths and the ER-membrane system, is reflected as a change in the surface charges on the protoplasts membrane this might result in altered electrophoretic mobility.

A simple electrophoretic system has therefore been developed which makes it possible to run protoplasts electrophoretically under simulated microgravity conditions on a fast clinostat.

The effect of a simulated microgravity environment on the initial cell divisions, growth and differentiation of tobacco plants from protoplasts has also been followed. The results indicate that protoplasts may be used as a model system to study regeneration processes under weightlessness and the radiation environment in space.

Acknowledgements. The author would like to thank Carolyn Baggerud and Geir Slupphaug for valuable contributions to the present study. Financial support from The Norwegian Research Council for Science and the Humanities (no. D.75. 49.052) is gratefully acknowledged.

References

Crowder AJ, Landgren CR, Rockwood LL (1979) Cultivar differences in starch content and protoplast yields from root cortical explants of *Pisum sativum*. Physiol Plant 46:85–88

Gamborg OL, Miller RA, Ojima K (1968) Nutrient requirements of suspension cultures of soybean root cells. Exp Cell Res 50:151–158

Haberlandt G (1900) Über die Perzeption des geotropischen Reizes. Ber Dtsch Bot Ges 18: 261–272

Harms CT, Potrykus I (1978) Fractionation of plant protoplast types by iso-osmotic density gradient centrifugation. Theor Appl Genet 53:57–63

Iversen TH (1969) Elimination of geotropic responsiveness in roots of cress *(Lepidium sativum)* by removal of statolith starch. Physiol Plant 22:1251–1262

Iversen TH (1970) The morphology, occurrence, and distribution of dilated cisternae of the endoplasmic reticulum in root cells of *Lepidium sativum* seedlings. Protoplasma 71:467–477

Iversen TH (1974) The roles of statoliths, auxin transport, and auxin metabolism in root geotropism. K Nor Vidensk Selsk Mus Misc 15:1–216

Iversen TH (1982) The effect of gravity on plant cells. The Physiologist Suppl 25:No 6

Iversen TH, Baggerud C (1980) Myrosinase activity in differentiated and undifferentiated plants of Brassiaceae. Z Pflanzenphysiol 97:399–407

Iversen TH, Larsen P (1971) The starch statolith hypothesis and the optimum angle of geotropic stimulation. Physiol Plant 25:23–27

Iversen TH, Larsen P (1973) Movement of amyloplasts in the statocytes of geotropically stimulated roots. The pre-inversion effect. Physiol Plant 28:172–181

Larkin PJ (1976) Purification and viability determinations of plant protoplasts. Planta (Berl) 128:213–216

Murashige T, Skoog F (1962) A revised medium for rapid growth and bioassays with tobacco tissue cultures. Physiol Plant 15:473–497

Němeč B (1900) Über die Art Wahrnehmung des Schwerkraftreizes bei den Pflanzen. Ber Dtsch Bot Ges 18:241–245.

Sack FD, Leopold AC (1982) Characteristics of statoliths from rootcaps and coleoptiles. Proc IUPS 33rd Annual Fall Meeting San Diego, California

Schatz A, Briegleb W (1982) Physikalische Grundlagen und Anwendung der 0-g Simulation nach H.J. Muller. Deutsche Luft- und Raumfahrt, Forschungsbericht 73–15:169–182

Seed LJ (1980) The preparation of fragile protoplasts for electron microscopy. J Microscopy (Oxf) 120:109–112

Senn A, Pilet P-E (1980) Isolation and some morphological properties of maize root protoplasts. Z Pflanzenphysiol 100:299–310

Senn A, Pilet P-E (1981) Electrophoretic mobility, zeta potential and surface charges of maize root protoplasts. Z Pflanzenphysiol 102:19–32

Skou JC (1959) Studies on the influence of the degree of unfolding and the orientation of the side chains on the activity of a surface-spread enzyme. Biochim Biophys Acta 31:1–10

Volkmann D, Sievers A (1979) Graviperception in multicellular organs. In: Haupt W, Feinleb ME (eds) Encycl plant physiol, vol 7. Physiology of movements. Springer, Berlin Heidelberg New York, p 573

Zimmermann U, Beckers F (1978) Generation of action potentials in *Chara corrallina* by turgor pressure changes. Planta (Berl) 138:173–179

Proton Extrusion in Protoplasts: Fusicoccin and Cytokinin Effects

D.C. ELLIOTT

Introduction

Among the advantages in using protoplasts for studies on the action of growth regulators are greater sensitivity in response, freedom from the buffering influences of cell walls and the barrier imposed by the cuticle, and from the complex interactions of adjacent cells. On the other hand, damage to the plasma membrane by the digestive enzymes used and the stress imposed by hypertonic media (Fleck et al. 1982, Rubinstein 1982) introduce possible disadvantages.

The question of whether protoplasts provide a valid model for the study of physiological processes in plants has been addressed by a number of workers (Galun 1981). On the one hand there is pronounced modulation of fatty acid patterns in protoplasts derived from some species. This appears to be correlated with reduced regenerative capacity of the protoplasts (Hock and Hartmann 1983). Uptake of adenine and of cytokinins is much retarded in *Brassica campestris* protoplasts as compared to root sections (Dekhuijzen et al. 1978). On the other hand, protoplasts appear to be very suitable for the study of other transport phenonema, e.g. sugar, amino acid and ion transport across the plasmalemma of *Pisum sativum, Nicotiana glutinosa, Zea mays* and *Beta vulgaris* (Guy et al. 1978, Mettler and Leonard 1979a, b; Briskin and Leonard 1979; Lin 1980, 1981; Schmidt and Poole 1980; Gronwald and Leonard 1982). Related parameters (swelling and contracting of cells) measured in guard cells from *Allium cepa* show responses to abscisic acid and fusicoccin (FC) correlated with, but four times as rapid as, the in vivo response of the stomata (Schnabl et al. 1978).

Protoplasts are potentially very valuable where particular cell types have specific functions and it is desirable to study these functions in the absence of surrounding cells which may interfere. The aim of the present work was to exploit the possibility of obtaining specific cell types as had been done with isolation of guard cells (Weyers et al. 1983) to investigate the mechanism of action of FC and cytokinin in the induction of betacyanin synthesis in *Amaranthus tricolor* cotyledons. Cytokinins in the dark (or red light alone) can be shown to regulate the induction of betacyanin synthesis in *A. tricolor* seedlings in specific cells. The target cells for pigment accumulation are the lower epidermis of the cotyledons and the endodermis of the hypocotyl (Elliott 1983a). FC and N^6-benzyladenine (BA) (Elliott 1979e) and red light and BA (Elliott 1979f) are synergistic in increasing betacyanin accumulation.

Abbreviations: IC_{50}, concentration giving 50% inhibition

The Physiological Properties of Plant Protoplasts
Edited by P. E. Pilet

Potential for cytokinin action is enhanced by pretreatments of half-seedlings (cotyledons plus top 4–5 mm hypocotyls), such as water stress, heat or aging in water (Elliott 1979b, 1979c, 1979d). Red-light and FC-dependent synthesis is also increased by the heat-aging pretreatment (Elliott 1979f). Inhibition of synthesis of membrane components inhibits both the stress-dependent potential for cytokinin action and the cytokinin-dependent induction itself (Elliott 1982a). Increases in membrane components occur during betacyanin induction, the levels being higher in the presence of BA (Elliott 1982b). These results show that new membrane synthesis and possibly increased membrane synthesis are required for potentiation of the cytokinin response and are part of the cytokinin response itself.

Some idea of the membrane function involved has been gained from the following experiments. Betacyanin accumulation is stimulated 50 to 100% by K^+ in the exogenous medium, the best induction being given by a K^+/Na^+ mixture with phosphate as the anion (Elliott 1979a). Preliminary results with proton extrusion (Batchelor and Elliott 1981), together with inhibitory effects of uncouplers used to destroy proton gradients and of inhibitors of membrane. ATPases (Elliott 1979e) led to the conclusion that cytokinins affect ion fluxes, particularly H^+/K^+ exchange.

Evidence of Ca^{2+} involvement (Elliott 1977, 1979a, 1983b) suggested that there was some inter-relation between cytokinin and Ca^{2+} also in this system. Hence, in view of the widespread involvement of calmodulin in many Ca^{2+}-regulated events, studies were undertaken to see if certain drugs which inhibit calmodulin-dependent processes have any effect on cytokinin regulation of betacyanin synthesis, cell divison or cell expansion. Activation at very low concentrations and inhibition by neuroleptic drugs [such as trifluoperazine (TFP)] and local anaesthetics [such as tetracaine (TC)] at IC_{50} (concentration giving 50% inhibition) values compatible with inhibition of calmodulin-dependent enzymes and processes suggested the involvement of a calcium-dependent regulator protein in cytokinin action (Elliott 1980, 1983b; Elliott et al. 1983). An alternative interpretation, that a CA^{2+}-dependent phospholipid-activation of an enzyme is involved, was also considered.

As a result of these studies a hypothesis has been formulated (Elliott 1983b) to explain cytokinin action as a two-fold event with (a) the hormone (and FC) acting through a membrane function (specifically K^+ and/or Ca^{2+} fluxes linked to proton extrusion in some way) to control events such as cell expansion. Control of developmental processes (such as betacyanin accumulation) by cytokinins would require in addition (b) a further action involving formation of the participating membrane structures being stimulated by stress pretreatment and by the hormone itself. Isolation and study of the "target cells" from *A. tricolor* cotyledons as protoplasts was seen as one way to test this hypothesis further.

Materials and Methods

Amaranthus tricolor seeds were germinated in the dark at 25 ± 1°C for 84 h under standard conditions (Elliott 1979b). The roots were removed and the remaining cotyledons plus hypocotyl were placed in distilled water for 1.5 h at 40 ± 1°C and a further 1.5 h at 25 ± 1°C (Elliott 1979b). Cotyledons were excised and protoplasts prepared as previously reported (Batchelor and Elliott 1983). The separation method produced a mixed population of which 25% were lower epidermal cells as judged by colour when the procedure was followed with cotyledons which had been pre-induced by incubation with 10 µM BA for 24 h. Details of the enzyme digestion medium and the suspension medium for measuring proton extrusion are summarized in Table 3. Measurement of apparent net H^+ efflux was carried out in a small perspex well fitted with a magnetic stirrer and pH electrode. pH changes were amplified and recorded continously on a moving chart recorder.

Chemicals

Cellulase R10 and macerozyme R10 were from All-Japan Biochemical Co., Mishinomya, Japan. Rohament P (Rohm and Haas Co., Philadelphia, USA) was purified by gel filtration on Sephadex G-25. All enzyme solutions were filter sterilised through 0.22 µm filters (Millipore Filter Corporation, Bedford, Mass, USA). Trifluoperazine was a gift from Smith, Kline and French Laboratories. The aqueous stock solution was 10^{-2} M, pH 6.2 (with NaOH). Fusicoccin was a generous gift from Professors A. Ballio (Rome) and E. Marrè (Milan). Benzyladenine (Sigma) and fusicoccin were dissolved in DMSO which had a negligible effect on H^+ extrusion at 0.1%. EtOH at this concentration inhibited about 50%.

Results

Protoplasts from *A. tricolor* cotyledons have been used to demonstrate apparent net proton efflux and stimulation of this efflux by 0.5 µM FC and 5 µM BA (60 and 40% respectively). The lag times for effects of FC and BA are 1–2 min and 7–9 min respectively (Batchelor and Elliott 1983). The maximal concentration for the BA effect is 10 µM but, using sub-optimal concentrations of 5 µM BA and 0.5 µM FC, synergism can be shown when these compounds are added sequentially (Batchelor and Elliott 1983).

As with regulator-induced effects in whole cotyledons (fresh weight increase and proton extrusion, Batchelor and Elliott 1983; betacyanin induction, Elliott 1979a), K^+ ions promote H^+ extrusion in protoplasts and Ca^{2+} ions (3 mM) inhibit (Table 1). The presence of K^+ ions appeared to be essential for BA-induced H^+ extrusion with no stimulation occurring in its absence. No study was made of the interrelation between K^+ ions and FC in the induction of H^+ extrusion. However, Cleland and Lomax (1977) showed the FC-induced H^+ intrusion in oat protoplasts did not have an obligatory counter movement of K^+ions, although the rate was enhanced in the presence of 1 mM KCl.

Inhibition of H^+ extrusion, both endogenous and BA-induced, was given by 3 mM $CaNO_3$ (Table 1). In previous work Ca^{2+} inhibition of BA effects has been interpreted in terms of the need to maintain a low (10^{-8} – 10^{-5} M) concentration of Ca^{2+} in the cytoplasm, and the subsequent inhibitory effect of high concentrations (Elliott 1983b). In view of such interaction between Ca^{2+} and BA the effect of

Table 1. Ionic regulation of BA-induced H^+ extrusion (Batchelor 1984)

		nmol H^+ min^{-1} /10^5 protoplasts	Control
	K^+ effect		
Expt. 1	Control ($-K^+$)	2.10	100
	BA (10 μM)	1.31	62
	BA (10 μM)+ KCl (10 mM)	4.43	211
Expt. 2	Control ($+K^+$)	2.27	100
	BA (10 μM)	4.17	183
	Ca^{2+} effect		
	Control	1.55	100
	BA (10 μM)	2.40	154
	$Ca(NO_3)_2$ (3 mM)	− 0.84	

Table 2. Effect of calmodulin-binding drug on BA-induced H^+ extrusion (Batchelor 1984)

		nmol H^+ min^{-1} /10^5 protoplasts	% Control
	Standard response		
	Control	1.55	100
	BA (10 μM)	2.40	154
	Trifluoperazine inhibition		
Expt. 1	Control	2.29	100
	TFP (20 μM)	2.18	95
	TFP (100 μM)	2.37	103
Expt. 2	Control	1.44	100
	BA (10 μM) + TFP (20 μM)	1.26	87

the calmodulin binding drug TFP was tested in protoplast H^+ extrusion studies (Table 2). TFP gave 50% inhibition at 20 μM. This is a lower concentration than the IC_{50} of this compound on other responses in whole cotyledons (50 μM on BA-dependent cotyledon fresh weight and 150 μM on betacyanin induction, Elliott et al. 1983).

Table 3. Comparison of rates of H^+ efflux by protoplasts

Plant tissue	Protoplast digestion medium	H^+ extrusion medium	Rate of H^+ extrusion	Reference
Avena sativa, leaves from 7-day-old light-grown seedlings	0.37 M mannitol 0.3 M sorbitol 1% cellulysin or 2% cellulase SS B5 salts (Gamborg 1970) pH 5.7	0.6 M mannitol 1 mM KCl	–FC, pH 6.1 → 6.0 in 15'. + 10 μM FC, pH 6.1 – 5.6 in 15'. Lag < 2 min	Cleland and Lomax (1977)
Parthenocissus tricuspidata, suspension cultures	13% mannitol 2% cellulase R-10 2% macerozyme R-10	13% mannitol 10 mM KCl	–FC, pH 5.96 → 5.84 in 2 h. +20 μM FC, pH 5.96 → 5.60 in 2 h.	Rollo et al. (1977)
Nicotiana glutinosa, suspension cultures	0.7 M mannitol 1% cellulysin 0.2% macerase	0.7 M mannitol 1 mM KCl Adj.pH 6.9 with NaOH	–FC, pH 6.9 → 6.1 in 40 min. +10 μM FC, pH 6.9 → 6.0 in 40 min	Mettler and Leonard (1979a)
Zea mays, roots from 3-day-old etiolated seedlings	0.6 M mannitol 2% cellulysin 1% hemicellulase 0.5% pectinase 0.2 mM $CaCl_2$ pH 5.5	0.7 M mannitol 0.2 mM $CaCl_2$ 1 mM Mes, pH 6.0	–FC, 0.701 nmol H^+ $min^{-1}/10^6$ protoplasts. +20 μM FC, 1.441 nmol H^+ $min^{-1}/10^6$ protoplasts	Lin (1980)
Amaranthus tricolor, cotyledons from 3.5–day-old etiolated seedlings	0.5 M mannitol 0.75 cellulase R-10 0.25% macerozyme 0.75% rohament 10 mM $CaCl_2$ 1 mM KNO_3 0.2 mM KH_2PO_4 1.4 mM $MgSO_4$ 0.9 μM KImol l 0.1 μM $CuSO_4$ pH 6.0	0.4 M mannitol 10 mM KCl 0.5 mM K_2HPO_4 3 mM $MgCl_2$ 0.5 mM Mes, pH 6.0	–FC, 1.5 nmol H^+ $min^{-1}/10^5$ protoplasts. +2 μM FC, 3.0 nmol H^+ $min^{-1}/10^5$ protoplasts. Lag 1–2 min	Batchelor and Elliott (1981, 1983)
Lunularia cruciata, thalli	0.5 M mannitol 5% cellulase 2.5% hemicellulase 0.75% pectinase	0.5 M mannitol	+100 μM FC, pH 5.8 → 4.8 in 30 s /2x10^5 protoplasts.	Thomas and Silcox (1983)

Discussion

Protoplasts of *A. tricolor* cotyledons have been used succesfully in this study to measure the apparent net proton efflux in response to FC and to BA, and correlation with data obtained from excised cotyledons can be made. Acidification of the surrounding medium by both protoplasts and cotyledons is increased by K^+ and dedreased by high (3 mM) Ca^{2+}. Both respond to FC with increased H^+ extrusion. In protoplasts the response is almost immediate (lag 1–2 min) while with cotyledons change in pH of the medium cannot be detected for 2–4 h. With BA there is a lag of 7–9 min before increased H^+ extrusion from protoplasts is measurable. BA-FC energism is seen with both materials.

Comparison of rates of H^+ extrusion in *A. tricolor* protoplasts with data from other sources is seen in Table 3. Increased aciddification of the medium in response to FC has been reported using protoplasts from leaves of 7-day-old light-grown seedlings of *Avena sativa* (Cleland and Lomax 1977), from suspension cultures of *Parthenocissus tricuspidata* (Rollo et al. 1977) and of *Nicotiana glutinosa* (Mettler and Leonard 1979a), from roots of 3-day-old etiolated seedlings of *Zea mays* (Lin 1980) and from haploid thalli of *Lunularia cruciata* (Thomas and Silcox 1983). The rate of H^+ extrusion by *A. tricolor* protoplasts is less than that reported for *L. cruciata* protoplasts, and about 10 times that given by *Z. mays* protoplasts. There is insufficient data to compare with other protoplast preparations.

Increased acidification of medium by protoplasts in response to cytokinin has not been reported previously although some reports of the effect of auxin on pH of medium have appeared.

To revert to the twofold hypothesis of cytokinin action outlined in the introduction, the use of protoplasts should facilitate the study of the ion transport events, particularly in view of the stimulation by K^+ and inhibition by high Ca^{2+}. The absence of the cell-wall, a major reservoir of cations, especially Ca^{2+}, will enable the ionic composition of the outside medium to be more accurately known. The effect of FC and BA on the induction of betacyanin accumulation has not yet been achieved in protoplasts. However, the demonstration of inhibitory effects of TFP on H^+ extrusion at the slightly lower concentration than that at which this drug inhibits other FC and BA dependent events is suggestive that at least one of the effects of this drug in inhibiting the developmental process of betacyanin induction can be studied in protoplasts. The results are consistent with a hypothesis that the activation of a plasma membrane ATPase having selective effects on monovalent ion transport may involve Ca^{2+}. Whether TFP is inhibiting by reacting with a Ca^{2+}-phospholipid-dependent (Schatzman et al. 1981), or a Ca^{2+}-calmodulin-dependent (Weiss et al. 1980) process would not affect this general conclusion.

References

Batchelor SM (1984) Ph D thesis, Flinders University of SA

Batchelor SM, Elliott DC (1981) The isolation of protoplasts from *Amaranthus tricolor* and their use in proton extrusion studies. Proc Aust Soc Biochem 14:117

Batchelor SM, Elliott DC (1983) The advantages of isolated protoplasts for plant growth regulation studies. Protoplasts 1983, Poster Proc 6th Int Protoplast Symp Basel. Birkhäuser, Basel, pp 178–179

Briskin DP, Leonard RT (1979) Ion transport in isolated protoplasts from tobacco suspension cells III. Membrane potential. Plant Physiol (Bethesda) 64:959–962

Cleland RE, Lomax T (1977) Hormonal control of H^+ excretion from oat cells. In: Ciferri O, Marrè E (eds) Regulation of cell membrane activities in plants. Elsevier, North Holland, New York, pp 161–171

Dekhuijzen HM, Bossen ME, Vonk CR (1978) The uptake of cytokinins by protoplasts and root sections of *Brassica campestris.* Physiol Plant 43:142–147

Elliott DC (1977) Action of transport anitbiotics on cytokinin-dependent betacyanin synthesis. In: Ciferri O, Marrè E (eds) Regulation of cell membrane activities in plants. Elsevier, North Holland, New York, pp 317–323

Elliott DC (1979a) Ionic regulation for cytokinin-dependent betacyanin synthesis in *Amaranthus* seedlings. Plant Physiol (Bethesda) 63:264–268

Elliott DC (1979b) Analysis of variability in the *Amaranthus* bioassay for cytokinins: effect of water stress on benzyladenine- and fusicoccin-dependent responses. Plant Physiol (Bethesda) 63:269–273

Elliott DC (1979c) Analysis of variability in the *Amaranthus* betacyanin assay for cytokinins: effects of "aging" excised cotyledons. Plant Physiol (Bethesda) 63:274–276

Elliott DC (1979d) Temperature-dependent expression of betacyanin synthesis in *Amaranthus* seedlings. Plant Physiol (bethesda) 63:277–279

Elliott DC (1979e) Differential effects on fusicoccin- and cytokinin-stimulated betacyanin synthesis by ATPase inhibitors and uncouplers. Plant Sci Lett 15:251–264

Elliott DC (1979f) Temperature-sensitive responses of red light-dependent betacyanin synthesis. Plant Physiol (Bethesda) 64:521–524

Elliott DC (1980) Calmodulin inhibitor prevents plant hormone response. Biochem Int 1:290–294

Elliott DC (1982a) Inhibition of cytokinin action and of heat/aging-induced potential for cytokinin action by inhibitors of membrane synthesis and function. Plant Physiol (Bethesda) 69:1169–1172

Elliott DC (1982b) Levels of membrane components regulated by cytokinins and by temperature and aging pretreatments of cotyledons. Plant Sci Lett 26:311–323

Elliott DC (1983a) Accumulation of cytokinin-induced betacyanin in specific cells of *Amaranthus tricolor* seedlings. J Exp Bot 34:67–73

Elliott DC (1983b) Inhibition of cytokinin-regulated responses by calmodulin-binding compounds. Plant Physiol (Bethesda) 71:215–218

Elliott DC, Batchelor SM, Cassar RA, Marinos NG (1983) Calmodulin-binding drugs affect responses to cytokinin, auxin and gibberellic acid. Plant Physiol (Bethesda) 71:219–224

Fleck J, Durr A, Fritsch C, Vernet T, Hirth L (1982) Osmotic-shock "stress proteins" in protoplasts of *Nicotiana sylvestris*. Plant Sci Lett 26:159–165

Galun E (1981) Plant protoplasts as physiological tools. Annu Rev Plant Physiol 32:237–266

Gamborg OL (1970) The effects of amino acids and ammonium on the growth of plant cells in suspension culture. Plant Physiol (Bethesda) 45:372–375

Gronwald JW, Leonard RT (1982) Isolation and transport properties of protoplasts from cortical cells of corn roots. Plant Physiol (Bethesda) 70:1391–1395

Guy M, Reinhold L, Laties GG (1978) Membrane transport of sugars and amino acids in isolated protoplasts. Plant Physiol (Bethesda) 61:593–596

Hock K, Hartmann E (1983) Modulation of fatty acid patterns during protoplast isolation. Protoplasts 1983, Poster Proc 6th Int Protoplast Symp Basel, Birkhäuser, Basel, pp 206–207

Lin W (1980) Corn root protoplasts: isolation and general characteristics of ion transport. Plant Physiol (Bethesda) 66:550–554

Lin W (1981) Inhibition of anion transport in corn root protoplasts. Plant Physiol (Bethesda) 68:435–438

Mettler IJ, Leonard RT (1979a) Ion transport in isolated protoplasts from tobacco suspension cells. I. General characteristics. Plant Physiol (Bethesda) 63:183–190

Mettler IJ, Leonard RT (1979b) Ion transport in isolated protoplasts from tobacco suspension cells II. Selectivity and kinetics. Plant Physiol (Bethesda) 63:191–194

Rollo F, Nielsen E, Sala F, Cella R (1977) Effect of fusicoccin on plant cell cultures and protoplasts. Planta (Berl) 135:199–202

Rubinstein B (1982) Regulation of H^+ excretion: role of protein released by osmotic shock. Plant Physiol (Bethesda) 69:945–949

Schatzman RC, Wise BC, Kuo JF (1981) Phospholipid-sensitive calcium-dependent protein kinase: inhibition by antiphsychotic drugs. Biochem Biophys Res Commun 98:669–676

Schmidt R, Poole RJ (1980) Isolation of protoplasts and vacuoles from storage tissue of red heat. Plant Physiol (Bethesda) 66:25–28

Schnabl H, Bornman H, Ziegler H (1978) Studies on isolated starch containing *Vicia faba* and starch deficient *Allium cepa* guard cell protoplasts. Planta (Berl) 143:33–40

Thomas RJ, Silcox KR (1983) Proton efflux and cell lysis induced by fusicoccin in haploid protoplasts of *Lunularia cruciata*. Plant Sci Lett 29:169–173

Weiss B, Prozialeck W, Cimino M, Barnette MS, Wallace TL (1980) Pharmacological regulation of calmodulin. Ann NY Acad Sci 356:319–345

Weyers JDB, Fitzsimons PJ, Mansey GM, Martin ES (1983) Guard cell protoplasts – aspects of work with an important new research tool. Physiol Plant 58:331–339

Protoplast Growth and Photoregulation

G. DE MARCH and A. TRÉMOLIÈRES

Introduction

Light is an essential factor in the control of plant development. Red light and blue light are the most effective spectrum regions and it is now a well established fact that they play a dual role in plant responses to the environment: first, they act through photosynthesis to provide a source of energy; secondly, they can modulate physiological responses through the action of two photoreceptors (phytochrome and the blue light pigment). Several reports have shown that phytochrome is responsible for perception of light quality while the blue-light pigment detects light quantity (Gaba and Black 1979, Thomas and Dickinson 1979, Ritter et al. 1981, Smith 1981). However, the mode of action of the pigments has not yet been clearly established. Membranes have been proposed as possible primary sites of action of phytochrome, the best known photoreceptor (Marmé 1977). Several studies have demonstrated that one of the primary events occurring just after light perception by phytochrome involves the transport of ions across membranes and particularly calcium ions (Satter et al. 1970; Tezuka and Yamamoto 1975; Weisenseel and Ruppert 1977; Brownlee and Kendrick 1979a, b; Dreyer and Weisenseel 1979). Studies to elucidate the mechanism of phytochrome action have focused primarily on physicochemical properties of the photoreceptor. It is now well known that the pigment exists in two interconvertible molecular forms. The P_{fr} form, absorbing in the far-red region of the spectrum, is induced by red light (R) and can be converted to the P_r form, absorbing in the red region, by far-red light (FR). It is assumed that P_{fr}, only, binds to a specific membrane receptor, resulting in modifications of membrane permeability and membrane potential (Marmé et al. 1973; Pratt and Marmé 1976; Marmé 1977; Löppert et al. 1978). However it is much more difficult to explain the relationship between membrane effects and morphogenetic responses such as stem or leaf growth, chloroplast development, etc. Although an increasing number of studies have attempted to elucidate the mechanisms of photoregulation, the sequence of events leading to the physiological responses still remains unknown. An explanation relates to the difficulty of studying the events occurring at the cellular level, either on whole plants or on excised tissue. Intact tissue constitutes a complex system where cuticle and cell cohesion hamper the study of solute transport (ions, hormones) as well as homogeneous incorporation of metabolic precursors. Moreover, cells integrated in a whole tissue are surrounded by a buffered and complex medium. In such a natural system, environmental parameters, which may be implicated in physiological processes, are not easily controlled. For these reasons plant physiologists have been

The Physiological Properties of Plant Protoplasts
Edited by P. E. Pilet

led to propose a simpler and more homogeneous system. Protoplasts appear to be a convenient tool for that purpose. They cannot only be maintained in a perfectly controlled medium but may also be readily subjected to regulatory changes. However the question arises as to whether protoplasts, which are free of tissue correlations, deprived of a cell-wall and subjected to an unusual osmotic stress, could be used to better understand physiological processes. In this report, we offer some answers to this question through a few examples concerning photoregulation of cellular processes. First, we present a synopsis of current knowledge about protoplast sensitivity to light quality or light quantity, via a photoreceptor. Second, we attempt to establish whether or not protoplasts may be used to understand physiological processes occurring in whole plants.

Protoplast Photoregulation

Several reports have shown the usefulness of protoplasts for studying photosynthesis at the cellular level (Nishimura and Akazawa 1975; Rathnam and Zilinskas 1977; Kaiser and Heber 1983). This aspect will not be detailed here but it should be stressed at this point that light perception via photosynthesis may cause modifications in ion uptake (Leurs et al. 1982) or protons efflux (Kelly 1983).

Protoplast Responses to Light Perception

Hale and Roux (1980) have shown that light perception in oat coleoptile protoplasts induced either an efflux of Ca^{2+} when protoplasts were subjected to red light, or an influx when they were subjected to far-red light. This Ca^{2+} flux was photoreversible by R or FR: a red irradiation following a far-red irradiation reversed the effect of FR and vice-versa. Far-red/red reversibility of Ca^{2+} movement across plasmalemma indicated that phytochrome was responsible for perception of light quality in these protoplasts. Moreover, the presence of phytochrome was confirmed by spectrophotometric measurements of phytochrome absorption according to its two interconvertible molecular forms, P_r and P_{fr}.

Recently, involvement of phytochrome has also been demonstrated in the swelling of etiolated wheat leaf protoplasts. Blakeley et al. (1983) showed that a 2 min red light irradiation, in presence of 1 mM KCl in the culture medium, was sufficient to cause protoplast swelling and could be reversed by FR. They assumed that phytochrome acts at the membrane level to regulate ion uptake and water movement. These rapid effects would be responsible for changes in protoplast size. Conversely, in a study on guard cell onion, Zeiger and Hepler (1977) reported that protoplast swelling was sensitive to blue light. As a working hypothesis, these authors proposed that a blue light photoreceptor, bound to plasmalemma or to the tonoplast, or both, might initiate an electron flow when energized with blue light; a resulting separation of OH^- and H^+ across the membrane would then establish a pH gradient capable of driving the uptake of K^+.

These various reports clearly demonstrated the action of both phytochrome and the blue-light-absorbing pigment on ion transport across the membrane (and therefore membrane permeability) in protoplasts. Thus, photoregulation of at least early events occurring just after light perception could be obtained. In contrast, subsequent events such as photoregulation of enzyme activities have not been actively studied in protoplasts (McClure 1981). Most investigations concerning photoregulation of metabolic pathways at the cellular level have been done on cell suspension culture and are well documented (Wellmann and Schopfer 1975; Gough et al. 1981; Duell-Pfaff and Wellmann 1982). However, protoplasts may also constitute a useful tool to investigate the complex physiological processes involved in the photoregulation of enzyme activity or metabolic pathways.

Protoplast Growth Photoregulation: A Case Study

Our laboratory has undertaken the study of growth photoregulation in *Vigna unguiculata= V. sinensis* green internode. There is conclusive evidence that in whole plants both phytochrome and the blue-light pigment are involved in the photocontrol of epicotyl elongation (Lecharny and Jacques 1980, 1982). In such a system, a positive correlation was established between the elongation rate and specific incorporation of oleic acid in two phospholipids: phosphatidylcholine (PC) and phosphatidylethanolamine (PE) (Lecharny et al. 1981; De March et al. 1984). These phospholipids are known to be mainly (for PC) and exclusively (for PE) located in extrachloroplastic compartments (endoplasmic reticulum or plasmalemma) specifically implicated in cell growth (Mazliak 1977).

Thus, the occurrence of oleic acid metabolism as a biochemical marker of epicotyl growth photoregulation provides a good tool for investigating growth photoregulation at the cellular level, particularly in a system like protoplasts where elongation cannot be measured.

Viable protoplasts from *Vigna unguiculata* epicotyls were obtained in high yield according to De March et al. (1983). Protoplasts released from the macerating medium were analyzed for their lipid composition. Total lipids were extracted following Bligh and Dyer (1959) and fatty acid methyl esters were prepared follow-

Table 1. Fatty acid content of intact epicotyls and epicotyl protoplasts from *Vigna unguiculata*. Results are expressed in percentage of total fatty acids. Each value is the mean of 3 replicates ± SD of (1) 750,000 protoplasts released in a simple salt macerating medium – KCl 330 mM, $CaCl_2$ $2H_2O$ 14 mM, pH 5.55-supplemented with enzymes (2) 10 excised epicotyls kept in the same salt medium, without enzymes. Lipid analysis was carried out after 17 h of incubation. 16:0, palmitic acid; 18:0, stearic acid; 18:1, oleic acid; 18:2, linoleic acid; 18:3, α-linolenic acid

Plant material	Fatty acid content (%)				
	16:0	18:0	18:1	18:2	18:3
Excised epicotyl	28.1 ± 0.3	5.3 ± 0.2	7.9 ± 0.3	19.4 ± 0.2	39.2 ± 0.2
Epicotyl protoplasts	27.4 ± 0.7	4.5 ± 0.3	6.6 ± 0.3	21.2 ± 0.5	40.3 ± 1.1

ing Metcalfe et al. (1966) for subsequent analysis by capillary gas chromatography. In our macerating conditions, fatty acids of freshly obtained protoplasts were found in similar proportions to those in intact epicotyls (Table 1). Lipid pattern was also not modified. These results indicate that major structural alterations of membranes did not occur during protoplast obtention, and that fatty acid metabolism did not seem to be affected either. This last point is consistent with results obtained by Haas et al. (1979), but is in contrast with that observed by Hock and Hartmann (1983) with particular plant species.

Photoregulation of oleic acid metabolism was investigated using labelling experiments with [1-^{14}C]-oleate incorporation. We observed that epicotyl protoplasts placed in particular environmental conditions were sensitive to various light/dark treatments (De March and Trémolières, in preparation). Incubation of protoplasts was done in a simple salt medium (KCl 250 mM, $CaCl_2$ $2H_2O$ 1.4 mM, MES 10 mM, pH 5.55) without growth substances or ose-alcohol osmoticum, but supplemented with Murashige and Skoog's (1962) macroelements at half strength, Heller's (1953) microelements, Morel and Whetmore's (1957) vitamins and 20 g l^{-1} of sucrose.

Figure 1 shows the label distribution of [1-^{14}C]-oleate within various lipid classes after a light/dark treatment for 24h. Results in Table 2 indicate that dark treatments (with or without a 20 min FR pretreatment) greatly enhanced [1-^{14}C]-oleate incorporation into PC and PE. After 1h incorporation, oleic acid was mainly incorporated in phosphatidic acid (PA) when protoplasts were subjected to

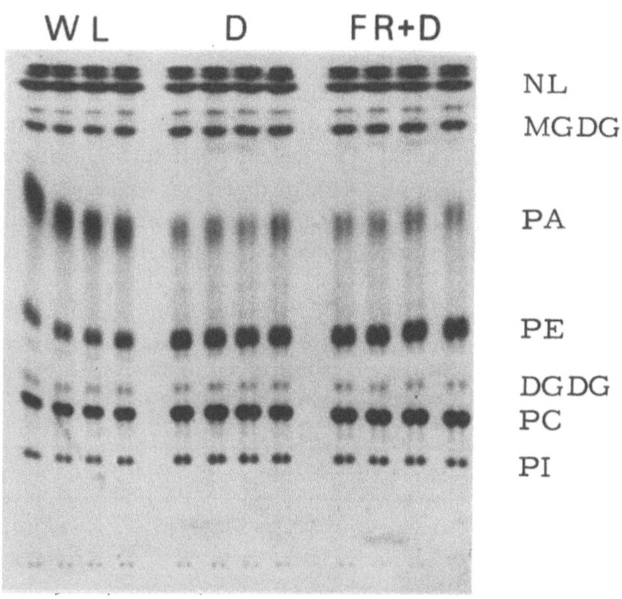

Fig. 1. Autoradiochromatogram showing labelling distribution of [1-^{14}C]-oleate into lipids of internode protoplasts. Protoplasts were incubated with [1-^{14}C]-oleate for 24h in white light *(WL)* at high fluence rate (60 μ mol m^{-2} s^{-1}), darkness *(D)*, or darkness after a FR pretreatment *(FR+D)*. *DGDG* diacyldigalactosylglycerol; *MGDG* diacylgalactosylglycerol; *NL* neutral lipids; *PA* phosphatidic acid: *PC* phosphatidylcholine; *PE* phosphatidylethanolamine; *PI* phosphatidylinositol

white light (60 μmol m^{-2} s^{-1}); in darkness, labelling was much less important in phosphatidic acid (the immediate precursor of PC and PE) but enhanced significantly in phosphatidylcholine and phosphatidylethanolamine (Table 2A). When protoplasts were subjected for 24h to darkness a significative enhancement of [1-^{14}C]-oleate incorporation into PC and PE was also observed. A weak stimulating effect of FR pretreatment could be noted when compared to dark treatment alone (Table 2B).

These experiments allow us to establish that photoregulation of lipid metabolism, linked to the growth process, occurs in a similar way in internode protoplasts and in the internodes of whole plants: dark, with or without a FR pretreatment, which stimulates epicotyl growth on whole plants compared to WL at high fluence rate, enhanced oleyl-PC and oleyl-PE biosynthesis in both systems. This result indicates that the early biochemical events occurring before cell elongation can be photoregulated on protoplasts, although whole plant correlations are suppressed. It may be assumed from these observations that the mode of action of the photoreceptors involved in cell elongation is probably not modified in protoplasts. However, it should be mentioned that even if photoregulation of lipid metabolism occurred in a similar way in internode protoplasts and in the internodes of whole plants, differences in photoresponses can be observed. The most important difference is that the FR promoting effect on PC and PE biosynthesis observed in whole plants, in comparison to dark effect alone, is less evident in protoplasts. The question arises as to whether the loss of the FR effect is due to the procedure by which protoplasts are obtained (Duke et al. 1981) or whether transduction of the light signal via phytochrome requires the pluricellular and complex organization of a whole plant. Indeed, it has often been reported that leaves may be involved in stem elongation via exportation of hormones or energetic sources (Garcia-Martinez and Rappaport 1982). A suitable medium, perhaps more complex, may eventually restore the FR effect in protoplasts (Tanada 1968a). The second point to be mentioned is that differences observed between light and dark treatment on whole plants are much more important in protoplasts. This may be due to a more homogeneous precursor incorporation in protoplast suspension and therefore to a more syn-

Table 2. Incorporation of [1-^{14}C]-oleate into lipids of epicotyl protoplasts. Protoplasts were incubated with [1-^{14}C]-oleate for 1h (A) or 24h (B) in various light or dark conditions: WL (60 μmol $m^{-2}s^{-1}$), darkness, or darkness after a FR treatment. Means ± SD were obtained from four replicates of 700,000 protoplasts each

Treatment		[pmol (μg of total fatty acids)$^{-1}$]					
		Total lipids	NL	PA	PC	PE	MGDG
A.	WL	76.5	54.0	13.2	2.0	3.2	1.1
	Dark	84.9	52.4	2.9	14.0	5.0	1.2
B.	WL	182 ± 23	59.6 ± 6.4	22.7 ± 5.9	21.7 ± 4.6	10.6 ± 1.7	7.9 ± 2.0
	Dark	208 ± 29	51.4 ± 4.3	8.1 ± 0.5	58.8 ± 8.2	22.7 ± 3.5	10.0 ± 1.4
	FR+dark	208 ± 14	49.4 ± 2.0	7.7 ± 0.4	63.2 ± 5.5	23.8 ± 1.7	10.4 ± 1.0

chronous response. Moreover if each cell in the internode is a potential "target cell", the physiological response may be amplified in protoplasts.

In conclusion, we may say that even if the FR effect is not observable under our experimental conditions, the effect of light and dark treatments on PC and PE biosynthesis is well expressed in protoplasts. At this stage in our work, we are not yet able to ascertain whether the blue-light pigment is involved in this photoresponse. Further studies are needed to determine whether the photoreceptor implicated in this particular response is the same as the one in whole plants.

Concluding Remarks

Our preliminary study concerning photoregulation of lipid metabolism in protoplasts has shown that a complex physiological response occurring in whole plants is observable at the cellular level. Even if photobiological studies on protoplasts are not sufficiently conclusive, we present evidence that protoplast photoregulation offers a powerful framework to investigate the molecular action of photoreceptors. One major reason to assume that protoplasts may provide a useful system for such studies is that the environmental conditions (ion concentrations, hormonal balance) can be adequately controlled. Thus, whether physiological responses to light perception are expressed in protoplasts or not, the factor(s) modulating or restoring these responses may be determined (Tanada 1968b). Such studies might be accompanied by immunological assays to ensure the photoreceptor presence in protoplasts. In wheat leaf protoplasts, Blakeley et al. (1983) have shown that gibberellic acid can replace the action of R and they suggested that this hormone might act as an intermediate in phytochrome-regulated protoplast swelling. Moreover, they assumed that these effects were related to wheat leaf unrolling, controlled by phytochrome. In this case, gibberellins also appeared to act as intermediates between light perception and the changes in cell volume. However, the interaction between light and hormones, with regard to a given physiological phenomenon, has never been demonstrated in protoplasts. Recently, Hasezawa and Syono (1983) obtained a spectacular elongation of tobacco protoplasts between the 4th and the 7th day of culture by balancing hormonal composition of the medium, but the effect of light on this system was not investigated. In regard to phytochrome action in oat coleoptile protoplasts (Hale and Roux 1980), regulation of Ca^{2+} movement has not been related to any physiological action of light in vivo and the physiological significance of this photoregulation cannot be conclusively shown. Nevertheless, this report can be related to other photobiological studies in higher plants pointing out the importance of Ca^{2+} as a second messenger in phytochrome-controlled morphogenesis (Weinsenseel and Ruppert 1977; Dreyer and Weinsenseel 1979). These results taken in connection with those obtained in our case study indicate that light-dependent events leading to the physiological response are worthy of investigation in protoplasts. Work on the mechanism of action of photoreceptors should focus on two major problems:

1. Upon phototransformation of photoreceptors, the often emphasized photoreceptor interaction with membranes must be elucidated. Protoplasts would then provide a useful tool to study subcellular localization of phytochrome which still remains controversial. Problems of identification and localization of the blue-light pigment could also be addressed by protoplast fractionation. Easily disrupted, protoplasts present as compared to homogenates the great advantage of releasing undamaged organelles (Haas and Heinz 1980).

2. Characterization of the primary response (ion fluxes across membranes, changes in membrane potential) sould be taken into account with subsequent events leading to physiological responses.

The effect of FR light on a Ca^{2+} pump, probably located in the plasma membrane, has been investigated by Dieter and Marmé (1983). These authors showed that FR light lowered the Ca^{2+} transport activity and thus induced an increase of the cytoplasmic, free Ca^{2+} concentrations. As a result, Ca^{2+} bound to calmodulin and thus activated Ca^{2+} calmodulin-dependent enzymes (Marmé 1982; Salimath and Marmé 1983). But regulation of plant enzyme activity by phytochrome or the blue-light pigment may result from either activation or de novo synthesis of enzymes. Further research using immunological techniques and deuterium labelling experiments will determine whether light is effective at the transcriptional or translational level or both (Schopfer 1977; Sharma and Schopfer 1982).

In conclusion, we must stress that in protoplasts, photoregulation studies and particularly the photoregulation study of metabolic processes are just beginning. Nevertheless, first results reveal that protoplasts are a very promising system through which mechanisms of photocontrol may be explained.

Acknowledgments. We thank Dr. Roger Jacques for helpful discussions and for reviewing the manuscript.

References

Blakeley SD, Thomas B, Hall JL, Vince-Prue D (1983) Regulation of swelling of etiolated-wheat-leaf protoplasts by phytochrome and gibberellic acid. Planta (Berl) 158:416–421

Bligh EG, Dyer WJ (1959) A rapid method of total lipid extraction and purification. Can J Biochem Physiol 37:911–917

Brownlee C, Kendrick RE (1979a) Ion fluxes and phytochrome protons in mung bean hypocotyl segments I. Fluxes of potassium. Plant Physiol (Bethesda) 64:206–210

Brownlee C, Kendrick RE (1979b) Ion fluxes and phytochrome protons in mung bean hypocotyl segments II. Fluxes of chloride, protons, and orthophosphate in apical and subhook segments. Plant Physiol (Bethesda) 64:211–213

De March G, Chupeau Y, Trémolières A (1983) Oleic acid as a biochemical marker to study growth photoregulation of internode protoplasts. In: Potrykus I, Harms CT, Hinnen A, Hütter R, King PJ, Shillito RD (eds) Protoplasts 1983, 6th Int Protoplast Symp Basel, Poster Proc Birkhäuser, Basel, pp 184–185

De March G, Trémolières A, Lecharny A (1984) Oleic acid metabolism: a biochemical marker for studying photoregulation of epicotyl elongation in *Vigna unguiculata*. Physiol Plant 60:401–408

Dieter P, Marmé D (1983) The effect of calmodulin and far-red light on the kinetic properties of the mitochondrial and microsomal calcium-ion transport system from corn. Planta (Berl) 159:277–281

Dreyer EM, Weisenseel MH (1979) Phytochrome-mediated uptake of calcium in *Mougeotia* cells. Planta (Berl) 146:31–39

Duell-Pfaff N, Wellmann E (1982) Involvement of phytochrome and a blue light photoreceptor in UV-B induced flavonoid synthesis in parsley (*Petroselinum hortense* Hoffm) cell suspension cultures. Planta (Berl) 156:213–217

Duke SO, Netzley DH, McClure JW (1981) Tissue localization of phytochrome in dark-grown barley leaves. Phytochemistry (Oxf) 20:2327–2328

Gaba V, Black M (1979) Two separate photoreceptors control hypocotyl growth in green seedlings. Nature (Lond) 278:51–54

Garcia-Martinez JL, Rappaport L (1982) Contribution of the leaves to gibberellin-induced epicotyl elongation in cowpea. J Plant Growth Reg 1:129–137

Gough SP, Girnth C, Kannangara CG (1981) α Aminolevulinate synthesis in greening barley I. Regulation. In: Akoyunoglou GA (ed) Photosynthesis V. Chloroplast development. Balaban International Science Servies, Philadelphia, Pa, pp 107–116

Haas R, Heinz E (1980) Use of mesophyll protoplasts to study lipid metabolism in cell organelles. In: Mazliak P, Benveniste P, Costes C, Douce R (eds) Biogenesis and function of plant lipids. Elsevier North-Holland Biomedical, Amsterdam, pp 19–28

Haas R, Heinz E, Popovici G, Weissenböck G (1979) Protoplasts from oat primary leaves as tools for experiments on the compartmentation in lipid and flavonoid metabolism. Z Naturforsch 34:854–864

Hale CC, Roux SJ (1980) Photoreversible calcium fluxes induced by phytochrome in oat coleoptile cells. Plant Physiol (Bethesda) 65:658–662

Hasezawa S, Syōno K (1983) Hormonal control of elongation of tobacco cells derived from protoplasts. Plant Cell Physiol 24:127–132

Hock K, Hartmann E (1983) Modulation of fatty acid patterns during protoplast isolation. In: Potrykus I, Harms CT, Hinnen A, Hütter R, King PJ, Shillito RD (eds) Protoplasts 1983, Poster Proc 6th Int Protoplast Symp Basel. Birkhäuser, Basel, pp 206–207

Kaiser G, Heber U (1983) Photosynthesis of leaf cell protoplasts and permeability of the plasmalemma to some solutes. Planta (Berl) 157:462–470

Kelly BM (1983) Light-stimulated changes in the acidity of suspensions of oat protoplasts. Plant Physiol (Bethesda) 72:351–355

Lecharny A, Jacques R (1980) Light inhibition of internode elongation in green plants. A kinetic study with *Vigna sinensis* L. Planta (Berl) 149:384–388

Lecharny A, Jacques R (1982) Photo-inhibition of internode elongation rate in light-grown *Vigna sinensis* L. Control by light quality. Plant Cell Environ 5:31–36

Lecharny A, Trémolières A, Jacques R (1981) Photomodulation of growth and specific changes in fatty acid amounts in internodes of green *Vigna sinensis* L. Planta (Berl) 152:487–490

Leurs CJ, Winter H, Wiersema PK, Helder RJ (1982) Light-dependent rubidium uptake into isolated mesophyll protoplasts from leaves of *Pisum sativum*. Physiol Plant 56:339–342

Löppert H, Kronberger W, Kandeler R (1978) Phytochrome-mediated changes in the membrane potential of subepidermal cells of *Lemna paucicostata* 6746. Planta (Berl) 138:133–136

Marmé D (1977) Phytochrome: membranes as possible sites of primary action. Annu Rev Plant Physiol 28:173–198

Marmé D (1982) The role of Ca^{2+} and calmodulin in plants. What's New in Plant Physiol 13: 37–40

Marmé D, Boisard J, Briggs WR (1973) Binding properties in vitro of phytochrome to a membrane fraction. Proc Natl Acad Sci USA 70:3861–3865

Mazliak P (1977) Glyco and phospholipids of biomembranes in higher plants. In: Tevini M, Lichtenthaler HK (eds) Lipids and lipid polymers in higher plants. Springer, Berlin Heidelberg New York, pp 48–74

McClure JW, Netzley DG, Duke SO (1981) Tissue localization of phytochrome and of secondary phenolic metabolism in barley primary leaves. In: Meet 13th Int Bot Congr, August 21–28 Sydney, NSW, Australia, p 27

Metcalfe LD, Schmitz AA, Pelka JR (1966) Rapid preparation of fatty acids esters from lipids for gas chromatographic analysis. Anal Chem 38:514–515

Nishimura M, Akazawa T (1975) Photosynthetic activities of spinach leaf protoplasts. Plant Physiol (Bethesda) 55:712–716

Pratt LH, Marmé D (1976) Red light-enhanced phytochrome pelletability; a reexamination and further characterization. Plant Physiol (Bethesda) 58:686–692

Rathnam CKM, Zilinskas BA (1977) Reversal of 3-(3,4-Dichlorophenyl)-1-1-Dimethylurea inhibition of carbon dioxide fixation in spinach chloroplasts and protoplasts by dicarboxylic acids. Plant Physiol (Bethesda) 60:51–53

Ritter A, Wagner E, Holmes MG (1981) Light quantity and quality interactions in the control of elongation growth in light-grown *Chenopodium rubrum* L. seedlings. Planta (Berl) 153: 556–560

Salimath BP, Marmé D (1983) Protein phosphorylation and its regulation by calcium and calmodulin in membrane fractions from zucchini hypocotyls. Planta (Berl) 158:560–568

Satter RL, Marinoff P, Galston AW (1970) Phytochrome controlled nyctinasty in *Albizzia julibrissin* II. Potassium flux as a basis for leaflet movement. Am J Bot 57:916–926

Schopfer P (1977) Phytochrome control of enzymes. Annu Rev Plant Physiol 28:223–252

Sharma R, Schopfer P (1982) Sequential control of phytochrome-mediated synthesis de novo of β-amylase in the cotyledons of mustard (*Sinapis alba* L) seedlings. Planta (Berl) 155: 183–189

Smith H (1981) Function, evolution and action of plant photosensors. In: Smith H (ed) Plants and the daylight spectrum. Academic, London, pp 499–508

Tanada T (1968a) A rapid photoreverisble response of barley root tips in presence of 3-indoleacetic acid. Proc Natl Acad Sci USA 59:376–380

Tanada T (1968b) Substances essential for a red, far-red light reversible attachment of mung bean root tips to glass. Plant Physiol (Bethesda) 43:2070–2071

Tezuka T, Yamamoto Y (1975) Control of ion absorption by phytochrome. Planta (Berl) 122: 239–244

Thomas B, Dickinson HG (1979) Evidence for two photoreceptors controlling growth in deetiolated seedlings. Planta (Berl) 146:545–550

Weisenseel MH, Ruppert HK (1977) Phytochrome and calcium ions are involved in light-induced depolarization in *Nitella*. Planta (Berl) 137:225–229

Wellmann E, Schopfer P (1975) Phytochrome-mediated de novo synthesis of phenylalanine ammonia-lyase in cell suspension cultures of parsley. Plant Physiol (Bethesda) 55:822–827

Zeiger E, Hepler PK (1977) Light and stomatal function: blue light stimulates swelling of guard cell protoplasts. Science (Wash DC) 196:887–889

Photorespiratory Metabolism in Protoplasts

G.E. EDWARDS and R. SCOTT

Introduction

The application of protoplasts as a tool for the study of photosynthesis was first done with C_4 plants in examining the intercellular and intracellular compartmentation of photosynthetic enzymes (Kanai and Edwards 1973; Gutierrez et al. 1975). Over the last decade, protoplasts have been isolated from C_3, C_4 and CAM plants to study various aspects of photosynthetic metabolism (Edwards and Walker 1983; Kobza and Edwards 1984). In the present review we discuss ways in which protoplasts have been used in the study of photorespiration, and the potential of this method for future research on this topic.

Materials and Methods

The first step in using protoplasts to study photorespiratory metabolism is to isolate protoplasts which have a good rate of photosynthesis (at least 50 to 100 μmol CO_2 fixed mg^{-1} chlorophyll h^{-1}). Factors which are critical in obtaining photosynthetically active protoplasts have been discussed previously (Edwards and Walker 1983; Kobza and Edwards 1984), and thus will not be covered in detail here. Condition of the parent plant, including leaf age, and length of digestion with macerating enzymes are important. For example, with mesophyll protoplasts of wheat, both the yield and photosynthetic capacity decrease with leaf age (Table 1). The yield and photosynthetic activity will depend on a combination of factors, including source

Table 1. Effect of leaf age on yield and photosynthetic activity of mesophyll protoplasts of wheat. (Huber and Edwards 1975a)

Plant age days	Yield mg chl g^{-1} fr. wt.	CO_2 fixation μmol mg^{-1} chl h^{-1}
6	1.16	148
8	0.97	173
10	0.48	29
13	0.50	20
21	0.13	16

The Physiological Properties of Plant Protoplasts
Edited by P. E. Pilet

of digestive enzymes, the species used and condition of leaf material. Therefore, in each case the investigator will need to establish optimum conditions.

When photosynthesis and photorespiratory properties of intact protoplasts are evaluated, stability can be achieved by use of an osmoticum (e.g. sorbitol at 0.5 M), $CaCl_2$ (1 to 5 mM) and a buffer (between pH 7 and 8). Protoplasts are stable for short periods in the presence of EDTA (Edwards et al. 1978), but the calcium salt tends to stabilize the protoplasts (Edwards et al. 1976) and, at the same time, will prevent photosynthesis of any chloroplasts which are released due to protoplast breakage (divalent cations inhibit photosynthesis of isolated chloroplasts). When assaying photosynthesis of isolated chloroplasts, chelation (e.g. either EDTA or pyrophosphate) and an optimum level of inorganic phosphate is required (Edwards and Walker 1983). When assaying photosynthesis with protoplasts, they should be stirred gently during assay to prevent breakage.

A method has been developed for isolation and purification of intact peroxisomes from protoplasts (Schmitt and Edwards 1982). The peroxisomes are purified by removal of the chloroplasts by differential centrifugation, and then separated from the mitochondria through use of a step gradient of Percoll, with the peroxisomes partitioning at the 30/45% interphase. Although functional mitochondria have been isolated from protoplasts of C_3 plants (Nishimura et al. 1982), compared to the use of conventional methods of preparing mitochondria this approach has not been utilized to any extent for studying mitochondrial functions in C_3 plants relative to photorespiration (glycine oxidation). An important application of the protoplast technique for studying photorespiration is through the recent isolation of mitochondria from mesophyll and bundle sheath protoplasts (Ohnishi and Kanai 1983, see results and discussion).

Results and Discussion

C_3 Plants

Brief Overview of Photorespiration

A simplified scheme of photorespiration in C_3 plants is illustrated in Fig. 1. O_2 inhibits photosynthesis by competitively inhibiting CO_2 fixation through RuBP carboxylase. This competitve inhibition by O_2 is associated with it reacting with RuBP as an oxygenase. The products of this reaction are P-glycolate and 3-phosphoglycerate. The P-glycolate is further metabolized through the glycolate pathway with a loss of CO_2 in the process. Thus, O_2 inhibits photosynthesis by two means, first as a competitive inhibitor of RuBP carboxylase, and secondly due to loss of CO_2 in glycolate metabolism. The pathway of metabolism in photorespiration is illustrated in Fig. 2. When 2 O_2 react with 2 RuBP, the 2 P-glycolate formed are converted to 2 glycolate in the chloroplast through P-glycolate phosphatase. The two glycolate are transported from the chloroplasts and metabolized to glycerate + CO_2 through the mitochondria and peroxisomes. The ammonia released during glycine decarboxy-

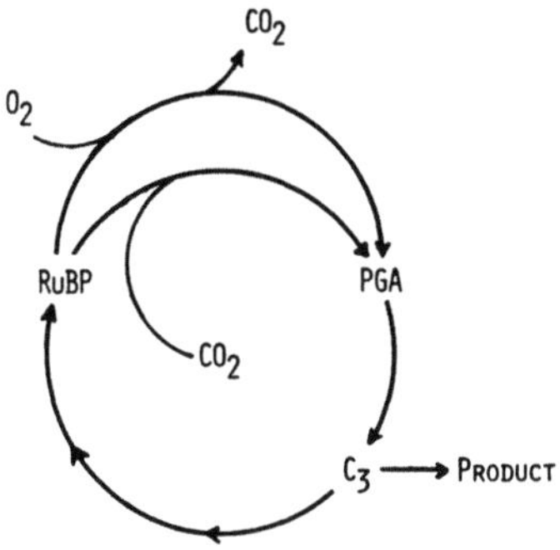

Fig. 1. Simplified scheme illustrating O_2 inhibition of photosynthesis in C_3 plants. Reaction of CO_2 with RuBP leads to CO_2 fixation through the RPP pathway. Reaction of O_2 with RuBP leads to photorespiration through the glycolate pathway

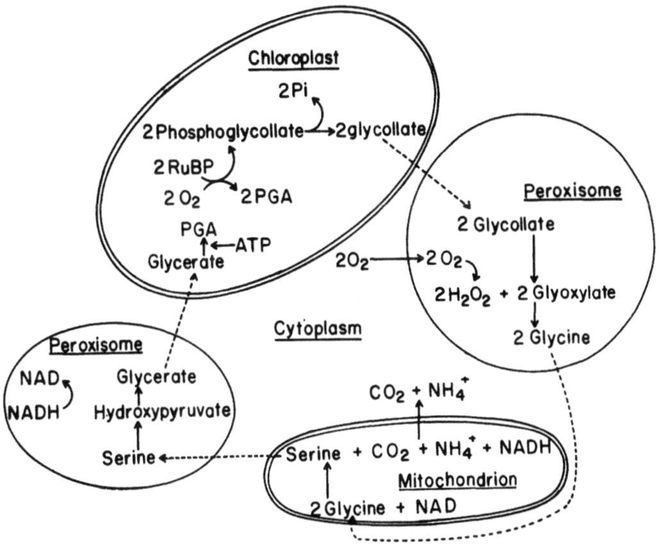

Fig. 2. Origin and metabolism of glycolate during photorespiration in a C_3 plant

lation is reassimilated through glutamine synthetase in the chloroplast. The glycerate is taken up by the chloroplast and converted to 3-PGA through glycerate kinase. Either high levels of CO_2 (approximately two fold above atmospheric concentrations) or low levels of O_2 (e.g. 2%) can largely prevent photorespiration and the associated O_2 inhibition of photosynthesis.

Use of Intact Protoplasts in the Study of Photorespiration

If higher plant cells can be isolated and used for studying photorespiration, including the interaction of O_2, CO_2, light, and temperature, this can offer some distinct advantages over the use of whole leaves. With protoplasts or isolated cells the equipment needed is less elaborate (following photosynthesis as O_2 evolution with an O_2 electrode or as $^{14}CO_2$ fixation) compared to that needed with whole leaves (use of an infrared gas analysis system). Control of temperature is relatively easy with isolated protoplasts compared to leaves. With isolated protoplasts a homogeneous preparation is obtained which can be used for various treatments, whereas with whole plants one must contend with leaf to leaf variations.

Either intact protoplasts or intact cells may be used with equal effectiveness for certain studies on photorespiration. There are many species, including C_3 and C_4 monocots, where mesophyll protoplasts can be isolated but mesophyll cells have not been isolated either mechanically or through use of commercially available enzymes. Obviously in these cases, e.g. with wheat and barley, the isolation of protoplasts is of considerable value. As will be discussed later, protoplasts are clearly of value when comparisons of photosynthetic or photorespiratory functions are made between the intact cell and various organelles and when protoplasts are used as a means to study the compartmentation of enzymes.

To date, there have been only a few studies of photorespiration on whole protoplasts of C_3 plants. O_2 inhibits photosynthesis of spinach protoplasts (C_3), with a Ki of about 0.3 mM, which is similar to that of the Ki(O_2) for RuBP carboxylase (Nishimura et al. 1975). However, the rate of photosynthesis by the protoplasts, even under nonphotorespiring conditions, was quite low (4 to 6 μmol mg^{-1} chlorophyll h^{-1}). Glycine, a metabolite of the glycolate pathway, was detected as a major product of photosynthesis under O_2 but not under anaerobic conditions. With protoplast extracts of barley, O_2 inhibited photosynthesis under low levels of bicarbonate (0.4 mM, pH 8.2), whereas under higher levels of bicarbonate (4.8 mM, pH 8.2) there was no difference in photosynthesis in the presence of 2 versus 21% O_2. The maximum rate of photosynthesis by the protoplast extracts was 137 μmol mg^{-1} chlorophyll h^{-1} (Huber and Edwards 1975b). Whether the glycolate pathway is functioning in the protoplast extracts has not been evaluated. With wheat protoplasts, increasing levels of O_2 up to 100% inhibits CO_2 fixation, while 4 mM bicarbonate was sufficient to prevent inhibition by O_2 (Edwards et al. 1976). These results indicate that high levels of CO_2 prevent O_2 inhibition of photosynthesis and that the protoplast system exhibits the competitive interaction between O_2 and CO_2 similar to that of intact leaves. Interestingly, with mesophyll protoplast extracts of C_4 plants, O_2 can *stimulate* pyruvate dependent assimilation of CO_2 through PEP carboxylase, which is considered to be due to an O_2 dependent production of ATP through pseudocyclic photophosphorylation (Huber and Edwards 1975b).

The effect of inhibition of the glycolate pathway on photosynthesis by wheat protoplasts was investigated by Usuda et al. (1980). They found that aminoacetonitrile (AAN), is an inhibitor of the glycolate pathway which blocks glycine oxidation by mitochondria. AAN inhibited CO_2 assimilation of isolated protoplasts under photorespiring conditions (low bicarbonate, 0.3 mM) but not under nonphotorespiring conditions (high bicarbonate, 10 mM). The compound had no effect on photosynthesis of isolated chloroplasts (where glycolate is a product of photosynthesis and not metabolized through the glycolate pathway) under either high or low bicarbonate. With protoplasts photosynthesizing under low levels of $^{14}CO_2$, AAN caused a large increase in the labelling of glycine and prevented the labelling of serine, which indicates glycine decarboxylation is inhibited. AAN also inhibited the decarboxylation of glycine added exogenously to the protoplasts. These results show that blockage of the glycolate pathway under photorespiring conditions leads to an inhibition of photosynthesis. Either prevention of return of carbon to the chloroplast in the form of glycerate, or feedback inhibition of photosynthesis by product accumulation in the glycolate pathway may account for the inhibitory effect. These results are consistent with other studies in which the glycolate pathway has been blocked either chemically or by mutation causing photosynthesis to be inhibited under photorespiring conditions (Sommerville and Ogren 1982; Edwards and Walker 1983).

Rathnam and Chollet (1980) studied photorespiration in mesophyll protoplasts from diploid and tetraploid cultivars of ryegrass. O_2 was shown to be a competitive inhibitor of CO_2 fixation by the protoplasts, and the protoplasts respired exogenously added glycolate and glycine with the release of CO_2. Although the $K_m(CO_2)$ was reported to be lower for the tetraploid than the diploid cultivar from measurements on CO_2 fixation (Rathnam and Chollet 1980), no difference has been found in the kinetic properties of RuBP carboxylase/oxagenase from studies on the isolated enzyme (see McNeil et al. 1981).

Use of Protoplasts to Study Compartmentation of Enzymes of Photorespiratory Metabolism

Considerable information on the compartmentation of enzymes in photorespiratory metabolism has been obtained using mechanical procedures to isolate organelles. Whether an enzyme is present in a given organelle can be determined by this method even though the percentage intactness is low. For example, with peroxisomes isolated mechanically from leaf tissue an intactness of only 5 to 10% is normally expected. Thus, with mechanical procedures it is difficult to account for the total intracellular distribution of a given enzyme, whereas this is possible using protoplasts. By gently breaking protoplasts a high percentage of intactness of organelles can be obtained (80 approaching 100% with mitochondria and chloroplasts, and 70 to 80% with peroxisomes). In a number of cases the question arises as to whether an enzyme is located both within an organelle and in the cytosol. One example is with the enzyme glycerate kinase. This enzyme catalyzes the terminal step of carbon metabolism in photorespiration whereby 3-PGA, a metabolite of the reductive pentose-P pathway is synthesized. If this enzyme is located in part in the cytosol it would suggest that carbon in the glycolate pathway could be directly metabolized to carbohydrate outside the chloroplast. In earlier studies using mechanical procedures it was suggested this enzyme is located in both the chloroplast and cytosol (Heber et al. 1974). However, in studies on compartmentation of glycerate kinase using protoplasts from several C_3 species, there is evidence that the enzyme is localized only in the chloroplast. Therefore, carbon in the glycolate pathway which is converted to glycerate must reenter the RPP pathway in the chloroplast. This indicates that there is an obligate link of carbon metabolism in photorespiration with photosynthesis and it is appropriate to refer to photorespiration as a photooxidative shunt of carbon around the RPP pathway (Schmitt and Edwards 1983a, Usuda and Edwards 1980a).

Another matter of importance in enzyme compartmentation is the site of reassimilation of the ammonia which is released from glycine decarboxylation in the mitochondria. Using protoplasts to study the compartmentation of glutamine synthetase, which is considered the enzyme responsible for ammonia assimilation, Wallsgrove et al. (1979) showed in pea *(Pisum sativum)* that about 40% of the enzyme is cytosolic and about 60% is chloroplastic. Therefore, in some species at least part of the ammonia from photorespiration may be assimilated in the cytosol. In other species, namely spinach, the enzyme is exclusively located in the chloroplast (Hirel et al. 1982), in which case all of the ammonia may be assimilated within this organelle. It is of interest to further establish the intracellular compartmentation of glutamine synthetase in various C_3 and C_4 plants in relation to the pathway, manner of provision of energy, and efficiency of nitrogen assimilation. There are other relevant studies on enzyme compartmentation which have been considered using protoplasts (for example the localization of NAD malate deyhdrogenase and aspartate aminotransferase within the cell relative to potential for intracellular shuttle of reductive power, Schmitt and Edwards 1983b), which cannot be discussed in detail in this review.

Use of Protoplasts to Study Functions of Organelles in Photorespiratory Metabolism

Since intact organelles can be isolated from protoplasts of various species, this allows one to study the metabolic function of the organelles and regulation of metabolism. Although mitochondria can be isolated from protoplasts of C_3 plants, this method has not been adopted over conventional mechanical methods of preparing mitochondria.

Intact peroxisomes have been isolated from protoplasts and are being used to study peroxisomal metabolism. A major question relative to function of peroxisomes in the glycolate pathway is how reductive power is provided for the hydroxypyruvate reductase for converting hydroxypyruvate to glycerate. Although a shuttle of reductive power to the peroxisomes was schematically presented more than a decade ago as occurring through a dicarboxylate shuttle, in a critical evaluation with isolated peroxisomes no evidence was found to support such a shuttle (Schmitt and Edwards 1983b), and it is possible that pyridine nucleotides are transported directly. While the predominant source of CO_2 release in the glycolate pathway is considered to occur through glycine decarboxylation, it has been suggested that some decarboxylation may occur through attack of peroxide on glyoxylate. Previous studies in vitro in this respect were with peroxisomes of questionable integrity (i.e. broken due to method of preparation). Through use of intact peroxisomes isolated from protoplasts evidence is presented that glycolate is not decarboxylated. Only under non-physiological conditions, by chemically inhibiting catalase activity or by its dilution using broken peroxisomes, is there evidence of glycolate decarboxylation through attack of peroxide on glyoxylate (Schmitt et al. 1984).

With respect to photorespiration, it is of interest to compare the effect of O_2 on CO_2 assimilation of the protoplasts versus the isolated chloroplast. With the isolated protoplast, O_2 presumably inhibits carbon assimilation for the two reasons already discussed, first, competitive inhibition with respect to CO_2 in RuBP carboxylase/oxygenase and second, inhibition due to loss of CO_2 in glycolate metabolism. In principle, one might expect with the isolated chloroplast that the competitive component of O_2 inhibition could be studied independent of the carbon loss from glycolate metabolism since glycolate will be exported from the chloroplast as an end product of photosynthesis. However, in addition to the competitive component of O_2 inhibition, under certain conditions O_2 can inhibit photosynthesis of the isolated Chloroplast for another reason. In order for photosynthesis to continue at a maximum rate, excessive export of carbon from the chloroplast must be prevented (in order that RuBP can be regenerated), an optimum level of Pi is required (in order to make ATP), and, or course, CO_2 and light must be available. It is well known that photosynthesis of the isolated chloroplasts is inhibited by a high level of Pi which excessive export of carbon in the form of triose-P and a limitation on regeneration of the acceptor RuBP (Edwards and Walker 1983). Under conditions favorable for photorepiration (presence of O_2, low CO_2), O_2 strongly inhibits photosynthesis at high Pi levels due to the increased export of carbon in the form of glycolate (Fig. 3). However, under low Pi levels O_2 can actually stimulate photosynthesis. When Pi is limiting, O_2 allows glycolate to be synthesized as a photosynthetic product with the isolated chloroplasts, and Pi is released from organic phosphate through P-glycolate phosphatase. The overall stoichiometry of photosynthesis under this condition if glycolate is the only product, is:

$2\,CO_2 + 2\,H_2O \longrightarrow 1.5\,O_2$ + glycolate (Usuda and Edwards 1982)

In vivo, while chloroplasts export glycolate during photorespiration they are thought to import glycerate. Isolated chloroplasts of the C_3 plant wheat metabolize glycerate to triose-P in the light at rates adequate to account for the rate of carbon flux through the glycolate pathway in vivo (Edwards and Walker 1984). Glycerate

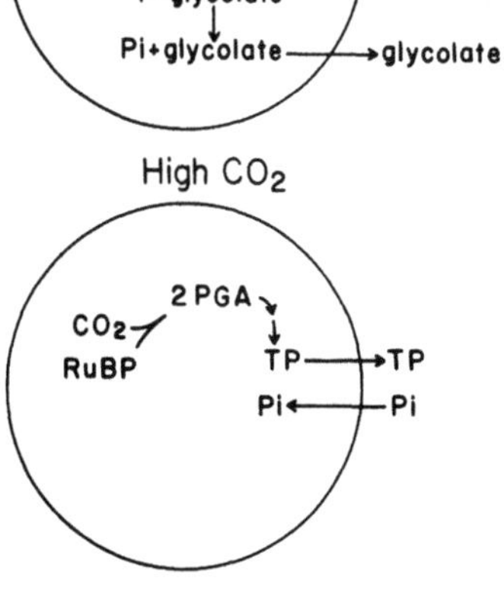

Fig. 3. Major forms of carbon exported from the chloroplasts of C_3 plants in the presence of Pi under photorespiratory conditions *(top)* and under nonphotorespiratory conditions *(bottom)*

also has a marked effect on CO_2 fixation by the isolated chloroplasts, in that relatively low concentrations (e.g. 100 to 200 μM) effectively prevent Pi inhibition of photosynthesis and shorten the induction period of photosynthesis. Thus, while export of glycolate from the isolated chloroplasts results in an increased sensitivity to Pi, glycerate prevents Pi inhibition of photosynthesis. When the glycolate pathway is blocked under photorespiratory conditions in vivo and return of carbon to the chloroplasts in the form of glycerate is prevented, photosynthesis is inhibited. In an analogous manner, with the isolated chloroplast under photorespiratory conditions, inhibition of photosynthesis under moderate to high levels of Pi can be relieved if glycerate is present to return carbon to the chloroplast.

In the future, it should be possible to study the competitive component of O_2 inhibition of photosynthesis with the isolated chloroplasts using conditions where excessive export of carbon is prevented and adequate Pi is available. Such conditions may be established with chloroplasts photosynthesizing in the presence of glycerate.

C_4 Plants

Under normal conditions photorespiration is considered to occur in C_4 plants, but at a rather low rate compared to the rate of photosynthesis, such that photorespiration is not apparent in whole plant measurements (i.e. atmospheric levels of O_2 do not inhibit photosynthesis). This is due to the CO_2 concentrating mechanism of C_4 photosynthesis, in which a relatively high CO_2/O_2 ratio exists in bundle-sheath cells where CO_2 is fixed through RuBP carboxylase (Edwards and Walker 1983).

In C_4 plants the enzymatic isolation of mesophyll protoplasts and bundle sheath strands or bundle-sheath protoplasts has allowed the intercellular compartmentation of enzymes of both photosynthesis and photorespiration to be established (see Edwards and Walker 1983; Ku and Edwards 1975; Usuda and Edwards 1980a; Ohnishi and Kanai 1983). RuBP carboxylase/oxygenase is located in bundle-sheath cells, and thus glycolate is thought to be synthesized in these cells at relatively low rates in vivo. Glycolate oxidase, catalase, glycine decarboxylase, serine hydromethyl transferase, and hydroxypyruvate reductase are located predominantly in the bundle-sheath cells (except for catalase 90% or more of the activity). Interestingly, glycerate kinase is located in mesophyll protoplasts of C_4 plants and, as discussed below, glycerate synthesis and metabolism in C_4 plants occur in different cell types.

In *Digitaria sanguinalis,* glycolate is decarboxylated by isolated bundle-sheath strands at a relatively high rate compared to mesophyll cells (Liu and Black 1972). Bundle-sheath cells of the C_4 plant *Panicum capillare* synthesize glycerate during assimilation of $^{14}CO_2$ under low bicarbonate. Glycerate synthesis is also known to be inhibited by AAN or by photosynthesis under high bicarbonate concentrations (Usuda and Edwards 1980b). These results indicate that the glycolate pathway can function in bundle-sheath cells converting glycolate to glycerate. Glycerate is not metabolized in isolated bundle-sheath cells, while in mesophyll protoplasts glycerate is converted to triose-P in the light (Usuda and Edwards 1980b). Thus, in C_4 plants, glycerate, the product of metabolism through the glycolate pathway, is not obligately linked to the C_3 pathway in bundle-sheath chloroplasts. It is uncertain why there is this differential compartmentation of the enzymes of glycerate synthesis and glycerate metabolism to triose-P. Perhaps under water stress, when CO_2 may be limited, C_4 plants photorespire and differential compartmentation of glycerate synthesis and metabolism may allow energy generated photochemically to be utilized in both mesophyll and bundle-sheath chloroplasts.

In C_4 plants, intact and functional mitochondria have been prepared from mesophyll and bundle-sheath protoplasts and used for studying photorespiratory functions (Ohnishi and Kanai 1983). These researchers showed that serine hydroxymethyl transferase and glycine decarboxylase, the two enzymes involved in glycine metabolism in mitochondria, were only localized in the bundle-sheath mitochondria. The bundle-sheath mitochondria oxidized glycine similar to mitochondria of C_3 plant leaves, whereas the mesophyll mitochondria did not oxidize glycine. Thus, the mitochondria of the photosynthetic cell types of C_4 plants have a differential expression of enzymes relative to photorespiratory metabolism.

In summary, in both C_3 and C_4 plants protoplasts are being employed to study photorespiratory metabolism through direct utilization of protoplasts, and through the fractionation of protoplasts to study enzyme compartmentation and organelle function. Continued studies at this level must be combined with whole plant investigations to obtain a full understanding of the pathway of photorespiration and its regulation. There are conditions where the O_2 inhibition of photosynthesis is variable at the whole plant level, for example with varying light, temperature, or under conditions where the utilization of triose-P may be limited. The use of protoplasts for studies on the mechanism of this regulation at the cellular level may prove to be particularly fruitful.

Acknowledgments. Appreciation is extended to John Kobza for the preparation of the figures. Supported by USDA Competitive Grant 83–CRCR–1–1321

References

Edwards GE, Walker DA (1983) C_3, C_4: mechanisms and cellular and environmental regulation of photosynthesis. Blackwell Scientific, Oxford

Edwards GE, Walker DA (1984) Influence of glycerate on photosynthesis of wheat chloroplasts. Arch Biochem Biophys 23:124–135

Edwards GE, Huber SC, Gutierrez M (1976) Photosynthetic properties of plant protoplasts. In: Peberdy JF, Rose AH, Rogers HJ, Cocking EC (eds) Microbial and plant protoplasts. Academic New York, pp 229–322

Edwards GE, Robinson SP, Tyler NJC, Walker DA (1978) Photosynthesis by isolated protoplasts, protoplast extracts and chloroplasts of wheat. Plant Physiol (Bethesda) 62:313–319

Gutierrez M, Huber SC, Ku SB, Kanai R, Edwards GE (1975) Intracellular carbon metabolism in mesophyll cells of C_4 plants. In: Avron M (ed) 3rd Int Congr on photosynthesis research. Elsevier Scientific, pp 1219–1230

Heber U, Kirk MR, Gimmler H, Schafer G (1974) Uptake and reduction of glycerate by isolated chloroplasts. Planta (Berl) 120:31–46

Hirel B, Perrot-Rechenmann C, Suzuki A, Vidal J, Gadal P (1982) Glutamine synthetase in spinach leaves. Immunological studies and immunocytochemical localization. Plant Physiol (Bethesda) 69:983–987

Huber S, Edwards G (1975a) An evaluation of some parameters required for the enzymatic isolation of cells and protoplasts with CO_2 fixation capacity from C_3 and C_4 grasses. Physiol Plant 35:203–209

Huber S, Edwards G (1975b) The effect of oxygen on CO_2 fixation by mesophyll protoplasts extracts of C_3 and C_4 plants. Biochem Biophys Res Commun 67:28–34

Kanai R, Edwards GE (1973) Separation of mesophyll protoplasts and bundle-sheath cells from maize leaves for photosynthetic studies. Plant Physiol (Bethesda) 51:1133–1137

Kobza J, Edwards GE (1984) Isolation of organelles: chloroplasts. In: Vasil IK (ed) Cell culture and somatic cell genetics of plants. A comprehensive treatise, vol 1. Laboratory techniques. Academic New York pp 471–482

Ku SB, Edwards GE (1975) Photosynthesis in mesophyll protoplasts and bundle-sheath cells of various types of C_4 plants. Z Pflanzenphysiol 77:16–32

Liu AY, Black CC (1972) Glycolate metabolism in mesophyll cells and bundle-sheath cells isolated from crabgrass, *Digitaria sanguinalis* (L) Scop, leaves. Arch Biochem Biophys 149: 269–280

McNeil PH, Foyer CH, Walker DA, Bird IF, Cornelius MJ, Keys AJ (1981) Similarity of RuBP carboxylase of isogenic diploid and tetraploid ryegrass (*Lolium perenne* L) cultivars. Plant Physiol (Bethesda) 67:530–534

Nishimura M, Douce R, Akazawa T (1982) Isolation and characterization of metabolically competent mitochondria from spinach leaf protoplasts. Plant Physiol (Bethesda) 69:916–920

Nishimura M, Graham P, Akazawa T (1975) Effect of oxygen on photosynthesis by spinach leaf protoplasts Plant Phsyiol. (Bethesda) 56:718–722

Ohnishi J, Kanai R (1983) Differentiation of photorespiratory activity between mesophyll and bundle-sheath cells of C_4 plants I. Glycine oxidation by mitochondria. Plant Cell Physiol 24:1411–1420

Rathnam CKM, Chollet R (1980) Photosynthetic and photorespiratory carbon metabolism in mesophyll protoplasts and chloroplasts isolated from isogenic diploid and tetraploid cultivars of ryegrass (*Lolium perenne* L). Plant Physiol (Bethesda) 65:489–494

Schmitt MR, Edwards GE (1982) Isolation and purification of intact peroxisomes from green leaf tissue. Plant Physiol (Bethesda) 70:1213–1217

Schmitt MR, Edwards GE (1983a) Glycerate kinase from leaves of C_3 plants. Arch Biochem Biophys 224:332–341

Schmitt MR, Edwards GE (1983b) Provision of reductant for the hydroxypyruvate to glycerate conversion in leaf peroxisomes. A critical evaluation of the proposed malate/aspartate shuttle. Plant Physiol (Bethesda) 72:728–734

Schmitt MR, Chestnut MH, Edwards GE (1984) Does glycolate decarboxylation occur in intact leaf peroxisomes? Plant Sci Lett 33:115–123

Somerville CR, Ogren WL (1982) Genetic modification of photorespiration. Trends Biochem Sci 7:171–174

Usuda H, Edwards GE (1980a) Localization of glycerate kinase and some enzymes for sucrose synthesis in C_3 and C_4 plants. Plant Physiol (Bethesda) 65:1017–1022

Usuda H, Edwards GE (1980b) Photosynthetic formation of glycerate in isolated bundle-sheath cells and its metabolism in mesophyll cells of the C_4 plant *Panicum capillare* L. Aust J Plant Physiol 7:655–662

Usuda H, Edwards GE (1982) Influence of varying CO_2 and orthophosphate concentrations on rates of photosynthesis, and synthesis of glycolate and dihydroxyacetone phosphate by wheat chloroplasts. Plant Physiol (Bethesda) 69:469–473

Usuda H, Arron GP, Edwards GE (1980) Inhibition of glycine decarboxylation by aminoacetonitrile and its effect on photosynthesis in wheat. J Exp Bot 31:1477–1483

Wallsgrove RM, Lea PJ, Miflin BJ (1979) Distribution of the enzymes of nitrogen assimilation within the pea leaf cell. Plant Physiol (Bethesda) 63:232–236

Subject Index

FSC
www.fsc.org
MIX
Papier aus verantwortungsvollen Quellen
Paper from responsible sources
FSC® C105338

If you have any concerns about our products, you can contact us on
ProductSafety@springernature.com

In case Publisher is established outside the EU, the EU authorized representative is:
Springer Nature Customer Service Center GmbH
Europaplatz 3, 69115 Heidelberg, Germany

Printed by Libri Plureos GmbH
in Hamburg, Germany